T0345271

Applications of Artificial Intelligence, Big Data and Internet of Things in Sustainable Development

This book focuses on different algorithms and models related to AI, big data and IoT used for various domains. It enables the reader to have a broader and deeper understanding of several perspectives regarding the dynamics, challenges, and opportunities for sustainable development using artificial intelligence, big data and IoT. *Applications of Artificial Intelligence, Big Data and Internet of Things (IoT) in Sustainable Development* focuses on IT-based advancements in multidisciplinary fields such as healthcare, finance, bioinformatics, industrial automation, and environmental science. The authors discuss the key issues of security, management, and the realization of possible solutions to hurdles in sustainable development. The reader will master basic concepts and deep insights of various algorithms and models for various applications such as healthcare, finance, education, smart cities, smart cars, etc. Finally, the book will also examine the applications and implementation of big data IoT, AI strategies to facilitate the sustainable development goals set by the United Nations by 2030. This book is intended to help researchers, academics, and policymakers to analyze the challenges and future aspects for maintaining sustainable development through IoT, big data, and AI.

Applications of Artificial Intelligence, Big Data and Internet of Things in Sustainable Development

Edited by
Sam Goundar
Archana Purwar
Ajmer Singh

CRC Press
Taylor & Francis Group
Boca Raton   London

CRC Press is an imprint of the
Taylor & Francis Group, an **informa** business

First edition published 2023
by CRC Press
6000 Broken Sound Parkway NW, Suite 300, Boca Raton, FL 33487–2742

and by CRC Press
4 Park Square, Milton Park, Abingdon, Oxon, OX14 4RN

CRC Press is an imprint of Taylor & Francis Group, LLC

© 2023 selection and editorial matter, Sam Goundar, Archana Purwar, Ajmer Singh; individual chapters, the contributors

Library of Congress Cataloging-in-Publication Data
Names: Goundar, Sam, 1967– editor. | Purwar, Archana, editor. | Singh, Ajmer, editor.
Title: Applications of artificial intelligence, big data and internet of things in
    sustainable development / edited by Sam Goundar, Archana Purwar, Ajmer Singh.
Description: Boca Raton : CRC Press, 2023. | Includes bibliographical references and
    index.
Identifiers: LCCN 2022014959 (print) | LCCN 2022014960 (ebook) |
    ISBN 9781032154022 (hardback) | ISBN 9781032157320 (paperback) |
    ISBN 9781003245469 (ebook)
Subjects: LCSH: Sustainable development—Technological innovations. | Internet
    of things—Environmental aspects. | Big data—Environmental aspects. | Artificial
    intelligence—Environmental aspects. | Internet of things—Economic aspects. | Big
    data—Economic aspects. | Artificial intelligence—Economic aspects.
Classification: LCC HC79.E5 A634 2023 (print) | LCC HC79.E5
    (ebook) | DDC 338.9/27—dc23/eng/20220406
LC record available at https://lccn.loc.gov/2022014959
LC ebook record available at https://lccn.loc.gov/2022014960

ISBN: 978-1-032-15402-2 (hbk)
ISBN: 978-1-032-15732-0 (pbk)
ISBN: 978-1-003-24546-9 (ebk)

DOI: 10.1201/9781003245469

Typeset in Times
by Apex CoVantage, LLC

Contents

Editors

Dr. Sam Goundar is an international academic having taught at twelve different universities in ten different countries. He is Editor-in-Chief of the *International Journal of Blockchains and Cryptocurrencies* (IJBC)—Inderscience Publishers, Editor-in-Chief of the *International Journal of Fog Computing* (IJFC)—IGI Publishers, Section Editor of the *Journal of Education and Information Technologies* (EAIT)—Springer, and Editor-in-Chief (Emeritus) of the *International Journal of Cloud Applications and Computing* (IJCAC)—IGI Publishers. He is also on the editorial review board of more than 20 high-impact factor journals. Professor Goundar is a senior member of IEEE; a member of ACS; a member of the IITP, New Zealand; Certification Administrator of ETA-I, USA; and past president of the South Pacific Computer Society. He also serves on the IEEE Technical Committee for Internet of Things, Cloud Communication and Networking, Big Data, Green ICT, Cybersecurity, Business Informatics and Systems, Learning Technology, and Smart Cities. He is a member of the IEEE Technical Society and a panelist with the IEEE Spectrum for Emerging Technologies.

Dr. Archana Purwar is currently working as Assistant Professor (Senior Grade) in Jaypee Institute of Information Technology, Noida (U.P.) and has been at the institute since 2011. Her areas of interest include database systems, information retrieval, and data mining, soft computing, and big data analytics.

Mr. Ajmer Singh earned B.Tech. and M.Tech. degrees from Kurukshetra University and his PhD from the Deenbandhu Chhoturam University of Science and Technology (DCRUST), Murthal. He is presently working as Assistant Professor in the Department of Computer Science and Engineering at DCRUST, Murthal, Sonepat, India. With more than 11 years of experience in academic and administrative affairs, he has published more than 20 research papers in various international/national journals and conferences. His research interests include software testing, information retrieval, and data sciences.

Contributors

Akshay Agarwal
University of Buffalo
State University of New York (SUNY)
Buffalo, New York, USA

Nidhi Agarwal
Department of Information Technology
KIET Group of Institutions
Ghaziabad, India

Imran Aslan
The Faculty of Health Science
Bingöl University
Bingöl, Turkey

Tanya Chauhan
Department of Computer Science and
    Engineering
DCRUST, Murthal
Haryana, India

Megha Chopra
Department of Computer science &
    Engineering and Information Technology
Jaypee Institute of Information Technology
Noida, India

Himanshi Garg
Amity Institute of Food Technology
NOIDA, U.P., India

Sam Goundar
School of Science, Engineering & Technology
RMIT University
Vietnam

Aishwarya Gupta
Jaypee Institute of Information Technology
Noida, India

Surbhi Gupta
Department of Computer Engineering,
    Model Institute of Engineering and
    Technology
Kot Bhalwal
J&K, India

Sardar M N Islam
ISILC & Decision Sciences and Modelling
Victoria University
Australia

Rachna Jain
Bhagwan Parshuram Institute of Technology
Delhi, India

Pankaj Khatak
Department of Mechanical Engineering
GJUST
Hisar (India)

Munish Kumar
Department of Mechanical Engineering
GJUST
Hisar (India)

Yogesh Kumar
Department of Computer science &
    Engineering
Indus University
Ahmedabad, India

Kumud Kundu
Department of Computer Science and
    Engineering
Inderprastha Engineering College
Ghaziabad, India

Sangeeta Lal
School of Computing and Mathematics
Keele University
UK

S. Mohanavel
Director
AJK Institute of Management
Coimbatore, India

Zarqua Neyaz
Computer Science and Engineering
    Department
Bhagwan Parshuram Institute of Technology
Delhi, India

Soumya Ranjan Purohit
Agricultural and Biosystems Engineering
 Department
South Dakota State University, USA

Archana Purwar
Department of Computer science
 & Engineering and Information
 Technology
Jaypee Institute of Information Technology
Noida, India

Ritu Rani
Department of Computer Science and
 Engineering
DCRUST
Murthal, India

Nalini Ratha
University of Buffalo
State University of New York (SUNY)
Buffalo, New York, USA

S. Gomathi alias Rohini
Department of Computer Applications
Sri Ramakrishna College of Arts and
 Science
Coimbatore, India

Deepanshu Sadhwani
Department of Electronics communication
 Engineering
Bharati Vidyapeeth's College of Engineering
New Delhi, India

Ovass Shafi
School of Computer Applications
Lovely Professional University
Punjab, India

Vasudha Sharma
Department of Food Technology
Jamia Hamdard, New Delhi, India

Tripti Sharma
Department of CSE (Data Science)
Inderprastha Engineering College
Ghaziabad, India

S. Jahangeer Sidiq
School of Computer Applications
Lovely Professional University
Punjab, India

Ajmer Singh
DCRUST/CSE/
Murthal, Haryana, India

Anjali Singhal
Department of Computer Science and
 Engineering
Inderprastha Engineering College
Ghaziabad, India

Nidhi Srivastava
Information Technology/Computer
 Science
Amity University
Lucknow, India

Tawseef Ahmed Teli
Department of Computer Applications
Amar Singh College
Srinagar, India

Narina Thakur
Computer Science and Engineering
Bharati Vidyapeeth's College of
 Engineering
New Delhi, India

Jeena Varghese
PG Scholar, Department of Computer
 Science and Engineering
Jyothi Engineering College
Cheruthuruthy, Thrissur, Kerala, India

Aswathy Wilson
Department of Computer Science and
 Engineering
Jyothi Engineering College
Cheruthuruthy, Thrissur, Kerala, India

Majid Zaman
Directorate of IT&SS
University of Kashmir
Srinagar, India

1 Introduction

Sam Goundar, Archana Purwar and Ajmer Singh

CONTENTS

1.1 WHAT IS SUSTAINABLE DEVELOPMENT?

Sustainable development (SD) is the continuous development by preserving necessary environmental resources, biological systems, protecting genetic diversity, and the unremitting use of species and resources. [1]. It is a worldwide challenge to preserve natural resources so future generations and the planet as a whole will also be able to flourish. The United Nations has set forth seventeen goals in 2030 agenda [1] for sustainable development. These SD goals (Global goals) [2–3] are integrated, development in one area will have an impact on other areas. Hence, sustainable development must balance various related aspects, such as society, economy, and environment. SD goals are designed so that all countries should work for global peace and prosperity and save resources on the earth for the wellbeing of all.

To facilitate the sustainable development in various countries, Information and Communications Technology (ICT) has emerged as a backbone for SD goals that helps to bring about their advancement toward meeting targets, partnerships, justice,

DOI: 10.1201/9781003245469-1

protecting natural resources such as forests, pollution-free atmosphere, water, and land. Countries should also provide fundamental amenities such as health services, schools and colleges, railways, roads, money, and energy [2, 4–5] along with techno-logical advancement. Moreover, the World Summit on Information Societies (WSIS) presents the largest annual meeting of ICT for the development community in the world. The main aim of this society is the improvement in the quality of living con-ditions by employing techniques and tools provides by ICTs.

1.2   TECHNOLOGIES IN SUSTAINABLE DEVELOPMENT

Nowadays, governments and businesses are leveraging progressed innovations to present modern trade models, advance sustainability, and move forward the common standard of living for today's worldwide citizens. There's an undiscovered opportu-nity to saddle modern advances to quicken advance on the Worldwide Objectives that will both broaden and develop current activity. Various technologies such as cloud computing, Internet of things (IoT), Artificial Intelligence (AI) and its sub domain machine learning, robotics, and fifth generation networks are a few advances that are being leveraged to realize the SDGs [2]. These innovations interface with citizens around the world, screen and track natural affect, and optimize mechanical wasteful aspects. Eventually, they will be changing customarily inefficient homes into feasible and productive places of operation.

The WSIS summit [3] in 2021 expressed that recent technologies are set to have a vital impact in our future. AI, Big Data, IoT, block chain, and many others are proving to hold immense potential in industries such as healthcare, education, and food as well as agriculture and many more. Frontier technological solutions such as startups will also address sustainable development challenges and help to facilitate innovation in a rapidly changing world.

The following sections focus on technologies, namely AI, big data, and IoT fields, meticulously for achieving informed decisions, platforms for online collaboration; social and economic progress; managing various resources such as energy, waste, water, and transport; and building cities in intelligent way.

1.2.1   ARTIFICIAL INTELLIGENCE

Vinuesa et al. [6] discussed that 79% of targets mentioned in the 2030 Agenda can benefit from AI. Mostly AI technology, applied in research and industry toward SD goals, is aimed to develop improved approaches to machine-learning techniques to forecast various diseases, minimum selling price, demand of load in electricity and other events. AI applications are right now focused toward SDG issues that are basically significant to those countries where most AI analysts reside and work. For example, numerous frameworks applying AI advances to agribusiness, e.g., to mechanize collecting or optimize its timing, are found in well-off countries.

AI facilitates machine learning from experience (training), works for prediction of new samples of test data set and carries out different tasks such as playing chess, on-line product recommendations, weather predictions, medical diagnosis systems, traffic monitoring systems. Machine learning, a branch of AI, is a way of analyzing data and making an automated model learn from data, detecting hidden patterns and insights with little human intervention. Broadly, machine learning can be studied in three types of learning as supervised learning, unsupervised learning and reinforcement learning. Moreover, deep learning is an emerging area of machine learning that establishes basic parameters of the data and trains a machine/computer to learn on its own by employing various layers during the training phase.

1.2.2 BIG DATA

The World Federation of Engineering Organizations acknowledged in the United Nations Educational, Scientific and Cultural Organization (UNESCO) Engineering Report released in 2021[7] that big data along with AI technology have promising strength for transformational and innovative engineering. It has great potential to inform and develop cutting-edge solutions to fulfil the Sustainable Development Goals.

Big data is majorly propounded with its characteristics like volume, variety and velocity [8]. The volume of data in the world is expanding at a tremendous rate due to mobile phones, social media, credit cards, and online activities in educational as well as other industries. Moreover, a COVID-19 pandemic has contributed a lot of growth in data all over the world.

The analysis of such a kind of data poses very interesting patterns, connections and recommendations in different applications that cannot be easily visualized. For illustration, ladies and young ladies, who frequently work within the casual division or at home, face challenges to their mobility and are suffering in both private and open decision-making. Most of the data is taken by companies for better decision making. Public-private associations are likely to be broader. The challenge will be guaranteeing they are maintainable over time, which clear systems are in place to clarify parts and desires for SD goals.

Big data, together with machine learning and deep learning, is used to analyze, process, and extract implicit patterns and information from extremely complex data. This data usually originated from various sources such as social media, satellite, videos, and sensors and is mainly described by its volume, velocity, and variety.

1.2.3 INTERNET OF THINGS (IoT)

The sustainable development areas like smart healthcare, smart agriculture, and smart transportation are met through next generation communication networks along with support of artificial intelligence [9].

Currently, interactive, compact, and synthesized solutions are being created for data-based decision making by exploiting the latest technology, IoT. It is one of the innovative advancements ordained to exponentially increase the network of different gadgets. The growth of IoT is affecting the daily life of people from smart watches to use of drones in various sectors [9]. The relentless development of IoT gadgets (right now more than 30 billion associated within the world) gives an inestimable sum of information that, in the event that it is accurately transmitted and handled (through cloud computing and other technologies), will turn into usable data [10]. This data, helpfully put away, envisioned, and dissected utilizing AI and big data analytics, will produce information that can produce positive input within the securing and preparing of the information and shrewd decision making based on profitable information [11]. Concurring with UNESCO, feasible and novel ideas and developed systems require proficient, straightforward, and non-static efforts [12]. The goals are to meet the need of industries and provide commercial innovative improvement and advancement that ensures the security and privacy of information. Further, it also promotes the open accessibility of collected data such that third parties can offer useful solutions, particularly innovative tools and techniques for academia and the research community [13].

IoT is an integrated system having many objects, namely things. It includes embedded software, sensors, devices, and technologies to exchange data with other devices and systems over the internet. This technology is exploring scalable and replicable models of business, investment, and collaboration across various organizations and with public authorities that can maximize their social value potential to meet SD goals.

1.3 APPLICATIONS

Various software development goals can be enabled by developing the application from AI-enabled healthcare systems/devices [12] to autonomous vehicles [11] and smart electrical grids [13] using technologies like AI, big data, and IoT. Some recently evolving areas of these technologies are highlighted in the following sections.

1.3.1 HEALTHCARE

Healthcare is one such prominent area where IOT, big data, and artificial intelligence are playing a significant role. Be it data analysis, telepathy, training, or visualization, these technologies are paving the way for better results, time saving, and convenience to the end users. A few of the developments in this field are depicted here in the following text.

1.3.2 IMPLANTABLE GLUCOSE MONITORING SYSTEMS

In traditional systems for diabetic treatment, the patient's blood sample is to be taken regularly and then analysis is performed. Based upon the outcome of analysis, the patient needs to control his glucose levels. But with the help of IOT devices like implantable glucose monitoring systems [14], patients can get reports of their glucose levels in real time and can control their diet accordingly.

1.3.3 HEART MONITORS

Nowdays, IOT in healthcare has become so popular that some health personnel call it Internet of Health things. Cardiovascular issues are becoming a matter of concern worldwide. Day by day increasing pressure and tensions in human life are making it even worse. Medical personnel are trying several contemporary products to diagnose and prevent cardiovascular issues. Different types of heart monitors systemsare elaborated on in [15].

1.3.4 MEDICAL ALERT SYSTEMS

Many individuals and elderly persons who are living alone can be assisted with the help of automatic medical alert systems [16]. Medical alert systems can be equipped with multiple monitoring capabilities. In case of some abnormality, these systems can automatically send alerts to a concerned caretaker or emergency department.

1.3.5 INGESTIBLE SENSORS

With the advent of technology now different monitoring systems can be easily installed in very minute devices. The size of these devices is smaller than the size of pills used for medication. These ingestible sensors can be easily ingested in a patient's body. Further, these sensors can communicate data to mobile apps of the patients and doctors, which can be further utilized for different purposes.

Many similar applications in the field of healthcare can be found in the literature.

1.3.6 TRAINING AND EDUCATION

Recently the world witnessed COVID-19 crises. Along with other sectors, education was badly affected due this pandemic. To minimize the spread of this pandemic, schools and universities were shut worldwide. During this testing time, several IOT- and AI-based platforms [18–19] have evolved to support the teaching and learning process. Nowadays these platforms have capabilities such as smart attendance marking, auto evaluation, etc. Also, many IOT- and AI-based simulators, virtual reality, and virtual augmentation platforms are being used to provide training to students and learners. These platforms not only reduce the risk as compared to actual systems, but also they are very convenient, economic, easily customizable, and reusable.

1.3.7 MANUFACTURING

Industry 4.0 is the center of attention nowadays. One of the main objectives of it is to imbibe sustainable technologies in different industrial processes. Many IOT- and AI-based approaches [20–22] are gaining popularity in providing sustainable solutions to many real-life issues. Smart cities, smart production, cyber-physical systems, and smart supply chain management are some of the rapidly growing fields.

1.3.8 COMMUNICATION

With the recent developments in communication technologies like 5G, integration of IOT and AI components has become possible. Communication between different IOT and AI devices have become new research domains. New protocols and standards are evolving to facilitate seamless integration of these technologies [23–24]. Edge AI and Edge computing in IOT are being investigated for better communication services.

1.3.9 MARKETING

Smart customer tracking systems automatic billing, chatting bots, transaction management, automatic reservations, etc. are some of the areas of marketing where these technologies are helping a lot [25].

1.3.10 TRANSPORTATION

IoT based solutions to the transportation system are becoming very popular nowadays. Accident-avoidance systems [26] and intelligent city transportation [27–28] are some of the future transport needs where IOT-based technologies will be playing a very significant role. New protocols and models like DSRC technology [29], intelligent traffic predictive model [30], and model for vehicle life integration [31] are some instances in literature that got the attention of researchers recently.

IOT and big data along with AI are delivering numerous solutions to our day-to-day problems. Besides the some of the key applications discussed earlier, there are a variety of fields like smart security systems for houses and organizations [32], smart agriculture [33], smart meters [34], smart grids [35], and smart homes [36].

1.4 SCOPE AND ORGANIZATION OF THE BOOK

This book aims to gather potential pieces of research of eminent academicians, researchers, and technicians working on various applications of artificial intelligence, big data and IoT to meet sustainable development goals. Accordingly, keeping in mind the gigantic scope of this book, chapters of the book will facilitate learners and researchers in having a broader and more profound understanding of several perspectives about the dynamics, challenges, and opportunities for the modern decade regarding sustainable development using artificial intelligence, big data, and IoT. The challenges in SD using major technological advancement such as artificial intelligence, big data, and IoT is expected to be a powerful ICT contribution to help us plan a common future for people. It will facilitate fulfillment of sustainable development goals by exploiting cutting edge recent technologies.

The remaining chapters of the book will present the various applications where machine learning/deep learning, which is a sub-field of artificial intelligence, is being used toward sustainable development across the world. Moreover, technologies such as big data and IoT are also included in different domains such as education, food recognition, and many more. Consequently, this book will enable researchers, academicians, and decision makers to analyze current state-of-the-art reviews, methods, and challenges/future aspects in the upcoming decades for approaching sustainable development goals across the globe.

REFERENCES

[1] International Union for Conservation of Nature, and World Wildlife Fund. *World conservation strategy: Living resource conservation for sustainable development*. Gland, Switzerland: IUCN, 1980.

[2] Tjoa, A. Min, and Simon Tjoa. "The role of ICT to achieve the UN sustainable development goals (SDG)." In *IFIP world information technology forum*, pp. 3–13. Cham: Springer, 2016.

[3] Vinuesa, Ricardo, Hossein Azizpour, IolandaLeite, Madeline Balaam, Virginia Dignum, Sami Domisch, Anna Felländer, Simone Daniela Langhans, Max Tegmark, and Francesco Fuso Nerini. "The role of artificial intelligence in achieving the sustainable development goals." *Nature Communications* 11, no. 1 (2020): 1–10.

[4] Sachs, Jeffrey D. "Achieving the sustainable development goals." *Journal of International Business Ethics* 8, no. 2 (2015): 53–62.

[5] Tyagi, Ruchi, Suresh Vishwakarma, Zubkov Sergey Alexandrovich, and Shariq Mohammmed. "ICT skills for sustainable development goal 4." *Quality Education* (2020): 435–442.

[6] Vinuesa, Ricardo, Hossein Azizpour, Iolanda Leite, Madeline Balaam, Virginia Dignum, Sami Domisch, Anna Felländer, Simone Daniela Langhans, Max Tegmark, and Francesco Fuso Nerini. "The role of artificial intelligence in achieving the sustainable development goals." *Nature Communications* 11, no. 1 (2020): 1–10.

[7] Guo, Huadong, Heide Hackmann, and Ke Gong. "Big data in support of the sustainable development goals: A celebration of the establishment of the international research center of big data for sustainable development goals (CBAS)." *Big Earth Data* 5, no. 3 (2021): 259–262.

[8] Purwar, Archana, and Indu Chawla. "Applications of big data in large and small systems: Soft computing." In *Applications of big data in large-and small-scale systems*, pp. 20–37. IGI Global, 2021.

[9] Maheswar, R., and G. R. Kanagachidambaresan. "Sustainable development through internet of things." *Wireless Networks* 26, no. 4 (2020): 2305–2306.

[10] Martínez, Ignacio, BelénZalba, Raquel Trillo-Lado, Teresa Blanco, David Cambra, and Roberto Casas. "Internet of things (IoT) as sustainable development goals (SDG) enabling technology towards smart readiness indicators (SRI) for university buildings." *Sustainability* 13, no. 14 (2021): 7647.

[11] Bonnefon, J.F., A. Shariff, and I. Rahwan. "The social dilemma of autonomous vehicles." *Science* 352 (2016): 1573–1576.

[12] De Fauw, J. et al. "Clinically applicable deep learning for diagnosis and referral in retinal disease." *Nature Medicine* 24 (2018): 1342–1350.

[13] International Energy Agency. *Digitalization & energy*. Paris: IEA, 2017. https://www.iea.org/reports/digitalisation-and-energy

[14] Joseph, J. I. "Review of the long-term implantable senseonics continuous glucose monitoring system and other continuous glucose monitoring systems." *Journal of Diabetes Science and Technology* 15, no. 1 (2021): 167–173.

[15] Marcus, A.G. Santos, Roberto Munoz, Rodrigo Olivares, Pedro P. Rebouças Filho, Javier Del Ser, and Victor Hugo C. de Albuquerque."Online heart monitoring systems on the internet of health things environments: A survey, a reference model and an outlook." *Information Fusion* 53 (2020): 22–239, ISSN 1566–2535.

[16] Nicholas, Wei Rong Ong, Andrew Fu Wah Ho, Bibhas Chakraborty, Stephanie Fook-Chong, Pasupathi Yogeswary, Sherman Lian, Xiaohui Xin, Juliana Poh, Kelvin Koon YeowChiew, and Marcus Eng Hock Ong. "Utility of a medical alert protection system compared to telephone follow-up only for home-alone elderly presenting to the ED—a randomized controlled trial." *The American Journal of Emergency Medicine* 36, no. 4 (2018): 594–601, ISSN 0735–6757.

[17] Flore, J. "Ingestible sensors, data, and pharmaceuticals: Subjectivity in the era of digital mental health." *New Media & Society* 23, no. 7 (2021): 2034–2051.

[18] Sultana, N., and M. Tamanna. "Evaluating the potentials and challenges of IoT in education and other sectors during pandemic: A case of Bangladesh." *Technology in Society* (2022): 101857.

[19] Pandey, D., N. Singh, V. Singh, and M. W. Khan. "Paradigms of smart education with IoT approach." In *Internet of things and its applications*, pp. 223–233. Cham: Springer, 2022.

[20] Oluyisola, O. E., S. Bhalla, F. Sgarbossa, and J. O. Strandhagen. "Designing and developing smart production planning and control systems in the industry 4.0 era: A methodology and case study." *Journal of Intelligent Manufacturing* 33, no. 1 (2022): 311–332.

[21] Dinesh, M., C. Arvind, S. S. Sreeja Mole, S. Kumar, P. Chandra Sekar, K. Somasundaram, and V. P. Sundramurthy. "An energy efficient architecture for furnace monitor and control in foundry based on industry 4.0 using IoT." *Scientific Programming* (2022).

[22] Sharma, A., V. Burman, and S. Aggarwal. "Role of IoT in industry 4.0." In *Advances in energy technology*, pp. 517–528. Singapore: Springer, 2022.

[23] Singh, A., S. C. Satapathy, A. Roy, and A. Gutub. "AI-Based mobile edge computing for IoT: Applications, challenges, and future scope." *Arabian Journal for Science and Engineering* (2022): 1–31.

[24] Singh, A., S. C. Satapathy, A. Roy, and A. Gutub. "AI-Based mobile edge computing for IoT: Applications, challenges, and future scope." *Arabian Journal for Science and Engineering* (2022): 1–31.

[25] Dwesar, R., and R. Kashyap. "IOT in marketing: Current applications and future opportunities." In *Internet of things and its applications*, pp. 539–553. Cham: Springer, 2022.

[26] Mohapatra, H., A. K. Rath, and N. Panda. "IoT infrastructure for the accident avoidance: An approach of smart transportation." *International Journal of Information Technology* (2022): 1–8.

[27] Beloualid, S., S. El Aidi, A. El Allali, A. Bajit, and A. Tamtaoui. "Applying advanced IoT network topologies to enhance intelligent city transportation cost based on a constrained and secured applicative IoT CoAP protocol." *Advances in Information, Communication and Cybersecurity: Proceedings of ICI2C'21* 357 (2022): 195.

[28] Spaho, E., and A. Koroveshi. "A low-cost solution for smart-city based on public bus transportation system using opportunistic IoT." In *International Conference on Emerging Internetworking, Data & Web Technologies*, pp. 175–182. Cham: Springer, 2022, February.

[29] Khan, A. R., M. F. Jamlos, N. Osman, M. I. Ishak, F. Dzaharudin, Y. K. Yeow, and K. A. Khairi. "DSRC technology in vehicle-to-vehicle (V2V) and vehicle-to-infrastructure (V2I) IoT system for intelligent transportation system (ITS): A review." *Recent Trends in Mechatronics Towards Industry 4.0* (2022): 97–106.

[30] Sathiyaraj, R., A. Bharathi, and B. Balusamy. *Advanced intelligent predictive models for urban transportation*. CRC Press, 2022.

[31] Iqbal, S., N. A. Zafar, T. Ali, and E. H. Alkhammash. "Efficient IoT-based formal model for vehicle-life interaction in VANETs using VDM-SL." *Energies* 15, no. 3 (2022): 1013.

[32] Khalid, M., N. Saunshi, P. Mehta, and M. N. Thippeswamy. "Smart college camera security system using IOT." In *Emerging research in computing, information, communication and applications*, pp. 295–309. Singapore: Springer, 2022.

[33] Mohapatra, B. N., R. V. Jadhav, and K. S. Kharat. "A prototype of smart agriculture system using internet of thing based on blynk application platform." *Journal of Electronics, Electromedical Engineering, and Medical Informatics* 4, no. 1 (2022): 24–28.

[34] Hasan, M. Y., and D. J. Kadhim. "Efficient energy management for a proposed integrated internet of things-electric smart meter (2IOT-ESM) system." *Journal of Engineering* 28, no. 1 (2022): 108–121.

[35] Mall, P., R. Amin, A. K. Das, M. T. Leung, and K. K. R. Choo. "PUF-based authentication and key agreement protocols for IoT, WSNs and smart grids: A comprehensive survey. *IEEE Internet of Things Journal* (2022).

[36] Qureshi, K. N., A. Alhudhaif, A. Hussain, S. Iqbal, and G. Jeon. "Trust aware energy management system for smart homes appliances." *Computers & Electrical Engineering* 97 (2022): 107641.

2 Artificial Intelligence for Sustainable Pedagogical Development

S. Gomathi alias Rohini and S. Mohanavel

CONTENTS

2.1 INTRODUCTION

Artificial Intelligence (AI) has been advancing through the use of Machine Learning (ML). Businesses, governments and start-ups pour billions of dollars into implementing AI in all the fields. AI is bringing evolutional changes in day-to-day life. The rapid development of AI and ML has affected many industries.

AI makes the learning environment more interactive. It adapts to individual student needs. Due to digital learning and personalized learning, owing to the COVID-19 pandemic, AI started to explore the education industry. Whether the policy makers, parents, institutions, teachers and students welcome it or not, AI is deployed in the education industry. AI can provide many positive changes in the learning environment and also helps the teachers and students to travel along with the speed of a digital era [1].

Educational technology companies leverage the benefits of AI to provide quality education online. AI connects educators and beneficiaries across the globe 24/7. It

DOI: 10.1201/9781003245469-2

11

cannot replace teachers, but some processes can be replaced by its tools and techniques. And teachers cannot block the involvement of AI in education [2].

India is facing difficulty in finding apt resources for the education sector. Hence it has to move toward AI solutions. India has shortage of teachers at schools, colleges and universities and requires qualified people to fill the vacancies. Experts say that the Indian education system needs a revolutionary technological intervention. AI might be a solution for the same [3]. AI holds incredible potential for improving education systems. With this tech-savvy generation of students, the education industry has transformed in the digitally driven world.

'Quality Education' is one of the 17 Sustainable Development Goals (SDG) of the United Nations Educational, Scientific and Cultural Organization (UNESCO) [4–5]. AI carries enormous potential for achievement of the SDG. The Compound Annual Growth Rate (CAGR) of the AI market in education is expected to be 48% in 2023 as per Technavio's market research [6].

AI market in education industry is predicted as a US$6B market in 2024. As per EDUCAUSE's survey, 54% of institutions use AI-based remote proctoring services. The overall market for online education is projected to reach US$350B by 2025. Price Waterhouse Coopers estimated that AI technologies would increase global GDP by US$15.7 trillion by 2030 [6].

2.2   RESEARCH METHODOLOGY

This chapter is descriptive in nature. This study describes the applications of AI for sustainable pedagogical development at present. It makes one understand and differentiate the characteristics of organizations that implement AI in education in the current scenario. The study allows thinking through the practices followed, offers ideas for probing and helps decision makers.

2.3   APPLICATIONS OF AI IN EDUCATION

AI in education can be used for developing smart content, creating personalized learning experiences, expanding the range of education, facilitating education management and intelligent tutoring and learning. AI applications in education cover training, content mapping, pedagogy, communications, assessment, administration and resource management [7]. It includes language conversion, visual aid, making decisions, pattern recognition etc.

AI systems make students learn from anywhere in the world 24/7. ML algorithms assist higher education institutions in marketing and estimating class sizes also. Robots or chat-boxes are used as colleagues or independent instructors. AI identifies when children are confused or bored to help them become engaged [8].

AI provides a wide view for students and empowers teaching with advanced tools and technologies. It also simplifies the administrative work among teachers i.e. grading assignments; tutorials, homework and test papers; attendance entry and calculation; communication with officials, other teachers, non-teaching and administrative staff and parents; documentation and correspondence; classroom management and related paperwork. Interactive Voice Response System (IVRS) is another application.

AI provides smart contents to students and improved and personalized learning experiences, not surrounded by boundaries, i.e., globally. Global learning includes individual discovery and reflection, team experience, learning circles, conferences and seminars and mentoring and virtual spaces like forums, blogs and resource sharing.

Owing to the lockdown, education has changed dramatically, where teaching is done remotely on digital platforms and examinations and assessments are conducted online. AI solutions provide online examination environments through retinal tracking, environment stimulus tracking and IP tracking for strict invigilation. Online enrollments and admission processes were also beneficiaries of AI's effective solutions during this pandemic.

Embedded systems, data mining, deep learning, image processing, computer vision, Bayesian K interface, face recognition, augmented reality and virtual reality are major AI technologies used in educational applications.

Many companies provide AI-As-A-Service (AAAS) platforms. AI is applied in school education, corporate training, language learning, higher education, quality training and reading. Tutorial applications are designed with the aim of supporting both teachers and students. To get the full benefit of online learning, video capabilities, collaboration tools and engagement methods are to be provided for inclusion, personalization and intelligence.

AI applications in education have three dimensions—system facing, student facing and teacher facing—and four categories—education management and delivery, learning and assessment, empowering teachers and enhancing teaching and lifelong learning [9].

The following are some of the major applications of AI in education [10–11].

- Personalized Learning: With the hyper-personalizationconcept, enabled through ML, AI is incorporated to design a customized learning profile for each individual student and tailor-make their training materials, considering the students' most comfortable learning mode, capability and experience individually. Teachers can split a lesson into small paragraphs or flashcards to help students to figure them out.
- Voice Assistants: Students can converse with educational materials without teachers using voice assistants. They can be employed anywhere for facilitating interaction with educational material or access to any extra learning assistance, e.g., Amazon's Alexa.
- Aiding in Administrative Tasks: AI systems deal with grading students and facilitating personalized responses, routine paperwork, logistics and personnel issues.
- AI Tutors: AI tutors like iTalk2Learn, Thirdspace Learning, DuolingoChatbot, Thinkster Math and EdTech Foundry provide accurate answers to questions and help students learn and help teachers plan lessons and monitor the progress of the students.
- Differentiated and Individualized Learning: AI-based intelligent instruction design and digital platforms offer learning, testing and feedback to students. They analyze the challenges they face, detect the knowledge gaps

and redirect them to suitable topics then and there. Companies like Content Technologies and Carnegie Learning develop these kinds of platforms. The Bill and Melinda Gates Foundation (BMGF) has invested more than US$120M in personalized learning.

- Smart Content: The learning experience of students at different levels can be enhanced by customized learning modes and digital content. The content becomes easy to grasp by separating and summarizing the main points. Through audio and video contents, students can easily gain access to all important materials, understand well and achieve their academic targets.
- Emotional Monitoring: Emotional monitoring is done through machine reading of facial expressions, body language and movements of students to determine important information like involvement during lessons and distractions.
- Attendance Marking: Attendance marking is achieved through fingerprint scanning.
- Automation of Grading: Teachers can automate grading for multiple-choice questions and fill-in-the-blank tests (e.g. BakPax).

Dialogue-based tutoring systems, exploratory learning environments, AI-driven discussion forum monitoring, AI-driven lifelong learning and AI-enabled records of lifelong learning achievements are the thrust areas on the pipeline [12].

2.4   AI TOOLS IN PEDAGOGICAL DEVELOPMENT

AI tools and devices aid in pedagogical development and making global classrooms accessible to all irrespective of their language or disabilities. Some of the top companies that create AI tools for pedagogical development are Brainly, Nuance, Querium and Century Tech.

Features of a few AI tools are given in the following list [13].

- Brainly's Math Solver assists users with finding solutions for the most complex mathematical problems and addresses the issue of understanding mathematical problems faced by some children.
- Nauance's Speech Recognition Software is useful to students who struggle with writing, spelling and word recognition. It dictates lectures and creates documents and email for teachers.
- Knewton'sAdaptive Learning Technology identifies gaps in a student's knowledge, provides relevant coursework, places them back on track and helps teachers teach at different educational levels.
- Cognii'sVirtual Learning Assistant uses conversational technology, guides improving students' critical thinking skills and provides real time feedback and one-to-one tutoring customized to each student's needs.
- Querium's Platform delivers customizable tutoring lessons to students and step-by-step tutoring assistance, gives insights into students' learning habits and designates areas in which the student could improve.

- Century Tech's Platform uses cognitive neuroscience and data analytics, tracks student progress, identifies knowledge gaps and offers personal study recommendations and feedback, creates personalized learning plans, gives access to resources and reduces time spent on planning, grading and managing homework for teachers.
- KidSense's Speech-to-Text Tool transfers speech to text to practice vocabulary and take notes and tests.
- Carnegie Learning's MATHia helps develop deeper conceptual understanding of math, learns students' habits and personalizes the learning experience.
- Kidaptive's Adaptive Learning Platform introduces and challenges students based on a student's strengths and weaknesses.
- Blippar's Webar SDK uses computer vision, intelligence technology and augmented reality to enhance students' learning with interactive visual materials.
- Thinkster Math uses human interaction with AI, provides custom programs, tracks work and helps students understand how they are correct or wrong.
- Volley's Knowledge Engine synthesizes course and test results and briefs to find knowledge gaps.
- Quizlet's Learn uses ML, provides adaptive plans, helps determine what to study and shows students the most relevant study materials.
- Altitude Learning focuses on each student's needs for their learning and progress and allows teachers to assign work to students individually or as a group.
- Gradescope enables instructors to grade exams, assignments, homework and projects and helps save their time for pedagogical development.
- Hugh's Library Assistant helps users quickly find any book in the library and takes them to the location.
- Ivy's Chatbotprovides answers about application forms, program details, tuition costs, deadlines, enrollment procedures, scholarships, grants, loans and other specific procedures.
- Knewton's Alta helps put achievement within reach for students through a personalized learning experience.
- Knowji's Vocabulary combines scientifically proven methodologies to help users learn fast and remember the words.
- EssayOnTimes' Plagiarism Checker helps students check their work for originality before submission.
- UKOU's OUAnalyse predicts student outcomes and identifies students at risk of failing.
- Swift E learning Services's SWIFT provides an insight into when and why the learner might be struggling or achieving [14].

AI tools have capabilities in analyzing multiple sources of data and comparing them with known patterns. This identifies the root causes of problems and drives toward more consistent outcomes across different classes, regardless of the experience of the teachers. AI provides analysis tracking and specific individual patterns for every student, which enhances his performance. In India, Andhra Pradesh Government uses AI to find patterns of school dropouts to solve their issues.

2.5   IMPACT OF COVID-19

Usage of AI tools has increased after the COVID-19 school closures. Having declared COVID-19 as a global pandemic, school systems are moving to remote or distance learning models in an effort to provide learning continuity [15–16]. There are many successful transitions as found in the following list.

- BYJU'S, a India based educational technology and online tutoring firm founded in 2011, has seen a 200% increase in the number of new students using its products. The effectiveness of online learning varies with respect to age. BYJU's observed that integration of games has demonstrated higher engagement and increased motivation toward learning especially among younger students.
- The largest online movement in the history of education took place with 7.3 lakhs or 81% of school students attending classes via Tencent K-12 Online School in China.
- Byte Dance of Lark, a Singapore-based company, offers teachers and students unlimited video conferencing time, auto-translation capabilities, editing of project work and smart calendar scheduling. Lark enables teachers to reach the students more efficiently and effectively through chat groups, video meetings, voting and document sharing. Students find it easier to communicate on Lark.
- Alibaba deployed more than one lakh cloud servers to expand its capacity largely for remote work for its educational tool DingTank. Zhejiang University had more than 5,000 courses online on DingTalk.
- The Los Angeles Unified School District partnered with PBS SoCal/KCET to offer local educational broadcasts.
- Bitesize Daily of BBC provides curriculum-based virtual learning for kids across the United Kingdom.
- Imperial College London offers a course—The Science of CORONA Virus on Coursera.

2.6   BENEFITS OF AI IN EDUCATION

AI and ML empower education to be individually driven. AI in education makes teaching, learning and related administration processes easier. It contributes in a number of ways to higher education. AI can help students and teachers get more out of the educational experience by assuming a number of roles. A new hybrid model of education will emerge with significant benefits. AI technology helps improve educational equity and quality in the developing world. Data analytics on education management improves the capacity to manage educational systems.

The following are the benefits of AI in students', teachers' and administrations' perspectives but are not limited to these [17–18].

2.6.1   BENEFITS TO STUDENTS

AI in education provides feedback on assignments to the students, makes the learning process comfortable, replies to the needs of students, improves their learning experiences and attention level, customizes/personalizes learning materials according to their needs and capabilities, lays greater emphasis on subjects, repeats the things that students haven't mastered, helps them to learn at their own speed and style, monitors students' attendance and feelings and appraises their progress.

AI suggests personalized learning models to the students, identifies students' level and interest, provides relevant coursework and value addition, places them on track, gives immersive and diverse experiences in doing research, offers continuous access to learning, reduces academic/social pressure of students significantly and the downtime due to fatigue/illness and makes interaction with teachers more comfortable and convenient.

It makes students who are not bold enough to ask questions in the class feel comfortable asking questions without the crowd; predicts students' moods, progress and learning to alter the module and assignments; improves learning experiences by providing early learning, adaptive learning, online learning and language learning; provides various personalized content based on students; makes learning something newer than with the teachers and improves learning outcomes.

Students recall 25–60% more lessons in online mode compared to 8–10% in a classroom, and e-learning requires 40–60% less time than learning in classroom.

2.6.2 BENEFITS TO TEACHERS

For teachers, implementing AI in education sets the syllabus and academic plan, identifies curriculum gaps, enhances human-centered approaches to pedagogy, maintains students' academic records, empowers and creates new dynamics with teachers, makes interactions with students more comfortable and convenient, identifies root causes of problems in pedagogy and drives more consistent outcomes.

It provides feedback to teachers and suggests improvements for an effective teaching-learning process, reduces teachers' burden so that they can concentrate on pedagogical development and improving; facilitates a kind of learning environment where students don't feel compared with their classmates; enables them to develop, manage and update content 24/7 from anywhere and gives recommendations on areas that require repeat or further explanation.

It also enables teachers to shape personalized learning styles for the students, analyses the students' learning speed and needs, helps reduce the administrative work, enables online proctoring during assessments, improves teachers' effectiveness and improves instructional quality and concerns about their approach to pedagogy.

Teachers may supplement AI lessons, assist students and provide human-computer interaction and in-person experiences. They can pay additional attention to places where students fall behind, monitor and analyze students' progress in real-time (i.e. they need not wait until they compile annual report sheets) and have good concentration on the subjects and the way of teaching.

2.6.3 BENEFITS TO ADMINISTRATION

Deploying AI in educational administration maintains students' records, provides new avenues for recruitment, strengthens learning management systems, aids budgeting, facilitates the management, improves efficiency of the institutions, arranges personal interaction with parents, and automates syllabus allocation, staff scheduling, substitute management, learners' profile management, grade management and school inspection and grading of multiple-choice and fill-in-the-blank tests (examinations done by thousands of students can be graded in days).

It supports online proctored assessments, evaluations, test preparations and analysis of student performances; reduces human bias, manpower cost, printing expenses, total operational cost and other overheads and breaks the barriers.

Institutions can personalize course outcomes based on students' strengths and weaknesses to enhance learning. Based on the students' learning speed and time, the courses can be tailor-made.

In general, commissioning AI in education demonstrates learning, critical thinking, and round the clock availability and problem-solving; enables everyone to learn and prosper and saves resources through use of digital assistants. It is faster, more economic and scalable in decision making.

2.7 CHALLENGES IN IMPLEMENTING AI IN EDUCATION

Though AI makes tremendous development in education sector, the realization of its potential is challenging mostly across the globe. The following are some of the challenges in implementing AI in education [15, 18, 3, 19].

AI in education needs policy support with equity and inclusion. AI requires infrastructure and an ecosystem of thriving innovators. Education leaders should be aided financially and ethically to focus on shaping learners who have the skills to thrive in the AI society. The technologies covered for capturing data might be costly for some countries. In data collection, storage and processing, data ethics like ownership, accountability, transparency, bias, privacy, security and quality must be maintained.

The accessibility of resources may be a determinant of the convenience of use. The cost of implementation, maintenance and training could be a financial burden for some students, teachers and institutions. There is resistance to change—most people would hold on to previously proven learning practices rather than adapt to AI solutions. AI software is highly vulnerable to cyber-attacks [20].

AI cannot develop a student's mind as a teacher can. Educators can offer multiple problem-solving methods, but not AI. AI cannot correct the errors. AI limits human interactions and connection with people. AI reduces learning of social skills. AI makes students highly dependable on computers, which may result in mental and physical health issues like headache, back pain, poor vision, stress, inferiority complex, anxiety etc.

AI increases the rate of cheating on assessments, i.e. poor students get higher marks than brilliant students. Special personalized education creates mental pressure on students. The unplanned and rapid move to online learning with no training, insufficient bandwidth and little preparation will result in poor user experience that is not conducive to sustained growth [21].

Broader ethical questions, digital divide, right to privacy, pedagogical choices, social and economic status, ethnic and cultural background, liberty and unhindered development, educational gap between economically rich and poor students, lack of electricity, ICT and Internet in underdeveloped countries, lack of language and culturally appropriate content and legal liability are other challenges.

2.8 SUGGESTIONS

UNESCO recommends establishing partnership between industry and academia to share material and financial resources, establishing partnership between universities and research institutes to foster collaborative research and creating international alliances to build the needed ICT infrastructure even in poor sectors of the world. Teachers must learn to use AI in pedagogy. AI developers must learn how teachers work and create solutions that are sustainable in real-life environmental growth [8, 21].

Teachers shall undergo training on research and data analytical skills, management skills and critical perspective on how AI affects human lives. AI technologies in pedagogical development should be based on immediate and long-term needs and grounded in benefit-risk analyses. Evidence for the teacher's efficacy and potential impact on their roles has to be measured. Pedagogy is required to develop human skills to help students learn how to live effectively in the AI-influenced world.

As institutions rush to move learning online, it is important to make sure good cyber security practices are in place. Protecting data privacy preferences and personal identifiable information is a must [20].

Weighing the costs carefully against the benefits (budgets have to be increased to cover the expenses), addressing certain societal and ethical concerns, encouraging data access for researchers without any block, government funding on AI research, promoting new models of digital education and AI workforce development, designing and developing new curriculum, technical and vocational education and training on AI and its applications and free online AI courses are other suggestions.

2.9 CONCLUSION AND FUTURE WORK

In recent years AI and ML have silently entered every aspect of life, transforming many trends with better experience, now moving into education [22–25]. AI has power to provide a new learning environment. But, since teachers remain at the frontline of education, creative and social-emotional aspects of teaching can never be neglected.

AI in education provides solutions to critical situations, enhances the performance and learning experience of the students and reduces the stress and workload of teachers. The system prepares learners for increased use of AI in all aspects of human activity. It must ensure ethical, inclusive and equitable use of AI by all citizens to live and work harmoniously.

Future enhancements may result in analyzing students reports, collaborative learning and setting up an open environment to learn anything students desire. Striving for quality education can be minimized through AI solutions. Classroom learning and e-learning can be done in parallel. The community will stick to AI tools even after the COVID-19 pandemic.

Smart applications of AI can be extended to co-curricular and other educational lines like drawing, dance, martial arts etc. Plans for using AI in education management, teaching, learning and assessment must be revised frequently to keep them up-to-date.

REFERENCES

[1] Hanania, Pierre-Adrien and Dabdoub, Farah. (2021). *How Can AI Help to Ensure Sustainability in Education?* Capgemini AI for Education.

[2] West, Darrell M. and Allen, John R. (2018). *How Artificial Intelligence Is Transforming the World.* www.brookings.edu.

[3] Dharmadhikari, Swapnil. (2021). *Artificial Intelligence (AI) in Indian Classrooms—A Need of the Hour!* ePravesh.

[4] Francesc, Pedro, Miguel, Subosa, Axel, Rivas and Paula, Valverde. (2019). *Artificial Intelligence in Education: Challenges and Opportunities for Sustainable Development.* UNESCO.

[5] UNESCO. (2021). *AI and Education Guidance for Policy Makers.* https://cit.bnu.edu.cn.

[6] Research and Markets. (2019). *Online Education Market Study 2019.* www.globenewswire.com/news-release.

[7] Goddard, William. (2020). AI in Education. *IT Chronicles.*

[8] Bozkurt, A., Karadeniz, A., Baneres, D., Guerrero-Roldán, A.E. and Rodríguez, M.E. (2021). Artificial Intelligence and Reflections from Educational Landscape: A Review of AI Studies in Half a Century. *Sustainability.* Vol. 13. No. 2.

[9] Ouyang, Fan and Jiao, Pengcheng. (2021). Artificial Intelligence in Education: The Three Paradigms. *Computers and Education: Artificial Intelligence.* Vol. 2.

[10] Rangaiah, Mallika. (2020). *6 Applications of AI in Education Sector.* www.analyticssteps.com.

[11] Larson, Sandra. (2020). *Artificial Intelligence in Classrooms: How Is Taking Over?* https://bigdata-madesimple.com.

[12] Eddington, Richard D. (2018). *Top 5 Artificial Intelligence Tutors in Education.* www.bigdata-madesimple.com.

[13] Schroer, Alyssa. (2018). *12 Companies Using AI in Education to Enhance the Classroom.* https://builtin.com.

[14] Gutierrez, Karla. (2019). *Facts and Stats That Reveal the Power of eLearning.* www.shiftelearning.com.

[15] Artiba. (2021). *Top 5 Challenges of Adopting AI in Education.* www.artiba.org.

[16] Li, Cathy and Lalani, Farah. (2020). *The COVID-19 Pandemic Has Changed Education Forever: This Is How.* World Economic Forum.

[17] Abraham, Priya. (2020). *The Challenges and Opportunities AI in Education Has to Offer.* https://wire19.com.

[18] Swain, Aaron. (2020). *Advantages and Challenges of AI in Education for Teachers and Schools.* www.robotlab.com.

[19] AnkitTewari, Shalini. (2020). *Sustainable Education in India through Artificial Intelligence: Challenges and Opportunities.* Association for Computing Machinery (ACM). WebSci'20 Companion.

[20] Mohanavel, S. and Gomathi alias Rohini, S. (2021). Prevention from Cyber Security Vulnerabilities in COVID-19 Pandemic Situation. *Research Trends in Multidisciplinary Subjects.* Vol. 1. pp. 225–229.

[21] Chakroun, B., Miao, F., Mendes, V., Crompton, H., Portales, P. and Wang, S. (2019). *Artificial Intelligence for Sustainable Development: Synthesis Report.* Mobile Learning Week.

[22] Zinchenko, Viktor and Hlushko, Tetiana. (2021). *Artificial Intelligence and Institutional Transformations of the Education System in the Context of the Sustainable Development Paradigm*. VIII International Scientific and Practical Conference on Current Problems of Social and Labour Relations.

[23] Marr, Bernard. (2021). *How Is AI Used in Education—Real World Examples of Today and a Peek into the Future*. https://bernardmarr.com.

[24] Chiu, Thomas K.F. and Chai, Ching. (2020). Sustainable Curriculum Planning for Artificial Intelligence Education: A Self-Determination Theory Perspective. *Sustainability*. Vol. 12.

[25] Editorial. (2020). *Top 6 AI Tools for Education—Learning Made Simple and Fun*. https://roboticsbiz.com.

3 Deep Learning in Computer Vision
Progress and Threats

Akshay Agarwal and Nalini Ratha

CONTENTS

3.1 INTRODUCTION

Deep learning-based algorithms are classified as automatic feature learning and classification algorithms. In order to effectively learn these features, these algorithms generally require a large amount of data and computational resources. The availability of large-scale databases and advancement in computing resources make it possible; hence, deep learning has shown tremendous success in solving complex tasks. In multiple tasks happening in constrained and semi-constrained environments, deep learning algorithms can surpass human-level performance as well. However, to avoid any misunderstanding, we want to highlight that the problem still is not completely solved and therefore needs further attention. In this chapter, we briefly describe the success of deep learning algorithms for two popular tasks: (i) object recognition and (ii) biometrics recognition. The object recognition tasks can be described as the classification of natural images such as dog, cat, and automobile into their corresponding classes, whereas biometrics recognition is the identification of a person based on physiological and behavioral characteristics such as face, fingerprint, signature, and gait. There are various deep learning algorithms presented in the literature; a few of them are: (i) Auto-Encoder (AE) for latent feature representation and dimensionality

DOI: 10.1201/9781003245469-3

reduction through nonlinear functions, (ii) Convolutional Neural Network (CNN) for features representation learning and classification, (iii) Boltzmann Machines (BM) used for pairwise layer learning and acting as a pre-trained network for future network training, and (iv) Recurrent Neural Network (RNN) used to handle the sequential and variable length input data such as time-series data. Among all these architectures, CNN and RNN are the two most popular deep learning architectures working for sequential data processing such as speech and fixed-size data processing such as images. Other than these, generative networks (GANs) and vision transformers have recently gained significant attention. GAN architectures are used for the generation of high-quality and high-resolution synthetic images, whereas the vision transformer aims to improve the capacity of the CNN in solving computer vision tasks.

3.1.1 DEEP LEARNING IN OBJECT RECOGNITION

In this section, we first described the publicly available and popular object recognition databases. Later, the algorithms that are developed to achieve state-of-the-art object recognition performances are discussed.

3.1.1.1 Databases

With the growing interest in object detection and object recognition so that a machine learning system can interact with those objects in the physical world, multiple databases are released by researchers. A few popular object recognition databases are:

1. **MNIST** [76]: It is a digit recognition database and was released in 1999 and is still popularly used for benchmarking the new algorithm. The images in the dataset are low-resolution images of size 28x28 and are single-channel gray-scale images. The dataset comes into two parts namely training and testing. The training set contains 60,000 images used for learning the parameters of the deep classifiers, whereas the testing set containing separate 10,000 images is used to report the performance.
2. **SVHN** [104]: In contrast to the MNIST, SVHN is an unconstrained digit recognition dataset consisting of color images. The resolution of the images is 32x32.
3. **CIFAR** [72]: It consists of two datasets of 10 and 100 classes. Each database contains 50,000 training images and 10,000 testing images. In contrast to the previous datasets, the images belong to several natural objects such as a car, cat, dog, and automobile. The images are of resolution 32x32 and are three-channel color images.
4. **ImageNet** [39]: It is one of the largest object recognition databases and the reason for such huge success of deep neural networks. It contains millions of annotated images that are helpful for object recognition. Due to the large-scale nature and complexity of the images, several object recognition competitions are also organized utilizing this dataset.
5. **Fashion-MNIST** [143]: It contains gray-scale images of fashion objects such as T-shirts/tops, trousers, and shirts. The resolution and number of images in the dataset are the same as the MNIST dataset.

Apart from these datasets, other popular object datasets can be found using the following repository [1].

3.1.1.2 Algorithms

The history of CNN is deep; however, the popularity of CNNs has been explored recently after 2012 architecture, namely AlexNet. In 1998, LeCunn presented the first ever CNN architecture known as LeNet [75]. LeNet is a conventional and shallow architecture consisting of only two convolutional layers followed by downsampling and feature extraction of fully connected layers. The architecture established the new benchmark in the domain of digit popularity and after that became popular for several other tasks such as object segmentation, detection, and classification. Later, in 2012, the first breakthrough network for object recognition known as the AlexNet [73] was proposed. The model consists of seven convolutional layers and is trained using multiple GPUs. The algorithm was able to beat the previous state-of-the-art handcrafted feature-based algorithm namely SIFT + FV by more than 12%. Since then, based on the potential of deep neural networks, several advanced architectures have been proposed to tackle the object recognition domain. Another popular architecture that came in 2014 is known as VGG [121] and comes in multiple forms containing 11, 16, and 19 layers. In contrast to the AlexNet, the authors of VGG have utilized the lower dimensional filters to train a deeper layer. The architecture was able to reduce the top-five error rate on the ImageNet dataset from 16.4% (Alexnet) to 7.4% (VGG-19). The concept of parallel implementation of various CNN operations such as pooling and convolution was proposed in the GoogleNet [130] architecture and won the large-scale object recognition challenge in 2014. The previous evolution of the architecture gives the feel that the increase in the depth of the network might be a better solution for higher performance. However, training of a deep network generally suffers from the gradient vanishing issue and, hence, the underfitting of the earlier layers of the network. To avoid such a problem, ResNet [62] introduces the concept of residual connection, which is connecting the earlier layers to later layers. The authors show the training of more than 150 layers of deep architecture and reduce the error rate to 3.57% on the ImageNet. Later, several variants of ResNet were proposed such as Inception-ResNet [129] and ResNeXt [145]. The recent success of vision transformers and attention modules gave rise to a different form of CNN architecture. Recently proposed ViT-G [149] and CoAtNet-7 [38] establish the new state-of-the-art results of 90.45% and 90.88% top-one classification accuracy on the ImageNet. ViT-G studied the relationship between various scale factors of data and models to analyze their effect on error rate and computation. CoAtNet ensembles the two domains, namely convolution and attention, through vertical convolution layers, depthwise convolution layers, and self-attention to improve the generalizability of the networks.

Figure 3.1 shows the trends in the top-one accuracy of the recent deep architectures on one of the challenging and benchmark object recognition dataset, namely ImageNet [39].

3.1.1.3 Challenges

While the performance of object recognition has shown tremendous success, the challenges that persist in the real world make the problem interesting and hard to

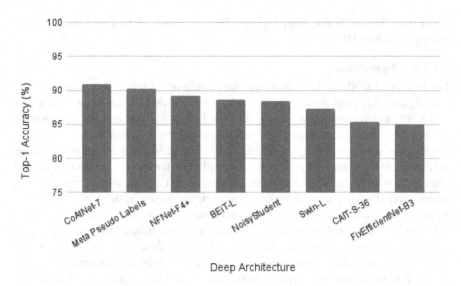

FIGURE 3.1   Top-one classification performance of the recent deep architectures on the ImageNet dataset. CoAtNet-7 [38], Meta Pseudo Labels [109], NFNet-F4+ [30], BEiT- L [26], NoisyStudent [144], Swin-L [84], CAIT-S-36 [133], FixEfficientNet-B3 [134].

solve. A few challenges toward the dataset sides are (i) orientation: it is the view angle between the object and camera that can significantly impact the availability of information for recognition; (ii) resolution, the distance between the camera and the object can either develop the final image in high resolution or low resolution, and identifying the objects in low resolution can significantly impact the performance; (iii) occlusion, sometimes in the natural world it might happen that the actual object might be hidden behind another object and hence make the classification complicated; (iv) illumination variation: captured images of indoor artificial light as compared to outdoor natural light drastically alter the pixel distribution of an image; (v) intra-class variation: while the variation between two objects can be a boon, the variation between the images of the same class hardens the learning of generalized decision hyperplanes. Apart from these challenges from the dataset side, the computational load of recent and state-of-the-work is also a major concern on green computing. For example, the SOTA architecture namely CoAtNet-7 and ViT-G discussed earlier consists of 2,440 and 1,843 million parameters. Therefore, not only is the optimization of these architectures difficult but also training of these networks requires hundreds of computing machines and days. Apart from these issues, an issue of bias also needs attention, which is the generalization capability of the model against the unseen distribution of the object images [27] and handling novel classes for open-set recognition [48].

3.1.1.4   Libraries

Several open-source deep learning libraries are also the possible reason for such a boom in the CNN architectures. A few popular deep learning libraries are: (i) Caffe

[66], (ii) Tensorflow [3], and (iii) PyTorch [108]. Other interesting libraries are explored are the Keras framework [36], MXNet [33], and CNTK [117]. Moreover, the release of machine learning platforms by several big-tech companies such as Colaboratory by Google, Azure by Microsoft, and web services by Amazon also leads to the training and testing of the machine learning algorithms by the resource-constrained institutes and individuals.

3.1.2 DEEP LEARNING FOR BIOMETRICS RECOGNITION

The popularity and success of deep learning are not limited to object recognition only but also exist for biometrics recognition as well. Biometrics modalities can be broadly grouped into two categories: (i) physiological: of which a human consists such as a face, iris, and fingerprint and (ii) behavioral: which is how the human interacts in the environment to prove its identities such as signature and walking pattern known as gait. Face and fingerprint are the most common biometric modalities for the identification of humans. The history of fingerprint recognition is ages old and first came into the picture around 1893. At that time, the fingerprint of a possible suspect in a murder case was found useful in establishing the proof of a crime [60]. Face recognition has significant attention due to its non-intrusive nature, which does not demand the cooperation of a user such as placement on a particular area or at a specific angle. However, the variation in the facial features with age and due to this non-intrusive nature poses several challenges. A few challenges are in the variation of the pose, illumination, and expression and facial accessories such as beards and eyeglasses. Similar to object recognition, the availability of a large-scale database made the training of deep networks an easy task. In this chapter, we briefly discuss the popular face recognition databases. At the end, the deep learning algorithms developed for face and fingerprint recognition are described.

3.1.2.1 Databases

First, the face recognition databases are listed. LFW [64] is one of the most popular datasets in the face recognition community to benchmark performance. The images of 5,749 identities in the LFW dataset are collected from the web. The dataset describes the face verification protocol on 6,000 pairs, and in total the dataset contains 13,233 images. Face verification represents the matching protocol where 1:1 matching is performed, i.e., at the time of testing the user with the face claims the identity against which the image should be matched. CASIA-Webface [148] introduces the semi-automatically collected face images containing 10,575 subjects and 494,414 images collected from the Internet. The dataset aimed to overcome the complexity of the face recognition task by introducing a large-scale dataset. The dataset comes with both face verification and identification protocol. Face identification refers to the matching of a test image with all the images of different identities present in the training dataset. The identity with which the maximum similarity is found is compared to the identity of the testing image. It refers to 1:n matching where a single image is matched with the images of n identities present in the dataset. IJB-B [142] is another large-scale unconstrained dataset again collected from the Internet. The dataset consists of 1,845 subjects with 11,754 images, 55,025 frames, and 7,011 videos. VGG-Face introduced by Parkhi et al. [107] consists of more than 2.5 million images of 2,622 identities. As mentioned earlier, age is one of the prime factors in the extreme change in the facial appearance of a person. Therefore, the matching of faces captured ages apart is a significant challenge. For that, several datasets are proposed to study the impact of age on face recognition. AgeDB [100] and Cross-age LFW (CALFW) [157] are the recently introduced face databases with an explicit focus on age-invariant face recognition. Another challenge in face recognition is the variability of the poses when captured in an unconstrained environment such as surveillance. The Cross-Pose LFW (CPLFW) [156] comes with an explicit focus on matching images with pose differences. Another dataset to tackle the pose variations is Celebrities in Frontal Profile in the Wild (CFP- FP) [118]. Other large-scale databases to enhance the capacity of deep networks are VGGFace2 [31], MegaFace [69], FaceScrub [105], MegaFace (R) [40], trillion-pairs for testing [2], Webface [159], and iQIYI-VID test set [83].

3.1.2.2 Algorithms

One of the early works for deep face recognition started with the introduction of DeepFace architecture [131]. The architecture shows the state-of-the-art performance on the LFW benchmark. The architecture shows the first-time-ever near-human performance on the unconstrained set of the dataset. Later, Sun et al. [126] proposed the DeepID2 architecture for joint training for identification and verification tasks. The architecture aims to decrease the similarity among the feature vectors belonging to a different class and increase the similarity among the features belonging to the same class. Contrary to CNN-based architectures, Majumdar et al. [89] introduce the sparse autoencoder for face verification. The supervisory signal is incorporated in the autoencoder with an assertion that the features that belong to the same class form a similar sparsity representation. The previous description shows that the problem of face recognition is challenging in separating the feature space belonging to different classes and bringing the feature space closer if it belongs to the same classes. For that, the development of an optimization function or loss function is an intuitive way to effectively train the deep CNN architectures. Therefore, in recent times, several deep face recognition architectures propose novel loss functions while utilizing the traditional object recognition deep architectures such as VGG and GoogleNet. Wang et al. [136] proposed the geometrically interpretable loss function termed additive margin Softmax to train a deep face recognition architecture. Other popular loss functions proposed around the same time are CosFace [138], ArcFace [40], Ring Loss [158], AdaCos [152], P2SGrad [153], UniformFace [45], and AdaptiveFace [80]. LarNet [147] proposed the pose robust recognition by estimating the rotation in the feature space corresponding to the image space. The improvement in training these deep architectures shows the path forward for highly accurate face recognition. Inspired by a human understanding of face recognition based on color, texture, and shape, Suri et al. [128] have proposed a novel dictionary learning technique. Other than the complexities mentioned in the database section such as pose, illumination, and expression, matching cross-domain face images is also a challenging task. For example, in an unconstrained and surveillance setting, the testing (also known as probe) images might be affected due to resolution and spectrum. Therefore, handling

TABLE 3.1

Face Recognition Results on Multiple Benchmark Datasets [29]

Algorithm	Training	LFW	AgeDB	CALFW	CPLFW	CFP-FP
GroupFace [70]	Clean MS1M	**99.85**	98.28	**96.20**	93.17	98.63
CircleLoss [127]	Clean MS1M	99.73	—	—	—	96.02
CurricularFace [65]	MS1MVV2	99.80	98.32	**96.20**	93.13	98.37
Dyn-arcFace [67]	Clean MS1M	99.80	97.76	—	—	94.25
MagFace [96]	MS1MVV2	99.83	98.17	96.15	92.87	98.46
Partial-FC-ArcFace [24]	MS1MVV2	99.83	98.20	96.18	93.00	98.45
Partial-FC-ArcFace [24]	MS1MVV2	99.83	98.03	96.20	93.10	98.51
ElasticFace-Arc [29]	MS1MVV2	99.80	**98.35**	96.17	93.27	**98.67**
ElasticFace-Cos [29]	MS1MVV2	99.82	98.27	96.03	93.17	98.61

these variations is also important in addition to proposing the face recognition algorithms working in the color RGB domain. Ghosh et al. [50] proposed the Subclass Heterogeneity Aware Loss (SHEAL) for cross-spectral and cross-resolution face matching. Dhamecha et al. [42] have presented an improvement over the ArcFace model using the discriminant analysis technique. The discriminant analysis aims to reduce the gap between the faces coming from a different distribution such as matching of cross-spectral face images and cross-modality face images.

Table 3.1 shows face recognition performance on multiple benchmark datasets achieved using recent deep learning algorithms.

Challenges: The challenges of face recognition lie in the heterogeneity of acquired face images due to external factors and inherent factors. The external factors that can affect face recognition performance are pose, illumination, and expression, whereas the inherent or natural factors such as age and facial characteristics such as beard significantly degrade the recognition performance of several SOTA algorithms. Apart from these challenges that occur due to variation in the distribution of training-testing images, bias is another critical weakness recently highlighted. The bias of face recognition algorithms can be described as the performance difference of an algorithm across different demographic entities such as ethnicity, gender, and race [101]. The first solution to tackle these bias issues is the development of large-scale databases where these characteristics are balanced. A few recent databases that are proposed with this aim are FairFace [68] and Racial faces in the wild [139]. FairFace contains 108,501 images from 7 different race groups labeled with race, gender, and age groups. Moreover, the databases alone are not sufficient, but the algorithm advancement [91, 102, 103] is also needed to tackle this important issue of doing extra favors for certain demographic entities as compared to others.

3.2   VULNERABILITIES OF DEEP LEARNING

In this section, the threats to deep learning algorithms are described. The most popular and effective attack against deep CNN is the adversarial perturbations. First, we discuss the adversarial perturbation, its generation, and possible defenses so far.

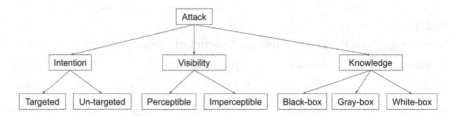

FIGURE 3.2   Taxonomy of the adversarial perturbations.

3.2.1   ADVERSARIAL PERTURBATIONS

In layman's terms, the adversarial perturbation can be defined as the addition of an intelligently crafted noise in the images to fool the deep classifiers.

Figure 3.2 shows the taxonomy of the adversarial perturbations. The perturbation can be divided based on the following three criteria: (i) intention, (ii) knowledge, and (iii) visibility.

The intention-based attack can further be divided into targeted and untargeted attacks. The targeted attack aims to misclassify the image in question into one of the desired target classes, whereas the untargeted attack fools the network into not predicting the original class of a testing image. The network can choose whatever random class in which it wants to classify, but an attacker does not pose any restriction on the selection of a class. The attribute defines the amount of information of a network an attacker utilizes while crafting the adversarial perturbation. In the black-box setting, no knowledge of the system should be used. In the gray-box setting, some information of the network such as decision probabilities and final selected class is utilized to manipulate the noise vector and increase its strength to achieve the attack goal. In the white box, full knowledge of the classifier is utilized such as its parameters and gradient flow to generate the effective noise vector. The white-box attacks are the strongest adversaries against any machine learning classifiers, including deep CNNs. The visibility of an adversarial attack is defined as whether the human examiners can identify whether an image is modified with the noise or not. The visibility of the noise vector in an image is termed perceptible attack, and non-visibility of noise is known as imperceptible modification of an image. While in this chapter we have used an image as input to the classifiers, a similar taxonomy of adversarial attack can be used against other inputs such as text and speech.

3.2.1.1   Attacks

The simplest and most effective attack on a deep classifier was proposed by Goodfellow et al. [55]. The authors have utilized the network gradient information and added back the sign of the gradient in the image itself intending to increase the loss of a classifier. The attack is popularly referred to as a fast gradient sign method (FGSM) attack. The attack applied the gradient in a single step and hence does not yield a strong adversary. Therefore, to increase its strength several advancements have been proposed such as iterative FGSM (IFGSM) [74], momentum iterative method (MIM) [44], and projected gradient descent (PGD) [88]. Trame'r et al. [135]

have utilized an ensemble of noises to fool the classifier effectively as compared to the single-step attacks. In the first step, a random noise pattern is added to the image followed by the computation of FGSM perturbation.

The aforementioned adversarial perturbation generation algorithm generates the noise vector against each input image, whereas a concept of a single noise vector was proposed by Moosavi-Dezfooli et al. [98]. The attack is referred to as a universal noise vector and is associated with the classifier rather than the input image. Therefore, it can misclassify multiple images containing the same noise by the classifier. Several enhancements of universal noise vectors have been proposed in recent times [41, 61, 99]. Another interesting adversarial attack that tries to minimize the number of pixel modifications is known as a one-pixel attack [125]. Agarwal et al. [14] showcase the impact of sign function used in the gradient-based attacks and demonstrate when the noise can be perceptible and imperceptible. In another contribution, the impact of gradient noise optimization further used as a defense is also reported. Anshumaan et al. [25] proposed an attack by decomposing the images into multiple frequency information through wavelet decomposition. The aim of an attack can be seen in that partial information of an image is being perturbed due to multiple entity decomposition obtained from the wavelet transformation. Most of the previously listed attacks come into the category of white-box attacks, which limits their deployment in the real world. Although it is seen that the adversarial perturbations are transferable against unseen classifiers, their success is limited. Agarwal et al. [22] proposed an interesting black-box attack based on the extraction of the inherent noise developed at the time of acquisition of an image. The inherent noise can be attributed to the fact that the environment is noisy or the camera processing steps incorporate different noise in the images. Agarwal et al. [9] proposed an adversarial attack on the tabular data common in the finance world. The authors have studied the impact of noise on several machine learning classifiers usually used for tabular datasets and show their vulnerability against minute perturbation of real or discrete feature variables. Goswami et al. [58] proposed the black-box on face recognition networks based on domain knowledge and image-specific attacks. The domain knowledge helps in creating face-specific attacks such as occlusion of the eye and beard region. Similarly, the image-specific attacks do not utilize any network information and either modify an entire image through some random lines or perturb the individual image bit planes. Another popular form of adversarial attack is the generation of a physical patch to deploy in the real world to fool the networks working directly on the camera-acquired images [63, 140].

3.2.1.2 Defenses

Similar to the adversarial perturbation algorithms, the adversarial defense algorithms can be broadly classified into three groups: (i) detection based: which aims to learn a binary classifier to predict whether an image in question is real or adversarially perturbed, (ii) mitigation algorithms try to remove the image to purify any incorporated adversarial noises, and (iii) robustness: it aims to train the robust classifier, which cannot be easily fooled by the adversarial perturbations.

Figure 3.3 gives the taxonomy of the adversarial defense algorithms. The mitigation algorithms are somewhat a two-step process wherein the first step the image is

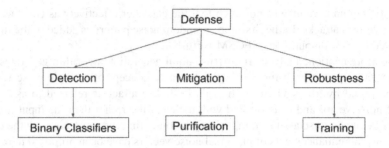

FIGURE 3.3   Taxonomy of the adversarial defense algorithms.

classified as real and adversarial and in the second step an adversarial image is sent for purification. The purification can be applied either on the image, such as image filtering or mapping an adversarial image to a corresponding clean image, or at the network level where the affected network components themselves are purified or dropped. Very limited work has been made toward mitigation, which might be due to these multi-step requirements. Goswami et al. [57] proposed a selective dropout algorithm to mitigate the effect of adversarial attacks on face recognition. The authors dropped the affected CNN filters and used the remaining filters for feature extraction and recognition. Song et al. [124] mapped the adversarial examples to their corresponding clean examples using a generative network.

In contrast to the mitigation defense, adversarial examples detection and training of a robust network gained significant attention and consists of highly rich literature. Further, we briefly describe the two defense algorithms; however, for detailed information, the readers can refer to the survey papers [32, 119]. Agarwal et al. [17] proposed one of the simplest adversarial example detection algorithms. The authors, through learning the distribution of real and adversarial images through principal component analysis (PCA), showcase that the detection of universal perturbation on face images might not be that difficult. Extensive experiments demonstrated the observation of the authors. Feinman et al. [46] proposed the training of a binary detector utilizing two different feature vectors such as kernel density estimates and Bayesian uncertainty estimates from the intermediate layers of a classifier. Goswami et al. [57, 58] proposed the detection of adversarial examples based on the statistical features of intermediate CNN filter maps. Safetynet quantized the ReLU activation outputs [85] and trained the non-linear support vector machine (SVM) classifier for binary classification. Apart from creating traditional classifiers, generative networks are also explored detection of the adversarial examples [49, 78]. The limitation of the existing detection and mitigation-based algorithms is the generalizability against unseen attacks, databases, and attack generation networks. To provide the defense in the wild setting, Agarwal et al. [5, 18] proposed the generalized adversarial examples detection algorithms. In the first algorithm, various image transformations are studied and demonstrated that the noise artifacts are attenuated in the transformed space [18]. The transformed space is then used for image feature extraction and SVM training. The score level ensemble of two transformations is used to detect the adversarial examples. In another multi-directional generalized detection algorithm [5], statistical features from the intermediate CNN maps along with non-linear latent encoding are used to train two SVM classifiers.

The most effective robustness defense is adversarial training or fine-tuning of the network through varying distributed images. In adversarial training, the adversarial images generated from the network itself are used for retraining the network with the hope to increase its robustness against the noise. Goodfellow et al. [55] proposed the training of the network based on their proposed FGSM attack and found that it can act as a better regularization technique. However, as the attack is not strong, the induced robustness was not very high against complex attacks and unseen attacks. As mentioned earlier, the majority of the method of the attack works in the white-box setting, and the gradient information is most critical for both the network update and adversarial examples generation. Therefore, to avoid the generation of adversarial examples, several research works have also proposed the regularization of the gradient flow while training the adversarially trained network [86, 113, 123]. In place of utilizing the adversarial examples of one network, the augmentation from multiple networks might create diversity among them and can act as a better regularizer. Trame'r et al.'s [165] ensemble of adversarial training is based on the augmentation of adversarial examples crafted from the target model and several pre-trained networks [135]. The major limitation of adversarial training methods are the computational complexity in generating the complex attack and then retraining of the target model, lack of generalizability against unseen examples, and loss of privacy of model due to the generation of adversarial examples from the network [95, 150]. Therefore, training of the networks using the images that might be generated outside of the network and in a computationally efficient manner might take time. In one such study, Agarwal et al. [20] proposed retraining of the networks based on cognitive data augmentation. It is known that deep networks require a huge amount of data, which might be a reason for overfitting toward training data distributions. To tackle that, the authors proposed the dropping of image information so that the model can learn to map the class to the input using a minimal amount of available information. The robustness is found against multiple architectures and attacks in a completely black-box setting. In another robustness study, Agarwal et al. [21] have proposed the data mix-up technique based on the corruption of images using Gaussian noise. The noisy images are mixed with the clean images, and a new image is generated and used for robustness enhancement of the CNN classifiers. Chabbra et al. [34] found that the amount of noise added in the image has a significant impact on its success in fooling the network. Based on this visual bound of the noise magnitude, the authors have proposed the generation of data and used that to increase the adversarial robustness. The CNN architectures have multiple hyper-parameters, and getting an ideal value of each of them is hard. Therefore, the correct optimization of which hyper-parameter might be important against adversarial robustness is also an interesting area of study. Agarwal et al. [19] proposed an adversarial robustness study based on the role of network optimization algorithms such as Adam and RMSProp.

The other branch of adversarial defense is based on the certification concept where it is defined that within a certain bound the network will be robust against the adversarial perturbations [37, 77] and reconfiguration of the network layers in such a way that the attack cannot correctly find the gradient flow [51, 52, 53].

3.2.1.3 Libraries

To advance the research in the adversarial learning domain, several libraries have also been released in the public domain. A few of them are: (i) ART by IBM [106], (ii) Smartbox by Goel et al. [54], (iii) foolbox [111], (iv) AdvPytorch [43], and (v) Advbox [56].

3.3 VULNERABILITIES OF FACE RECOGNITION

In the face recognition system, generation of deepfake video, face morphed images/videos, and presentation attacks are considered some of the stealthy threats. These attacks work both in the digital domain and physical domain. First, we discuss the presentation attacks and their countermeasure algorithm followed by the description of the face morphing attacks. In the end, deepfake attack generation and detection works are listed.

3.3.1 PRESENTATION ATTACKS

The presentation attacks on face recognition systems can be broadly divided into two categories: (i) 2D attacks and (ii) 3D attacks. The effective 2D attacks are printed photos, display attacks, and video replay on the electronic screen. While the creation of 2D attacks is easy and computationally efficient, the lack of 3D structure and face alike texture is a serious concern, whereas the 3D attacks are costlier as compared to the 2D attacks due to the development of sophisticated masks that can fit on the real face. Multiple types of face masks are available, and those used for presentation attacks are latex masks and silicone masks. In the literature, several face presentation attack databases are presented utilizing both 2D and 3D attacking instruments.

Table 3.2 shows a few challenging and benchmark face presentation attack databases. The databases vary in terms of the number of subjects, type of attacking

TABLE 3.2

A Few Popular Face Presentation Attack Databases

Database	Modality	Subjects	Videos	Attack
CASIA-FASD [155]]	RGB	50	600	2D
Replay-Attack [35]	RGB	50	1200	2D
MSU-MFSD [141]	RGB	35	440	2D
SMAD [92]	RGB	65	130	2D
MLFP [23]	RGB/IR/Thermal	10	1200	2D/3D
CASIA-SURF [151]	RGB/IR/Depth	1000	21000	2D
CelebA-Spoof [154]	RGB	10,177	625537*	2D/3D
OULU-NPU [28]	RGB	55	5940	2D
HKBU-MARs V1 [81]	RGB	12	180	3D

*Represents the number of images, not videos.

instruments, and modality in which the images/videos are captured. Initial face presentation attack databases such as CASIA-FASD, Replay-attack, and MSU-MFSD are limited in terms of the number of subjects and modality. However, as mentioned, the 2D attacks are less challenging to be performed in the real world; therefore, the first-ever silicone mask attack database from Internet videos was prepared by Manjani et al. [92] and termed SMAD. The face recognition systems are not limited to any modality such as RGB; its use-cases are wide such as in surveillance, and infra-red spectrum is highly utilized. Hence, the protection of these systems is also important. Keeping that in mind, MLFP [23] and CASIA-SURF [151] are prepared in multiple imaging spectra. The CASIA- SURF database is further limited in terms of attack medium, and the MLFP database contains a low number of subjects. The advantage of the MLFP database is not only the multi-modal video but also varying 3D mask instruments including latex and silicone. CelebA-spoof is one of the largest datasets in terms of subjects; however, it is limited in terms of modality. While the database contains nine different types of 2D attacking instruments, another limitation is that a single 3D attacking medium is used. It is clear from the previous discussion that no database so far has all the required characteristics to make it effective for large-scale in-the-wild real-world study, and hence research toward that is also needed.

The previously listed presentation attack database was found effective in not only fooling humans but also in degrading the performance of several face recognition algorithms including commercial and deep face recognition networks [23, 141]. Therefore, to protect the integrity of the face recognition systems, several attack detection algorithms are proposed. The proposed algorithms can be divided into handcrafted machine learning-based algorithms and deep CNN-based classifiers. In one of the first studies toward generalized face presentation attack detection, Agarwal et al. [12] proposed a feature fusion algorithm by decomposing the images into separate color channels followed by wavelet decomposition to attenuate the artifacts. The evaluation has been performed on multiple databases covering multiple attacking instruments and shows the state-of-the-art performance. Siddiqui et al. [120] have proposed the fusion of multiple handcrafted features for the training of an SVM classifier for multiple presentation attack detection. Based on the popularity and success of CNN architectures in object recognition, several research works have recently started exploring it for face presentation attack detection (PAD) [82, 87, 112, 137]. Manjani et al. [92] proposed the deep dictionary learning framework contrary to deep convolutional architecture and show state-of-the-art (SOTA) performance against multiple attacks. A novel feature extraction technique is proposed by utilizing the filters of a CNN architecture. A novel feature extraction algorithm coupled with an SVM classifier is found effective against silicone mask attacks and yields computationally efficient implementation. Inspired by the categorization of the presentation attack instruments, Sanghvi et al. [114] have proposed a multi-branch deep architecture to learn different feature representations for each attack type. Not only the face modality but also other modalities such as iris and fingerprint are subjected to presentation attacks and significant efforts are being made to tackle these attacks [59, 71, 79, 146]. While the development of an effective presentation attack detection algorithm is a necessity, evaluating their robustness against attacks is also critical.

Recently, it was found that face presentation attacks are vulnerable to feature transformation attacks and simple image intensity manipulations [10, 11].

3.3.2 FACE MORPHING AND SWAPPING ATTACKS

Another severe threat to face recognition is referred to as a face morphing or swapping attack. In 2014, Ferrara et al. [47] proposed the first face morphing attack database using various computer graphics commercial software such as GIMP. The best morphing image for the attack is selected based on the face matching score of the recognition algorithm. Later, one of the largest face morph or swap databases was prepared by Agarwal et al. [15, 16]. The authors used several mobile applications and Internet platforms for the generation of different types of face morphing and neural transformation alteration databases. Face morphing is done not only through digital software but also to make them effective for their deployment in the physical world; scanning of these morphed images is also performed [110]. It is shown through multiple face recognition experiments that face morphing can create a new identity that contains the facial features of identities used for its generation. Therefore, it is extremely important to detect these images so that no single person can claim two or more identities. To get the smooth and visually appealing face morphing images, most of the software or applications perform the blending of the swapped regions that hide the alteration. Hence, to properly detect the morphed images, highlighting the artifacts is important for discriminating feature learning. To keep that in mind, in the first-ever large-scale study, Agarwal et al. [15, 16] proposed a novel feature engineering algorithm to highlight the minute discrepancies developed due to the morphing of faces. Raja et al. [110] have used the deep features of multiple pre-trained CNN architectures and used them for both digital and scanned morphed face detection. Other effective and recent face morphing detection algorithms are either utilizing the noise features generated due to morphing [115] or deep neural networks [116]. In an interesting defense, Mehta et al. [94] proposed a novel loss function based on the cross-entropy and focal loss to develop a 'panoptic' face presentation attack detection algorithm. The novel loss function aims to learn a compact representation of the real class and multiple loss feature clustering of the attack classes. The panoptic defense is the first-ever work that tackles both the physical presentation attack and the digital face morphing attack.

3.3.3 DEEPFAKE

The last attack that we want to focus on in this chapter is deepfake. Deepfake can be broken into two pieces: deep + fake, which means the utilization of deep architectures for the generation of 'fake' images. Due to the tremendous success of deep generative architectures, the generation of fake videos and images became extremely easy. The generated videos are of high quality and real face similarity such that it is hard to detect these images by just looking at them.

Table 3.3 shows a few deepfake databases proposed in the literature. Due to the rich literature on deepfake, here we present a brief overview of the existing deepfake detection algorithms. Therefore, we refer the readers to the survey papers [97, 132].

TABLE 3.3
Face Deepfake Detection Databases

Database	Year	Videos		Attack
		Real	Fake	
FF++	2019	1,000	4,000	YouTube
DFDC	2019	1,131	4,1119	Actors
DF-1.0	2020	50,000	10,000	YouTube
Wild Deepfake	2020	3,805	3,509	YouTube
KoDF	2021	62,166	175,776	Actors
OpenForensics	2021	45,473	70325	Actors

Similar to any attack detection algorithms, deepfake detection algorithms range from traditional handcrafted features to deep CNN architectures. While the handcrafted features are computationally efficient, they sometimes lack generalizability due to lower learning capacity. There are deep architectures proven effective that are generally trained on large-scale databases and seen in multiple variations possible in the images. Agarwal et al. [15] proposed the handcrafted feature learning techniques boosted from the learned filters of the CNN networks. The ensemble of raw input images with their corresponding transformed images to learn better supervisory signals are found to be more effective as compared to training a classifier utilizing single RGB information [4]. The challenges that need to be tackled further in deepfake detection are the generalizability of the algorithms against an unseen database, unseen attack, and bias-free attribute against multiple ethnicities [90, 93]. We assert that the recent development in the large-scale database covering a variety of attacks and demographics will improve the deepfake detection performances. Other possible directions where the biometrics systems need improvement are sensor interoperability, mask-based face recognition, and drone surveillance [6, 7, 8, 13]. A brief overview of the threats to the biometrics recognition systems can also be found in the overview paper [122].

3.4 CONCLUSION

Deep learning architectures have shown tremendous success in solving a variety of computer vision tasks including object recognition and face recognition. The possible reasons for such successful performance are the availability of large databases, computing infrastructures, an increase of theoretical and practical knowledge, and open-source libraries. In this chapter, first, a comprehensive survey of deep learning algorithms used for object recognition and face recognition is discussed. Later, various vulnerabilities against the object recognition algorithms and face recognition algorithms are described to showcase that the world is not entirely fair, and these algorithms need protection as well. Not only does the chapter provide a survey of the existing algorithms, but also the open challenges in each entity are discussed to increase the research to further improve each field. We believe the availability of such a chapter can boost the understanding of new and experienced researchers in understanding the current research trends in such a broad field and can motivate them to contribute to and enrich the field further.

REFERENCES

[1] Datasets: 5,207 machine learning datasets. https://paperswithcode.com/datasets.
[2] Face feature test/trillion pairs. http://trillionpairs.deepglint.com/overview.
[3] Martin Abadi, Ashish Agarwal, Paul Barham, Eugene Brevdo, Zhifeng Chen, Craig Citro, Greg S. Corrado, Andy Davis, Jeffrey Dean, Matthieu Devin, et al. Tensorflow: Large-scale machine learning on heterogeneous distributed systems. *arXiv preprint arXiv:1603.04467*, 2016.
[4] Aayushi Agarwal, Akshay Agarwal, Sayan Sinha, Mayank Vatsa, and Richa Singh. Md-CSD network: Multi-domain cross stitched network for deepfake detection. *IEEE International Conference on Automatic Face and Gesture Recognition*, pages 1–8, 2021.
[5] Akshay Agarwal, Gaurav Goswami, Mayank Vatsa, Richa Singh, and Nalini K. Ratha. Damad: Database, attack, and model agnostic adversarial perturbation detector. *IEEE Transactions on Neural Networks and Learning Systems*, pages 1–13, 2021.
[6] Akshay Agarwal, Rohit Keshari, Manya Wadhwa, Mansi Vijh, Chandani Parmar, Richa Singh, and Mayank Vatsa. Iris sensor identification in a multi-camera environment. *Information Fusion*, volume 45, pages 333–345, 2019.
[7] Akshay Agarwal, Nalini Ratha, Mayank Vatsa, and Richa Singh. Impact of super-resolution and human detection in drone surveillance. *IEEE International Workshop on Information Forensics and Security*, pages 1–6, 2021.
[8] Akshay Agarwal, Nalini Ratha, Mayank Vatsa, and Richa Singh. When sketch face recognition meets mask obfuscation: Database and benchmark. *IEEE International Conference on Automatic Face and Gesture Recognition*, pages 1–5, 2021.
[9] Akshay Agarwal and Nalini K Ratha. Black-box adversarial entry in finance through credit card fraud detection. *International Conference on Information and Knowledge Management (CIKM) Workshop on Modelling Uncertainty in the Financial World*, volume 3052, pages 1–12, 2021.
[10] Akshay Agarwal, Akarsha Sehwag, Richa Singh, and Mayank Vatsa. Deceiving face presentation attack detection via image transforms. In *2019 IEEE Fifth International Conference on Multimedia Big Data (BigMM)*, pages 373–382. IEEE, 2019.
[11] Akshay Agarwal, Akarsha Sehwag, Mayank Vatsa, and Richa Singh. Deceiving the protector: Fooling face presentation attack detection algorithms. In *2019 International Conference on Biometrics (ICB)*, pages 1–6. IEEE, 2019.
[12] Akshay Agarwal, Richa Singh, and Mayank Vatsa. Face anti-spoofing using haralick features. In *2016 IEEE 8th International Conference on Biometrics Theory, Applications and Systems (BTAS)*, pages 1–6. IEEE, 2016.
[13] Akshay Agarwal, Richa Singh, and Mayank Vatsa. Fingerprint sensor classification via mélange of handcrafted features. In *2016 23rd International Conference on Pattern Recognition (ICPR)*, pages 3001–3006. IEEE, 2016.
[14] Akshay Agarwal, Richa Singh, and Mayank Vatsa. The role of 'sign' and 'direction' of gradient on the performance of cnn. In *Proceedings of the IEEE/CVF Conference on Computer Vision and Pattern Recognition Workshops*, pages 646–647, 2020.
[15] Akshay Agarwal, Richa Singh, Mayank Vatsa, and Afzel Noore. Magnet: Detecting digital presentation attacks on face recognition. *Frontiers in Artificial Intelligence*, page 136.
[16] Akshay Agarwal, Richa Singh, Mayank Vatsa, and Afzel Noore. Swapped! digital face presentation attack detection via weighted local magnitude pattern. In *2017 IEEE International Joint Conference on Biometrics (IJCB)*, pages 659–665. IEEE, 2017.

[17] Akshay Agarwal, Richa Singh, Mayank Vatsa, and Nalini Ratha. Are image-agnostic universal adversarial perturbations for face recognition difficult to detect? In *IEEE International Conference on Biometrics Theory, Applications and Systems (BTAS)*, pages 1–7, 2018.

[18] Akshay Agarwal, Richa Singh, Mayank Vatsa, and Nalini K Ratha. Image transformation based defense against adversarial perturbation on deep learning models. *IEEE Transactions on Dependable and Secure Computing*, volume 18, no. 5, pages 2106–2121, 2020.

[19] Akshay Agarwal, Mayank Vatsa, and Richa Singh. Role of optimizer on network fine-tuning for adversarial robustness (student abstract). *Proceedings of the AAAI Conference on Artificial Intelligence*, volume 35, pages 15745–15746, 2021.

[20] Akshay Agarwal, Mayank Vatsa, Richa Singh, and Nalini Ratha. Cognitive data augmentation for adversarial defense via pixel masking. *Pattern Recognition Letters*, volume 146, pages 244–251, 2021.

[21] Akshay Agarwal, Mayank Vatsa, Richa Singh, and Nalini Ratha. Intelligent and adaptive mixup technique for adversarial robustness. *2021 IEEE International Conference on Image Processing (ICIP)*, pages 824–828. IEEE, 2021.

[22] Akshay Agarwal, Mayank Vatsa, Richa Singh, and Nalini K Ratha. Noise is inside me! generating adversarial perturbations with noise derived from natural filters. *Proceedings of the IEEE/CVF Conference on Computer Vision and Pattern Recognition Workshops*, pages 774–775, 2020.

[23] Akshay Agarwal, Daksha Yadav, Naman Kohli, Richa Singh, Mayank Vatsa, and Afzel Noore. Face presentation attack with latex masks in multispectral videos. *Proceedings of the IEEE Conference on Computer Vision and Pattern Recognition Workshops*, pages 81–89, 2017.

[24] Xiang An, Xuhan Zhu, Yuan Gao, Yang Xiao, Yongle Zhao, Ziyong Feng, Lan Wu, Bin Qin, Ming Zhang, Debing Zhang, et al. Partial fc: Training 10 million identities on a single machine. *Proceedings of the IEEE/CVF International Conference on Computer Vision*, pages 1445–1449, 2021.

[25] Divyam Anshumaan, Akshay Agarwal, Mayank Vatsa, and Richa Singh. WaveTransform: Crafting adversarial examples via input decomposition. In *European Conference on Computer Vision*, pages 152–168. Springer, 2020.

[26] Hangbo Bao, Li Dong, and Furu Wei. Beit: Bert pre-training of image transformers. *arXiv preprint arXiv:2106.08254*, 2021.

[27] Andrei Barbu, David Mayo, Julian Alverio, William Luo, Christopher Wang, Danny Gutfreund, Joshua Tenenbaum, and Boris Katz. Objectnet: A large-scale bias-controlled dataset for pushing the limits of object recognition models. In *International Conference on Neural Information Processing Systems*, pages 9453–9463. MIT Press, 2019.

[28] Zinelabinde Boulkenafet, Jukka Komulainen, Lei Li, Xiaoyi Feng, and Abdenour Hadid. Oulu-npu: A mobile face presentation attack database with real-world variations. In *2017 12th IEEE International Conference on Automatic Face & Gesture Recognition (FG 2017)*, pages 612–618. IEEE, 2017.

[29] Fadi Boutros, Naser Damer, Florian Kirchbuchner, and Arjan Kuijper. Elasticface: Elastic margin loss for deep face recognition. *Proceedings of the IEEE/CVF Conference on Computer Vision and Pattern Recognition*, pages 1578–1587, 2022.

[30] Andrew Brock, Soham De, Samuel L Smith, and Karen Simonyan. High-performance large-scale image recognition without normalization. In *International Conference on Machine Learning*, pages 1059–1071. PMLR, 2021.

[31] Qiong Cao, Li Shen, Weidi Xie, Omkar M Parkhi, and Andrew Zisserman. Vggface2: A dataset for recognising faces across pose and age. In *2018 13th IEEE international conference on automatic face & gesture recognition (FG 2018)*, pages 67–74. IEEE, 2018.

[32] Kai Chen, Haoqi Zhu, Leiming Yan, and Jinwei Wang. A survey on adversarial examples in deep learning. *Journal on Big Data*, volume 2, no. 2, page 71, 2020.

[33] Tianqi Chen, Mu Li, Yutian Li, Min Lin, Naiyan Wang, Minjie Wang, Tianjun Xiao, Bing Xu, Chiyuan Zhang, and Zheng Zhang. Mxnet: A flexible and efficient machine learning library for heterogeneous distributed systems. *arXiv preprint arXiv:1512.01274*, 2015.

[34] Saheb Chhabra, Akshay Agarwal, Richa Singh, and Mayank Vatsa. Attack agnostic adversarial defense via visual imperceptible bound. In *2020 25th International Conference on Pattern Recognition (ICPR)*, pages 5302–5309. IEEE, 2021.

[35] Ivana Chingovska, André Anjos, and Sébastien Marcel. On the effectiveness of local binary patterns in face anti-spoofing. In *2012 BIOSIG-Proceedings of the International Conference of Biometrics Special Interest Group (BIOSIG)*, pages 1–7. IEEE, 2012.

[36] François Chollet et al. Keras: The python deep learning library. *Astrophysics Source Code Library*, pages ascl–1806, 2018.

[37] Jeremy Cohen, Elan Rosenfeld, and Zico Kolter. Certified adversarial robustness via randomized smoothing. In *International Conference on Machine Learning*, pages 1310–1320. PMLR, 2019.

[38] Zihang Dai, Hanxiao Liu, Quoc V. Le, and Mingxing Tan. Coatnet: Marrying convolution and attention for all data sizes. *arXiv preprint arXiv:2106.04803*, 2021.

[39] Jia Deng, Wei Dong, Richard Socher, Li-Jia Li, Kai Li, and Li Fei-Fei. Imagenet: A large-scale hierarchical image database. In *IEEE Conference on Computer Vision and Pattern Recognition*, pages 248–255, 2009.

[40] Jiankang Deng, Jia Guo, Niannan Xue, and Stefanos Zafeiriou. Arcface: Additive angular margin loss for deep face recognition. In *Proceedings of the IEEE/CVF Conference on Computer Vision and Pattern Recognition*, pages 4690–4699, 2019.

[41] Yingpeng Deng and Lina J. Karam. Universal adversarial attack via enhanced projected gradient descent. In *IEEE International Conference on Image Processing (ICIP)*, pages 1241–1245, 2020.

[42] Tejas Indulal Dhamecha, Soumyadeep Ghosh, Mayank Vatsa, and Richa Singh. Kernelized heterogeneity aware cross-view face recognition. *Frontiers in Artificial Intelligence*, volume 4, page 68, 2021.

[43] Gavin Weiguang Ding, Luyu Wang, and Xiaomeng Jin. Advertorch v0.1: An adversarial robustness toolbox based on pytorch. *arXiv preprint arXiv:1902.07623*, 2019.

[44] Yinpeng Dong, Fangzhou Liao, Tianyu Pang, Hang Su, Jun Zhu, Xiaolin Hu, and Jianguo Li. Boosting adversarial attacks with momentum. In *Proceedings of the IEEE Conference on Computer Vision and Pattern Recognition*, pages 9185–9193, 2018.

[45] Yueqi Duan, Jiwen Lu, and Jie Zhou. Uniformface: Learning deep equidistributed representation for face recognition. In *Proceedings of the IEEE/CVF Conference on Computer Vision and Pattern Recognition*, pages 3415–3424, 2019.

[46] Reuben Feinman, Ryan R Curtin, Saurabh Shintre, and Andrew B Gardner. Detecting adversarial samples from artifacts. *arXiv preprint arXiv:1703.00410*, 2017.

[47] Matteo Ferrara, Annalisa Franco, and Davide Maltoni. The magic passport. In *IEEE International Joint Conference on Biometrics*, pages 1–7. IEEE, 2014.

[48] Chuanxing Geng, Sheng-jun Huang, and Songcan Chen. Recent advances in open set recognition: A survey. In *IEEE Transactions on Pattern Analysis and Machine Intelligence*. IEEE, 2020.

[49] Partha Ghosh, Arpan Losalka, and Michael J. Black. Resisting adversarial attacks using gaussian mixture variational autoencoders. *Proceedings of the AAAI Conference on Artificial Intelligence*, volume 33, pages 541–548, 2019.

[50] Soumyadeep Ghosh, Richa Singh, and Mayank Vatsa. Subclass heterogeneity aware loss for cross-spectral cross-resolution face recognition. *IEEE Transactions on Biometrics, Behavior, and Identity Science*, volume 2, no. 3, pages 245–256, 2020.

[51] Akhil Goel, Akshay Agarwal, Mayank Vatsa, Richa Singh, and Nalini Ratha. Deepring: Protecting deep neural networks with blockchain. In *Proceedings of the IEEE/CVF Conference on Computer Vision and Pattern Recognition Workshops*. IEEE, 2019.

[52] Akhil Goel, Akshay Agarwal, Mayank Vatsa, Richa Singh, and Nalini Ratha. Securing cnn model and biometric template using blockchain. In *2019 IEEE 10th International Conference on Biometrics Theory, Applications and Systems (BTAS)*, pages 1–7. IEEE, 2019.

[53] Akhil Goel, Akshay Agarwal, Mayank Vatsa, Richa Singh, and Nalini K Ratha. Dndnet: Reconfiguring cnn for adversarial robustness. In *Proceedings of the IEEE/CVF Conference on Computer Vision and Pattern Recognition Workshops*, pages 22–23. IEEE, 2020.

[54] Akhil Goel, Anirudh Singh, Akshay Agarwal, Mayank Vatsa, and Richa Singh. Smartbox: Benchmarking adversarial detection and mitigation algorithms for face recognition. In *2018 IEEE 9th International Conference on Biometrics Theory, Applications and Systems (BTAS)*, pages 1–7. IEEE, 2018.

[55] Ian J. Goodfellow, Jonathon Shlens, and Christian Szegedy. Explaining and harnessing adversarial examples. *arXiv preprint arXiv:1412.6572*, 2014.

[56] Dou Goodman, Hao Xin, Wang Yang, Wu Yuesheng, Xiong Junfeng, and Zhang Huan. Advbox: A toolbox to generate adversarial examples that fool neural networks. *arXiv preprint arXiv:2001.05574*, 2020.

[57] Gaurav Goswami, Akshay Agarwal, Nalini Ratha, Richa Singh, and Mayank Vatsa. Detecting and mitigating adversarial perturbations for robust face recognition. *International Journal of Computer Vision*, volume 127, no. 6, pages 719–742, 2019.

[58] Gaurav Goswami, Nalini Ratha, Akshay Agarwal, Richa Singh, and Mayank Vatsa. Unravelling robustness of deep learning based face recognition against adversarial attacks. In *Proceedings of the AAAI Conference on Artificial Intelligence*, volume 32, 2018. AAAI.

[59] Mehak Gupta, Vishal Singh, Akshay Agarwal, Mayank Vatsa, and Richa Singh. Generalized iris presentation attack detection algorithm under cross-database settings. In *2020 25th International Conference on Pattern Recognition (ICPR)*, pages 5318–5325. IEEE, 2021.

[60] Mark Hawthorne. *Fingerprints: Analysis and Understanding*. CRC Press, 2017.

[61] Jamie Hayes and George Danezis. Learning universal adversarial perturbations with generative models. In *2018 IEEE Security and Privacy Workshops (SPW)*, pages 43–49. IEEE, 2018.

[62] Kaiming He, Xiangyu Zhang, Shaoqing Ren, and Jian Sun. Deep residual learning for image recognition. In *Proceedings of the IEEE Conference on Computer Vision and Pattern Recognition*, pages 770–778. IEEE, 2016.

[63] Yu-Chih-Tuan Hu, Bo-Han Kung, Daniel Stanley Tan, Jun-Cheng Chen, Kai-Lung Hua, and Wen-Huang Cheng. Naturalistic physical adversarial patch for object detectors. In *Proceedings of the IEEE/CVF International Conference on Computer Vision*, pages 7848–7857. IEEE, 2021.

[64] Gary B. Huang, Marwan Mattar, Tamara Berg, and Eric Learned-Miller. Labeled faces in the wild: A database for studying face recognition in unconstrained environments. In *Workshop on Faces in 'Real-Life' Images: Detection, Alignment, and Recognition*, pages 1276–1284, 2008.

[65] Yuge Huang, Yuhan Wang, Ying Tai, Xiaoming Liu, Pengcheng Shen, Shaoxin Li, Jilin Li, and Feiyue Huang. Curricularface: Adaptive curriculum learning loss for deep face recognition. In *Proceedings of the IEEE/CVF Conference on Computer Vision and Pattern Recognition*, pages 5901–5910, 2020.

[66] Yangqing Jia, Evan Shelhamer, Jeff Donahue, Sergey Karayev, Jonathan Long, Ross Girshick, Sergio Guadarrama, and Trevor Darrell. Caffe: Convolutional architecture for fast feature embedding. *Proceedings of the 22nd ACM International Conference on Multimedia*, pages 675–678, 2014.

[67] Jichao Jiao, Weilun Liu, Yaokai Mo, Jian Jiao, Zhongliang Deng, and Xinping Chen. Dyn-arcface: Dynamic additive angular margin loss for deep face recognition. *Multimedia Tools and Applications*, pages 1–16, 2021.

[68] Kimmo Karkkainen and Jungseock Joo. Fairface: Face attribute dataset for balanced race, gender, and age. *arXiv preprint arXiv:1908.04913*, 2019.

[69] Ira Kemelmacher-Shlizerman, Steven M Seitz, Daniel Miller, and Evan Brossard. The megaface benchmark: 1 million faces for recognition at scale. *Proceedings of the IEEE Conference on Computer Vision and Pattern Recognition*, pages 4873–4882, 2016.

[70] Yonghyun Kim, Wonpyo Park, Myung-Cheol Roh, and Jongju Shin. Groupface: Learning latent groups and constructing group-based representations for face recognition. *Proceedings of the IEEE/CVF Conference on Computer Vision and Pattern Recognition*, pages 5621–5630, 2020.

[71] Jascha Kolberg, Marcel Grimmer, Marta Gomez-Barrero, and Christoph Busch. Anomaly detection with convolutional autoencoders for fingerprint presentation attack detection. *IEEE Transactions on Biometrics, Behavior, and Identity Science*, volume 3, no. 2, pages 190–202, 2021.

[72] Alex Krizhevsky, Geoffrey Hinton et al. *Learning Multiple Layers of Features from Tiny Images*, Master's thesis, University of Toronto, 2009.

[73] Alex Krizhevsky, Ilya Sutskever, and Geoffrey E. Hinton. Imagenet classification with deep convolutional neural networks. *Advances in Neural Information Processing Systems*, volume 25, pages 1097–1105, 2012.

[74] Alexey Kurakin, Ian Goodfellow, Samy Bengio et al. Adversarial Examples in the Physical World. *Artificial intelligence safety and security*, pages 99–112. Chapman and Hall/CRC, 2018.

[75] Yann LeCun, Bernhard Boser, John S. Denker, Donnie Henderson, Richard E. Howard, Wayne Hubbard, and Lawrence D. Jackel. Backpropagation applied to handwritten zip code recognition. *Neural Computation*, volume 1, no. 4, pages 541–551, 1989.

[76] Yann LeCun, Leon Bottou, Yoshua Bengio, and Patrick Haffner. Gradient-based learning applied to document recognition. *Proceedings of the IEEE*, volume 86, no. 11, pages 2278–2324, 1998.

[77] Mathias Lecuyer, Vaggelis Atlidakis, Roxana Geambasu, Daniel Hsu, and Suman Jana. Certified robustness to adversarial examples with differential privacy. In *2019 IEEE Symposium on Security and Privacy (SP)*, pages 656–672. IEEE, 2019.

[78] Hyeungill Lee, Sungyeob Han, and Jungwoo Lee. Generative adversarial trainer: Defense to adversarial perturbations with gan. *arXiv preprint arXiv:1705.03387*, 2017.

[79] Feng Liu, Haozhe Liu, Wentian Zhang, Guojie Liu, and Linlin Shen. One-class fingerprint presentation attack detection using auto-encoder network. *IEEE Transactions on Image Processing*, volume 30, pages 2394–2407, 2021.

[80] Hao Liu, Xiangyu Zhu, Zhen Lei, and Stan Z Li. Adaptiveface: Adaptive margin and sampling for face recognition. In *Proceedings of the IEEE/CVF Conference on Computer Vision and Pattern Recognition*, pages 11947–11956. IEEE, 2019.

[81] Si-Qi Liu, Xiangyuan Lan, and Pong C. Yuen. Remote photoplethysmography correspondence feature for 3d mask face presentation attack detection. *Proceedings of the European Conference on Computer Vision (ECCV)*, pages 558–573, 2018.

[82] Si-Qi Liu, Xiangyuan Lan, and Pong C. Yuen. Multi-channel remote photoplethysmography correspondence feature for 3d mask face presentation attack detection. *IEEE Transactions on Information Forensics and Security*, volume 16, pages 2683–2696, 2021.

[83] Yuanliu Liu, Bo Peng, Peipei Shi, He Yan, Yong Zhou, Bing Han, Yi Zheng, Chao Lin, Jianbin Jiang, Yin Fan et al. Iqiyi-vid: A large dataset for multi-modal person identification. *arXiv preprint arXiv:1811.07548*, 2018.

[84] Ze Liu, Yutong Lin, Yue Cao, Han Hu, Yixuan Wei, Zheng Zhang, Stephen Lin, and Baining Guo. Swin transformer: Hierarchical vision transformer using shifted windows. *arXiv preprint arXiv:2103.14030*, 2021.

[85] Jiajun Lu, Theerasit Issaranon, and David Forsyth. Safetynet: Detecting and rejecting adversarial examples robustly. *Proceedings of the IEEE International Conference on Computer Vision*, pages 446–454, 2017.

[86] Chunchuan Lyu, Kaizhu Huang, and Hai-Ning Liang. A unified gradient regularization family for adversarial examples. In *2015 IEEE International Conference on Data Mining*, pages 301–309. IEEE, 2015.

[87] Yukun Ma, Lifang Wu, Zeyu Li, et al. A novel face presentation attack detection scheme based on multi-regional convolutional neural networks. *Pattern Recognition Letters*, volume 131, pages 261–267, 2020.

[88] Aleksander Madry, Aleksandar Makelov, Ludwig Schmidt, Dimitris Tsipras, and Adrian Vladu. Towards deep learning models resistant to adversarial attacks. *arXiv preprint arXiv:1706.06083*, 2017.

[89] Angshul Majumdar, Richa Singh, and Mayank Vatsa. Face verification via class sparsity based supervised encoding. *IEEE Transactions on Pattern Analysis and Machine Intelligence*, volume 39, no. 6, pages 1273–1280, 2017.

[90] Puspita Majumdar, Akshay Agarwal, Mayank Vatsa, and Richa Singh. Facial Retouching and Alteration Detection. *Handbook of Digital Face Manipulation and Detection,* pages 367–387. Springer, Cham, 2022.

[91] Puspita Majumdar, Saheb Chhabra, Richa Singh, and Mayank Vatsa. Subgroup invariant perturbation for unbiased pre-trained model prediction. *Frontiers in Big Data*, volume 3, pages 590–596, 2021.

[92] Ishan Manjani, Snigdha Tariyal, Mayank Vatsa, Richa Singh, and Angshul Majumdar. Detecting silicone mask-based presentation attack via deep dictionary learning. *IEEE Transactions on Information Forensics and Security*, volume 12, no. 7, pages 1713–1723, 2017.

[93] Aman Mehra, Akshay Agarwal, Mayank Vatsa, and Richa Singh. Detection of digital manipulation in facial images (student abstract). In *AAAI Conference on Artificial Intelligence*. AAAI, 2021.

[94] Suril Mehta, Anannya Uberoi, Akshay Agarwal, Mayank Vatsa, and Richa Singh. Crafting a panoptic face presentation attack detector. In *2019 International Conference on Biometrics (ICB)*, pages 1–6. IEEE, 2019.

[95] Felipe A. Mejia, Paul Gamble, Zigfried Hampel-Arias, Michael Lomnitz, Nina Lopatina, Lucas Tindall, and Maria Alejandra Barrios. Robust or private? Adversarial training makes models more vulnerable to privacy attacks. *arXiv preprint arXiv:1906.06449*, 2019.

[96] Qiang Meng, Shichao Zhao, Zhida Huang, and Feng Zhou. Magface: A universal representation for face recognition and quality assessment. *Proceedings of the IEEE/CVF Conference on Computer Vision and Pattern Recognition*, pages 14225–14234, 2021.

[97] Yisroel Mirsky and Wenke Lee. The creation and detection of deepfakes: A survey. *ACM Computing Surveys (CSUR)*, volume 54, no. 1, pages 1–41, 2021.

[98] Seyed-Mohsen Moosavi-Dezfooli, Alhussein Fawzi, Omar Fawzi, and Pascal Frossard. Universal adversarial perturbations. *Proceedings of the IEEE Conference on Computer Vision and Pattern Recognition*, pages 1765–1773, 2017.

[99] Konda Reddy Mopuri, Aditya Ganeshan, and R Venkatesh Babu. Generalizable data-free objective for crafting universal adversarial perturbations. *IEEE Transactions on Pattern Analysis and Machine Intelligence*, volume 41, no. 10, pages 2452–2465, 2018.

[100] Stylianos Moschoglou, Athanasios Papaioannou, Christos Sagonas, Jiankang Deng, Irene Kotsia, and Stefanos Zafeiriou. Agedb: The first manually collected, in-the-wild age database. *Proceedings of the IEEE Conference on Computer Vision and Pattern Recognition Workshops*, pages 51–59, 2017.

[101] Shruti Nagpal, Maneet Singh, Richa Singh, and Mayank Vatsa. Deep learning for face recognition: Pride or prejudiced? *arXiv preprint arXiv:1904.01219*, 2019.

[102] Shruti Nagpal, Maneet Singh, Richa Singh, and Mayank Vatsa. Attribute aware filter-drop for bias invariant classification. *Proceedings of the IEEE/CVF Conference on Computer Vision and Pattern Recognition Workshops*, pages 32–33, 2020.

[103] Shruti Nagpal, Maneet Singh, Richa Singh, and Mayank Vatsa. Diversity blocks for debiasing classification models. In *2020 IEEE International Joint Conference on Biometrics (IJCB)*, pages 1–9. IEEE, 2020.

[104] Yuval Netzer, Tao Wang, Adam Coates, Alessandro Bissacco, Bo Wu, and Andrew Y Ng. NIPS Workshop on Deep Learning and Unsupervised Feature Learning. *Reading Digits in Natural Images with Unsupervised Feature Learning*, 2011.

[105] Hong-Wei Ng and Stefan Winkler. A data-driven approach to cleaning large face data-sets. In *2014 IEEE International Conference on Image Processing (ICIP)*, pages 343–347. IEEE, 2014.

[106] Maria-Irina Nicolae, Mathieu Sinn, Minh Ngoc Tran, Beat Buesser, Ambrish Rawat, Martin Wistuba, Valentina Zantedeschi, Nathalie Baracaldo, Bryant Chen, Heiko Ludwig et al. Adversarial robustness toolbox v1.0.0. *arXiv preprint arXiv:1807.01069*, 2018.

[107] Omkar M. Parkhi, Andrea Vedaldi, and Andrew Zisserman. Deep Face Recognition. *Proceedings of the British Machine Vision Conference,* pages 41.1–41.12. University of Oxford, 2015.

[108] Adam Paszke, Sam Gross, Francisco Massa, Adam Lerer, James Bradbury, Gregory Chanan, Trevor Killeen, Zeming Lin, Natalia Gimelshein, Luca Antiga et al. Pytorch: An imperative style, high-performance deep learning library. *Advances in Neural Information Processing Systems*, volume 32, pages 8026–8037, 2019.

[109] Hieu Pham, Zihang Dai, Qizhe Xie, and Quoc V Le. Meta pseudo labels. *Proceedings of the IEEE/CVF Conference on Computer Vision and Pattern Recognition*, pages 11557–11568, 2021.

[110] Kiran Raja, Sushma Venkatesh, R.B. Christoph Busch et al. Transferable deep-CNN features for detecting digital and print-scanned morphed face images. *Proceedings of the IEEE Conference on Computer Vision and Pattern Recognition Workshops*, pages 10–18, 2017.

[111] Jonas Rauber, Roland Zimmermann, Matthias Bethge, and Wieland Brendel. Foolbox native: Fast adversarial attacks to benchmark the robustness of machine learning models in pytorch, tensorflow, and jax. *Journal of Open Source Software*, volume 5, no. 53, page 2607, 2020.

[112] Yasar Abbas Ur Rehman, Lai-Man Po, and Jukka Komulainen. Enhancing deep discriminative feature maps via perturbation for face presentation attack detection. *Image and Vision Computing*, volume 94, page 103858, 2020.

[113] Kevin Roth, Aurelien Lucchi, Sebastian Nowozin, and Thomas Hofmann. Adversarially robust training through structured gradient regularization. *arXiv preprint arXiv: 1805.08736*, 2018.

[114] Nilay Sanghvi, Sushant Kumar Singh, Akshay Agarwal, Mayank Vatsa, and Richa Singh. Mixnet for generalized face presentation attack detection. In *2020 25th International Conference on Pattern Recognition (ICPR)*, pages 5511–5518. IEEE, 2021.

[115] Ulrich Scherhag, Luca Debiasi, Christian Rathgeb, Christoph Busch, and Andreas Uhl. Detection of face morphing attacks based on prnu analysis. *IEEE Transactions on Biometrics, Behavior, and Identity Science*, volume 1, no. 4, pages 302–317, 2019.

[116] Ulrich Scherhag, Christian Rathgeb, Johannes Merkle, and Christoph Busch. Deep face representations for differential morphing attack detection. *IEEE Transactions on Information Forensics and Security*, volume 15, pages 3625–3639, 2020.

[117] Frank Seide and Amit Agarwal. CNTK: Microsoft's open-source deep-learning toolkit. *Proceedings of the 22nd ACM SIGKDD International Conference on Knowledge Discovery and Data Mining*, pages 2135–2135, 2016.

[118] Soumyadip Sengupta, Jun-Cheng Chen, Carlos Castillo, Vishal M. Patel, Rama Chellappa, and David W Jacobs. Frontal to profile face verification in the wild. In *2016 IEEE Winter Conference on Applications of Computer Vision (WACV)*, pages 1–9. IEEE, 2016.

[119] Alex Serban, Erik Poll, and Joost Visser. Adversarial examples on object recognition: A comprehensive survey. *ACM Computing Surveys (CSUR)*, volume 53, no. 3, pages 1–38, 2020.

[120] Talha Ahmad Siddiqui, Samarth Bharadwaj, Tejas I. Dhamecha, Akshay Agarwal, Mayank Vatsa, Richa Singh, and Nalini Ratha. Face anti-spoofing with multifeature videolet aggregation. In *2016 23rd International Conference on Pattern Recognition (ICPR)*, pages 1035–1040. IEEE, 2016.

[121] Karen Simonyan and Andrew Zisserman. Very deep convolutional networks for large-scale image recognition. *arXiv preprint arXiv:1409.1556*, 2014.

[122] Richa Singh, Akshay Agarwal, Maneet Singh, Shruti Nagpal, and Mayank Vatsa. On the robustness of face recognition algorithms against attacks and bias. *Proceedings of the AAAI Conference on Artificial Intelligence*, volume 34, pages 13583–13589, 2020.

[123] Ayan Tuhinendu Sinha, Andrew Rabinovich, Zhao Chen, and Vijay Badrinarayanan. *Gradient Adversarial Training of Neural Networks*. US Patent App. 17/051,982, June 24, 2021.

[124] Yang Song, Taesup Kim, Sebastian Nowozin, Stefano Ermon, and Nate Kushman. Pixeldefend: Leveraging generative models to understand and defend against adversarial examples. *arXiv preprint arXiv:1710.10766*, 2017.

[125] Jiawei Su, Danilo Vasconcellos Vargas, and Kouichi Sakurai. One pixel attack for fooling deep neural networks. *IEEE Transactions on Evolutionary Computation*, volume 23, no. 5, pages 828–841, 2019.

[126] Yi Sun. *Deep Learning Face Representation by Joint Identification-Verification*. The Chinese University of Hong Kong (Hong Kong), 2015.

[127] Yifan Sun, Changmao Cheng, Yuhan Zhang, Chi Zhang, Liang Zheng, Zhong- dao Wang, and Yichen Wei. Circle loss: A unified perspective of pair similarity optimization. *Proceedings of the IEEE/CVF Conference on Computer Vision and Pattern Recognition*, pages 6398–6407, 2020.

[128] Saksham Suri, Anush Sankaran, Mayank Vatsa, and Richa Singh. Improving face recognition performance using tecs2 dictionary. *Pattern Recognition Letters*, volume 145, pages 88–95, 2021.

[129] Christian Szegedy, Sergey Ioffe, Vincent Vanhoucke, and Alexander A. Alemi. *Inception-v4, Inception-Resnet and the Impact of Residual Connections on Learning*. Thirty-First AAAI Conference on Artificial Intelligence, 2017.

[130] Christian Szegedy, Wei Liu, Yangqing Jia, Pierre Sermanet, Scott Reed, Dragomir Anguelov, Dumitru Erhan, Vincent Vanhoucke, and Andrew Rabinovich. Going deeper with convolutions. *Proceedings of the IEEE Conference on Computer Vision and Pattern Recognition*, pages 1–9, 2015.

[131] Yaniv Taigman, Ming Yang, Marc'Aurelio Ranzato, and Lior Wolf. Deepface: Closing the gap to human-level performance in face verification. *Proceedings of the IEEE Conference on Computer Vision and Pattern Recognition*, pages 1701–1708, 2014.

[132] Ruben Tolosana, Ruben Vera-Rodriguez, Julian Fierrez, Aythami Morales, and Javier Ortega-Garcia. Deepfakes and beyond: A survey of face manipulation and fake detection. *Information Fusion*, volume 64, pages 131–148, 2020.

[133] Hugo Touvron, Matthieu Cord, Alexandre Sablayrolles, Gabriel Synnaeve, and Herve Jegou. Going deeper with image transformers. *arXiv preprint arXiv:2103.17239*, 2021.

[134] Hugo Touvron, Andrea Vedaldi, Matthijs Douze, and Herve Jegou. Fixing the train-test resolution discrepancy: Fixefficientnet. *arXiv preprint arXiv:2003.08237*, 2020.

[135] Florian Tram'er, Alexey Kurakin, Nicolas Papernot, Ian Goodfellow, Dan Boneh, and Patrick McDaniel. Ensemble adversarial training: Attacks and defenses. *arXiv preprint arXiv:1705.07204*, 2017.

[136] FengWang, Jian Cheng, Weiyang Liu, and Haijun Liu. Additive margin softmax for face verification. *IEEE Signal Processing Letters*, volume 25, no. 7, pages 926–930, 2018.

[137] Guoqing Wang, Hu Han, Shiguang Shan, and Xilin Chen. Cross-domain face presentation attack detection via multi-domain disentangled representation learning. *Proceedings of the IEEE/CVF Conference on Computer Vision and Pattern Recognition*, pages 6678–6687, 2020.

[138] Hao Wang, Yitong Wang, Zheng Zhou, Xing Ji, Dihong Gong, Jingchao Zhou, Zhifeng Li, and Wei Liu. Cosface: Large margin cosine loss for deep face recognition. *Proceedings of the IEEE Conference on Computer Vision and Pattern Recognition*, pages 5265–5274, 2018.

[139] Mei Wang, Weihong Deng, Jiani Hu, Xunqiang Tao, and Yaohai Huang. Racial faces in the wild: Reducing racial bias by information maximization adaptation network. *Proceedings of the IEEE/CVF International Conference on Computer Vision*, pages 692–702, 2019.

[140] Yajie Wang, Haoran Lv, Xiaohui Kuang, Gang Zhao, Yu-an Tan, Quanxin Zhang, and Jingjing Hu. Towards a physical-world adversarial patch for blinding object detection models. *Information Sciences*, volume 556, pages 459–471, 2021.

[141] Di Wen, Hu Han, and Anil K. Jain. Face spoof detection with image distortion analysis. *IEEE Transactions on Information Forensics and Security*, volume 10, no. 4, pages 746–761, 2015.

[142] Cameron Whitelam, Emma Taborsky, Austin Blanton, Brianna Maze, Jocelyn Adams, Tim Miller, Nathan Kalka, Anil K. Jain, James A Duncan, Kristen Allen et al. Iarpa janus benchmark-b face dataset. *Proceedings of the IEEE Conference on Computer Vision and Pattern Recognition Workshops*, pages 90–98, 2017.

[143] Han Xiao, Kashif Rasul, and Roland Vollgraf. Fashion-mnist: A novel image dataset for benchmarking machine learning algorithms. *arXiv preprint arXiv:1708.07747*, 2017.

[144] Qizhe Xie, Minh-Thang Luong, Eduard Hovy, and Quoc V. Le. Self-training with noisy student improves imagenet classification. *Proceedings of the IEEE/CVF Conference on Computer Vision and Pattern Recognition*, pages 10687–10698, 2020.

[145] Saining Xie, Ross Girshick, Piotr Dollár, Zhuowen Tu, and Kaiming He. Aggregated residual transformations for deep neural networks. *Proceedings of the IEEE Conference on Computer Vision and Pattern Recognition*, pages 1492–1500, 2017.

[146] Daksha Yadav, Naman Kohli, Akshay Agarwal, Mayank Vatsa, Richa Singh, and Afzel Noore. Fusion of handcrafted and deep learning features for large-scale multiple iris presentation attack detection. *Proceedings of the IEEE Conference on Computer Vision and Pattern Recognition Workshops*, pages 572–579, 2018.

[147] Xiaolong Yang, Xiaohong Jia, Dihong Gong, Dong-Ming Yan, Zhifeng Li, and Wei Liu. Larnet: Lie algebra residual network for face recognition. In *International Conference on Machine Learning*, pages 11738–11750. PMLR, 2021.

[148] Dong Yi, Zhen Lei, Shengcai Liao, and Stan Z. Li. Learning face representation from scratch. *arXiv preprint arXiv:1411.7923*, 2014.

[149] Xiaohua Zhai, Alexander Kolesnikov, Neil Houlsby, and Lucas Beyer. Scaling vision transformers. *arXiv preprint arXiv:2106.04560*, 2021.

[150] Huan Zhang, Hongge Chen, Zhao Song, Duane Boning, Inderjit S Dhillon, and Cho-Jui Hsieh. The limitations of adversarial training and the blind-spot attack. *arXiv preprint arXiv:1901.04684*, 2019.

[151] Shifeng Zhang, Xiaobo Wang, Ajian Liu, Chenxu Zhao, Jun Wan, Sergio Escalera, Hailin Shi, Zezheng Wang, and Stan Z Li. A dataset and benchmark for large-scale multimodal face anti-spoofing. *Proceedings of the IEEE/CVF Conference on Computer Vision and Pattern Recognition*, pages 919–928, 2019.

[152] Xiao Zhang, Rui Zhao, Yu Qiao, Xiaogang Wang, and Hongsheng Li. Adacos: Adaptively scaling cosine logits for effectively learning deep face representations. *Proceedings of the IEEE/CVF Conference on Computer Vision and Pattern Recognition*, pages 10823–10832, 2019.

[153] Xiao Zhang, Rui Zhao, Junjie Yan, Mengya Gao, Yu Qiao, Xiaogang Wang, and Hongsheng Li. P2sgrad: Refined gradients for optimizing deep face models. *Proceedings of the IEEE/CVF Conference on Computer Vision and Pattern Recognition*, pages 9906–9914, 2019.

[154] Yuanhan Zhang, ZhenFei Yin, Yidong Li, Guojun Yin, Junjie Yan, Jing Shao, and Ziwei Liu. Celeba-spoof: Large-scale face anti-spoofing dataset with rich annotations. In *European Conference on Computer Vision*, pages 70–85. Springer, 2020.

[155] Zhiwei Zhang, Junjie Yan, Sifei Liu, Zhen Lei, Dong Yi, and Stan Z Li. A face antispoofing database with diverse attacks. In *IAPR International Conference on Biometrics (ICB)*, pages 26–31. IEEE, 2012.

[156] Tianyue Zheng and Weihong Deng. Cross-pose lfw: A database for studying cross-pose face recognition in unconstrained environments. *Beijing University of Posts and Telecommunications, Tech. Rep*, volume 5, page 7, 2018.

[157] Tianyue Zheng, Weihong Deng, and Jiani Hu. Cross-age LFW: A database for studying cross-age face recognition in unconstrained environments. *arXiv preprint arXiv:1708.08197*, 2017.

[158] Yutong Zheng, Dipan K. Pal, and Marios Savvides. Ring loss: Convex feature normalization for face recognition. *Proceedings of the IEEE Conference on Computer Vision and Pattern Recognition*, pages 5089–5097, 2018.

[159] Zheng Zhu, Guan Huang, Jiankang Deng, Yun Ye, Junjie Huang, Xinze Chen, Jiagang Zhu, Tian Yang, Jiwen Lu, Dalong Du et al. Webface260m: A benchmark unveiling the power of million-scale deep face recognition. *Proceedings of the IEEE/CVF Conference on Computer Vision and Pattern Recognition*, pages 10492–10502, 2021.

4 Smart Parking System Using YOLOv3 Deep Learning Model

Narina Thakur, Sardar M N Islam, Zarqua Neyaz, Deepanshu Sadhwani and Rachna Jain

CONTENTS

4.1 INTRODUCTION

The main and most important requirement of smart cities is smart parking systems. A typical Indian driver spends 20 minutes [1] per day searching for the perfect car park. Living in a city can be challenging. Rapid population growth will always exist, and city traffic has become extremely congested. As per the United Nations Department of Economic and Social Affairs (UNDESA), cities comprise the majority of the world's population. The world population increase has been projected to reach 68 percent by 2050 [2]. Indian metropolises are undergoing mobility issues as a result of rapid population and economic growth. Cities today face a huge burden on parking lots, resulting in several issues such as traffic congestion, disproportionate supply and demand, as well as environmental pollution. India suffers from chaotic situations such as congested footpaths, illegal parking, and illicit behavior resulting

DOI: 10.1201/9781003245469-4

from inadequate scrutiny as a consequence of poor parking management and policy. Population growth is not only an important and urgent concern for the government, but it is also an everyday reality for the majority of citizens, which is a significant problem, especially in metro cities. On average, forty percent of cars are parked on the roads, causing traffic congestion. According to a 2018 BCG study, India loses 1.47 lac crore rupees per year due to heavy traffic in the four urban areas alone. If we simply double this figure to include the rest of the country and factor in fuel inflation, the estimated loss due to traffic congestion would exceed India's consolidated health-care and education expenditures. Further, the burden it puts on our urban planners to continuously create flyovers, subways, and other infrastructure is an altogether different challenge. The rapid pace of urbanization and modernization will result in an increase in parking demand.

An enticing smart city research study is expanding knowledge and understanding about vehicle parking systems in various metropolitan areas. As a result, academics and practitioners have obtained an interconnected automobile parking dataset. Sometimes, during peak hours, parking lot operators do not register detailed information or register erroneous details into the system, creating issues for vehicle owners while leaving the parking lot and posing a massive security threat. In this paper, a robust solution to the difficulties is proposed by designing a system called automatic number plate recognition (ANPR) using the YOLOv3 deep learning model. As more cities suffer from traffic congestion and inadequate parking, the vehicle parking sector is still evolving. Predicting the site of parking has been a critical challenge in our daily lives for so many years. This is one area of smart parking systems that has received considerable attention. The authors propose the use of a neural networks (NN) [3] model to predict vehicle parking space, with a particular emphasis on smart parking systems and ANPR using deep learning models. This chapter also highlights quality concerns and areas for improvement in smart parking solutions currently being used in smart cities. Even though parking and transportation are such significant characteristics of daily life, there is a huge demand for innovative and cost-effective solutions. ANPR has numerous applications, including stolen vehicle detection, parking management, traffic flow monitoring, etc. As a result, researchers from all around the world have been studying this topic in order to improve ANPR performance in everyday situations.

The existing ANPR algorithms perform admirably in bounded spaces; however, when confronted with complex scenarios, performance suffers. The three major steps in an ANPR method are license plate detection or localization, segmentation, and OCR. The most important step in selecting an appropriate object model is to utilize. We have reviewed a large number of models, each of which has its own set of advantages, disadvantages, and parameters. The two most common categories or models are YOLO and Visual Geometry Group with 16 layers (VGG16). The first step in the proposed smart parking system is license plate detection, which is performed by YOLOv3 and VGG16. By implementing the models with the chosen dataset, the prediction accuracy of the models was compared. For the detection of the vehicle plate, the most accurate model was chosen. YOLOv3 was found to outperform the other models. YOLO takes an image as input, runs it through a Neural

Network, and displays the predicted bounding box. Uncontrollable variables such as uneven lighting, weather, image distortion, image blurring, etc. are causing problems with automatic license plate recognition (ALPR). Because of variances in number plate design between countries and the lack of publicly available transnational number plate databases, ANPR is a difficult issue in a variety of diverse areas. The chapter proposes a cost-effective ALPR system employing accurate deep ALPR method using YOLOV3 that may be used on license plates from a range of places. The proposed algorithm for license plate detection is simple and may successfully categorize various license plate layouts. It can bring a number of benefits, such as traffic safety adherence, safety in the event of susceptibility, ease of use, and immediate access to information—compared to the phase of segmentation searching for registration details of vehicle ownership. The lighting, terminology, car shade, non-uniform plate size, characters on the plate, distinct font, and background color are factors that can affect ANPR results. The rest of the chapter organization is as follows: Section 2 presents various methodologies for identifying license plates, as well as their limitations. Section 3 introduces the proposed methodology. Section 4 shows the results of experiments and simulations, and Section 5 discusses the results as well as findings before drawing conclusions and highlighting future prospects in Section 6.

4.2 LITERATURE REVIEW

The ANPR process is primarily divided into three stages: number plate detection, segmentation, and alphanumeric recognition. Over the years, there has been a lot of interest in detecting license plates. Many studies have undertaken ANPR research using various algorithms, each of which has attempted to improve the performance of ANPR. There are two types of ANPR approaches: traditional image processing methodologies and deep learning techniques. The ANPR process is divided into three stages: number plate detection, segmentation, and detection and recognition. Over the years, there has been a lot of interest in detecting license plates. Many studies have undertaken ANPR research using various algorithms, each of which has attempted to improve the performance of ANPR. There are two types of ANPR approaches: traditional image processing methodologies and deep learning techniques. We have conducted an extensive literature review on license plate detection techniques that have been used in recent years, segmentation techniques and character recognition. This section also discusses some of the techniques' limitations. We also discussed how ANPR's performance has improved over the years with the use of diverse deep learning methods.

Pranav Adarsh et al. [4] presented a three deep multinational ALPR system that works exclusively with number plates from multiple countries, integrating a deep convolutional neural network with an image-based multinational number plate design identification technique. The focus of this article was to suggest a method to identify and recognize new objects in a single stage using an upgraded model. It also evaluated the performance of two different types of object detection techniques or detectors viz. two-stage detector algorithms encompassing RCNN, Fast

RCNN, Faster RCNN, the single-stage detector algorithms encompassing YOLOv1, YOLOv2, YOLOv3, and SSD. The two phase detectors are more concerned with accuracy, while one is more concerned with speed. This research paper proposed a new single-stage model for speeding up without losing precision. The latter is the best of two-stage detectors out of RCNN, Fast RCNN, and Faster RCNN, as per comparison results. The YOLO v3-Tiny increases object detection speed while ensuring the precise results in a stage detector of YOLOv1, YOLOv2, YOLOv3, and SSD. 320. A single end-to-end model, on the other hand, is efficient enough just to detect number plates from diverse locations.

Initially, the Darknet frame and ImageNet-1000 data packages were used in YOLOv1 for the model training. The image is divided into a grid of the cells of S×S. The YOLOv1 limits are based on objects in close proximity to the image. Small objects cannot be found if the objects appear to be in a cluster. YOLOv2 is a replacement for YOLO that provides high time-precision equilibrium. YOLOv2 introduces batch normalization, which helps to improve accuracy by adding that to every layer of convolution by 2 percent. The next advanced variant for YOLO is YOLOv3, which computes the target score using logistic regression. It displays the score for all targets in each boundary box. Since it employs the softmax function used in YOLOv2 for each class, YOLOv3 can perform multi-label classification. Authors have suggested the use of the Darknet-53 in YOLOv3; it has 53 convolution layers. These have more depth layers than YOLOv2's Darknet-19.

YOLOv3 has an advantage over YOLOv2 in that some changes are incorporated into the error function, and three scales are used for small to large-scale detections. The multi-class problem has become a multi-label issue, and the performance of small objects has improved. The single-shot sensor (SSD) achieves an excellent speed-to-accuracy ratio. To create the map, a CNN model was superposed. It also employs anchor boxes with varying characteristics, similar to faster RCNN, and it learns offset rather than box. Unlike YOLO, SSD does not divide the image into random grids. The offset of predefined anchor boxes for each function map location (default boxes). Every box has a definite structure, ratio, and position in relation to the corresponding element. The YOLOv3 Tiny is a lightweight YOLOv3 variant that requires less operating time and precision in YOLOv3 examination.

C. Henry et al. [5] proposed multinational platform recognition using general sequence detection as the application of the YOLOv3 model. This paper is mainly based on YOLO networks. The second step employs YOLOv3-SPP, with the YOLOv3 comprising a Space Pyramid Pooling (SPP) block. In particular, the smaller YOLOv3 was used for character identification throughout. The localized license plate is fed into YOLOv3 SPP. The character recognition network reverts the prediction character's bounding boxes but no information on the license plate number sequence. An incorrectly sequenced license plate number is not considered correct. As a result, they suggested a design algorithm for retrieving the correct sequence of plate number multinational license plates.

The CNN-based methodology for license plate detection [6–7] allows for an estimation of the license plate's location. This model generates a score for each sub region of the image, which allows prediction of the position of the detected license plate by integrating the results from sparsely overlapping regions. S. Zain Masood et al. [6] presented an improved CNN plate detector model, developing an output function that can bring together the results from a subset of pictorial subregions that are available for other object detection tasks and developing a challenging picture benchmark that is freely available for research applications. In the detection of license plate numbers, many other researchers also used CNN, such as in [7], where they could train their model with CNN.

A new OKM-CNN [8] technique has been presented to effectively detect and recognize license plates. Three major stages operate on the proposed OKM-CNN model. In the first phase, the location and detection processes on license plates are carried out using IBA and CCA models. Then the clustering technique based on OKM is implemented in the LP image segment and characters will be recognized using a CNN model in the license plate system. Classifiers are used in previous work on object detection. Object detection could perhaps be framed as a regression problem with spatially separated bounding boxes and associated class probabilities. A single NN predicts boundary boxes with class probabilities from the entire image in a single evaluation. As a result, YOLO [9] was employed to introduce a novel approach to object detection. YOLO is extremely fast when compared to previously introduced algorithms. In training, YOLO can process video streaming with less than 25-millisecond latency and outperforms other detection methods.

R. NarenBabu et al. [10] postulated an ALPR system based on YOLO object detectors. CNNs are trained and fine-tuned for each ALPR phase. They devised a two-stage strategic plan based on simple data-increase tricks such as Inverted License Plates and Flipped Characters. H. Bura et al. [11] trained ALPR system with a YOLO License Plate Locating Algorithm, as well as character recognition, which then sorted the recognized features to match the license platform from left to right.

Redmon et al. [12] proposed a YOLO-based ALPR system that utilizes advanced YOLO deep learning, distributed phones, edge computing, and data analysis algorithms to create a more adaptable and cost-effective smart parking system. YOLOv3 is an improved YOLO that detects an item using features learned by a deep CNN. The results indicate that 320 *320 YOLOv3 is as precise as SSD and yet three times faster in 22ms with a 28.2 mAP, and YOLOv3 was compared using the old IOU mAP (mean average precision) detection measurements. Titan X has 57:9 AP50 per 51 ms while RetinaNet has 57:5 AP50 per 8 ms, which is similar but with 3.8 times faster performance.

Chen et al. [13], proposed a single-class detector sliding window through a small YOLO CNN classifier. The paper addressed the issue of detecting car licenses by using a deep learning frame of YOLO-Darknet, where YOLO's uses 7 convolutionary layers to detect a single class. As the pattern can predict boundary boxes from data, it strives in new or unusual aspects or configurations to generalize objects. This restriction of YOLO has led researchers to investigate other object detectors. J. Yepezet al. [14] proposed network architecture to identify license plates with inverted residues using bottleneck-separable depth convolution. SSD architecture [14] builds the neural network used in the location of license plates. VGG-16 is the original extractor in SSD. VGG-16 is composed of 13 layers of convolution followed by 3 layers of fully integrated connection. They discovered that only by combining the versatility of profoundly separable convolutions with the fundamental ideas of relevant information extraction, abstraction, and accumulation of sequential constraints, can the parking system provide a precise and fast solution for the location of license plaques without compromising overall accuracy. VGG, on the other hand, has approximately 140 million parameters, making a system computer-complex and necessitating the use of a powerful Graphics processor unit (GPU) to function efficiently in an acceptable timeframe. This is the foremost restriction of the VGG-16 model. Template matching [15–16] is another proposed algorithm for detecting the license plate. Its purpose is to match the template scheme for the number plate of the vehicle. The car number plate must be placed first and foremost from the input of the car so that the application of template matching occurs before the morphology procedure starts.

M. Karakaya et al. [17] present a method for gathering information about parking lot availability on an embedded system by using a convolutional neural network (CNN) for image processing. The authors were able to use a very low-cost embedded system in comparison to the exiting methodologies on this topic by utilizing an efficient neural network model. Ghazali et al. [15, 18] highlighted that the use of neural networks for plate recognition has been very prevalent over the past several years. The priorities must be to obtain the boundary box of the plate number. The OCR was used to identify every character on the number plate of the car. Various artificial neural networks (ANNs) have been used and exploited by the various researchers in recent past. Different neural networks models such as ANN [19], probabilistic neural network (PNN) [20] and back propagation (BP) neural network [21] have been extensively employed in plate identification and recognition. In [19], an OCR algorithm based on feed-forward Artificial Neural Network (ANN) specifically for an ANPR system has been proposed and implemented and has validated the algorithm using MATLAB. The main objective of this study is the application for one FPGA of the entire ANPR scheme. Fikriye et al. [21] proposed BP Neural Network research on Vehicle Plate Recognition. A BP Neural Network is essentially a collection of samples from input and output that has been transformed into a nonlinear problem. It is a weight learning algorithm that employs a gradient algorithm. In [21], the number plate is located by the Otsu's threshold. For character segmentation, vertical and horizontal histograms are used. Last, Probabilistic Neural Networks recognizes the character.

C. Anantha et al. [22] proposed that symmetric wavelets and multi-level genetic algorithm be employed for locating license plates. These were the methods used by ANPR before profound models of learning. This paper employed multiple license plates in one single image using the genetic algorithm at many layers. This can identify and locate any number of license plates in a single picture. Symbols can be located on two-dimensional compound items using multi-level genetic techniques at excellent precision rates. V. Himani et al. [23] locate the data using symmetric wavelets and mathematical morphology. In the pre-processing technique the input images were first transformed into a grey-scale, followed by a mask. The pre-processing produced dominant values. The results have been evaluated using root mean square error (RMSE) and peak signal to noise ratio (PSNR) [23] evaluation metrics.

4.3 METHODOLOGY

This section discusses the methodology for the proposed YOLO-based smart parking system.

Figure 4.1 shows the flowchart of the proposed method.

The method proposed consisted of various processes such as number plate detection, identification of the characteristics of the number plate detected, and data storage on the database system. The YOLOv3 CNN has been used for license plate detection, taking the license plate images as input and returning the license plate location annotations. The recognized area of the vehicle license plate is then pre-processed and applied to OCR, which successfully acknowledges the number plate character and then saves data in the data frame for the same number plate.

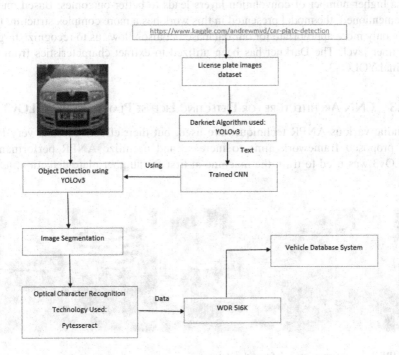

FIGURE 4.1   YOLOv3 based smart parking system flow chart.

4.3.1 Dataset

The dataset Car License Plate Detection [24] contains 433 images of bounding box annotations of the car license plates within the image.

Figure 4.2 depicts a few sample images from the dataset. The PASCAL VOC format is being used for annotations. There are 433 images and 433 annotations in the dataset. With the same name in the same directory the. xml files of every image contains the coordinates of the images as <x_center><y_center><width><height>.

4.3.2 Training

For number plate detection, the device was trained, and the program was written in Python. We used Darknet framework, an open source neural network framework, for training the detector. In the proposed work the detector is the YOLOv3 deep learning model. YOLO uses an image as an input, runs it via a neural network, and predicts the bounding boxes. The prediction for each bounding box includes five components, i.e. x, y, w, h, and confidence. (x, y) represents the center of the bounding box, while (w, h) represents the width and the height of the boxes, and confidence is the estimated accuracy of the object prediction. Training is done just by placing the YOLOv3 model as input and annotations as output on image data i.e. x, y, w, h, and confidence. Because of its speed and precision, the YOLOv3 model has been used in the proposed model, which has a variety of applications. The study showed that a higher number of convolution layers leads to better outcomes. Based on the aforementioned, the model presented in this work has a more complex structure that is not only more appropriate for our database but also allows us to recognize targets at a finer level. The Darknet has been utilized to extract characteristics from the original YOLOv3.

4.3.3 CNN Architecture for Detecting License Plates Using YOLOv3

In India, various ANPR techniques are used, but their effectiveness is very low. The proposed framework aims to increase and optimize ANPR performance. YOLOv3 was used to train the machine at first for number plate detection, using

FIGURE 4.2 Sample images from the dataset.

Convolutional Neural Network (CNN), which is capable of detecting objects and entities. CNN applies a filter on the input, which generates the feature map. Then in the input, the presence of detected features is summarized by the generated feature map. Within a given image, this model is in charge of locating and identifying a license plate. To train this network, we employ a collection of real-world license plate photos as well as license plate annotations. Since the training data we utilized covers a wide range of variations, the network is meant to adapt to different scenarios and be adaptable to regional changes in license plates. The last three convolutional layers, logical, 52*52, and 26*26, should produce the 104*104, 52*52, and 26*26 feature maps.

Since high-level characteristics have more semantic information, we still want to collect finer-grained level characteristics. The low-level features have more spatial details, on the other hand. The updated model adds a new convolution layer behind the 26*26 function plan to achieve a smaller feature map with a stride of 2. The 13*13 feature is then linked to the 26*26 function map, to perform the maximum pooling process. The results detect a high-level target. The results then can be updated and linked with the 26*26 feature map and 52*52 feature map to be used following a concatenation method for max-pooling. The findings are a medium-level goal. The intermediate-level function map samples and fits in with the 52*52 function map.

4.3.4 IMAGE SEGMENTATION

The next step is to segment the number plate out of the image after successfully detecting the number plate.

Figure 4.3 shows segmented number plate from the original image. The image segmentation has been implemented using OpenCV library, by cropping the number plate region and then saving it as the new image. Segmentation serves as a link between character recognition and number plate extraction. Boundary box analysis is another name for segmentation, and the characters are extracted by using this analysis.

4.3.5 OPTICAL CHARACTER RECOGNITION USING PYTESSERACT

Conversion of manually printed or printed text images into computer text is known as the OCR. There are a number of OCR engines available; the proposed work uses Python-Tesseract, also called Pytesseract. Python-Tesseract is a python-based OCR application. It can recognize and interpret text embedded in pictures as shown in Figure 4.4. It shows a picture of OCR using Pytesseract.

Tesseract contains a new neural network component that can recognize text lines. It is based on OCRopus' Python-based LSTM implementation; however it has been rewritten in C++ for Tesseract. Tesseract's neural network system predates Tensor Flow, but it is compatible with it because it uses the Variable Graph Specification Language as a network description language (VGSL). The Pytesseract

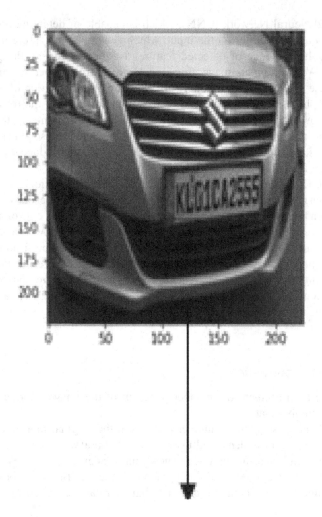

FIGURE 4.3   Segmented number plate from the original image.

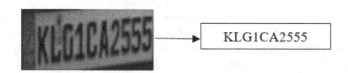

FIGURE 4.4   OCR using Pytesseract.

	License Number	Entry time	Exit time	Difference in minutes	Amount
0	/KAO3-MG- 2784	2021–06–16 18:40:42.866059	2021–06–16 18:44:43.005556	4.000000	30
1	DL7C N 5617	2021–06–16 18:40:43.124571	2021–06–16 19:44:18.931102	63.583333	80
2	SRV P	2021–06–16 18:40:43.322041 0		0.000000	0
3	COU\nEAGLE W	2021–06–16 18:40:43.523503 0		0.000000	0
4	JAG2 UAR	2021–06–16 18:40:43.735934	2021–06–16 21:16:23.689468	155.650000	100

FIGURE 4.5 Data entry in Database system.

OCR recognizes the segmented image as input and then recognizes the characters in the number plate image. The information gathered is then saved in a database or a data file.

4.3.6 STORING EXTRACTED DATA IN DATABASE SYSTEM

The characters extracted from the OCR of license plates will be stored with the date, entry and exit time, and vehicle number in the database system using the Pandas data frame. Figure 4.5 shows data entry in the database system.

Also, a unique User ID is generated for each vehicle entering in the parking lot. With the vehicle's entry and exit time in the parking lot, the total vehicle parked duration can be computed, which further can be used to compute the parking charges as per the parked time by the vehicle in the parking lot.

4.4 EXPERIMENTAL RESULTS

The experiments have been conducted on several features of vehicles with completely various shapes and dimensions, subject to different conditions in order to assess the precision. The OCR was applied for number plate recognition using Tesseract API available in python called Pytesseract. The segmentation approach did not produce the anticipated outcomes because the algorithm's accuracy was restricted by plate orientation to a certain degree and plates at the edge of the image. To be more efficient and effective, a proper camera angle configuration is necessary.

4.4.1 EVALUATION CRITERIA

The evaluation criteria considered are Accuracy, Mean Squared error.

A) Accuracy:

Accuracy is the fraction of number of correct prediction to the total number of predictions as in equation (1).

$$Accuracy = \frac{TP + TN}{TP + TN + FP + FN} X100\% \tag{1}$$

where TP stands for True Positives, TN stands for True Negatives, FP stands for False Positives, and FN stands for False Negatives.

B)   Mean squared error (MSE):

The MSE rates the accuracy of a predictor (a function that maps arbitrary inputs to a sample of values for a random variable) or an estimator i.e., a mathematical function mapping a sample of data to an estimate of a parameter of the population from which the data is sampled as in equation (2).

$$MSE = \sum_{i=1}^{n} \frac{\left(y(i) - \tilde{y}(i)\right)^2}{n} \tag{2}$$

where n is the number of data points, y are the observed values, and $\tilde{y}$ are the predicted values.

4.4.2   Result and Analysis

The literature review identified the fastest R-CNN model, VGG 16, YOLOv3, and Tiny-YOLOv3, as the most efficient and appropriate algorithms for detecting number plates in real-time. The training loss and accuracy performance of the YOLOv3 and VGG 16 algorithms were obtained and compared. The findings and evaluation plots and Tanning loss plots have been presented. The proposed method yielded the following outcomes:

Figure 4.6 demonstrates the evaluation scores of Smart Parking system using YOLOv3 detector. Score depicts the training data score, whereas score validation

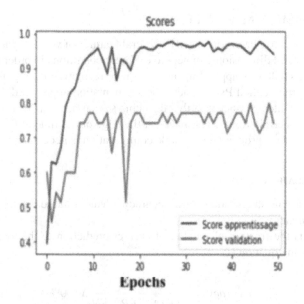

FIGURE 4.6   Evaluation scores of Smart Parking system using YOLOv3.

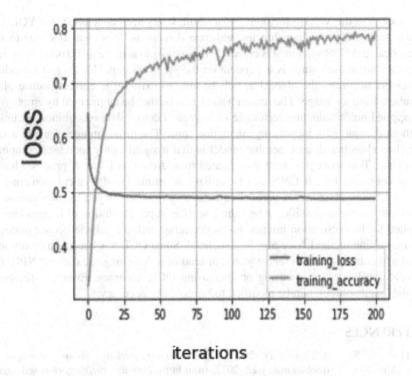

iterations

FIGURE 4.7   Loss function of Smart Parking system using VGG16.

depicts validation score (works as a part of test data from the same training dataset). It yielded an accuracy score on training data and on V = validation to be 94.2 percent and 80 percent respectively for 50 epochs.

Figure 4.7 shows the evaluation scores when the dataset was trained using a VGG16 detector. It yielded an accuracy score of 79.13 percent on 200 epochs. The YOLOv3 algorithms were found to have a higher accuracy score than the VGG16 algorithms. Hence YOLOv3 has been identified to be the optimum algorithm for real-time detections in an ANPR-based Smart Parking system.

4.5   CONCLUSION AND FUTURE WORK

An extensive literature review was conducted to understand various deep learning models capable of performing real-time number plate detection. The fastest R-CNN model, VGG 16, YOLOv3, and Tiny-YOLOv3 have been identified as the most efficient and appropriate algorithms for detecting number plates in real-time in a literature review. The proposed system was trained using the YOLOv3-Darknet framework. The model for license plate detection was trained using YOLOv3 with CNN, which is capable of detecting object and entities. The results of our method yielded an accuracy score on training data and on validation to be 94.2 percent and 80 percent respectively. The YOLOv3 algorithms have been proven to be more accurate than the VGG16 methods. As a result, it has been concluded that YOLOv3 is the most effective algorithm for real-time detections. It is clear that due to the complicated ANPR system, it is currently impossible to achieve a 100 percent overall accuracy since each stage is dependent on the previous step. However, if bounding boxes are accurate, the algorithm will be able to extract the correct license plate numbers from an image. The research work can further be augmented by employing an applied noise reduction technique to improve license plate recognition accuracy without dramatically increasing calculation time. The disadvantage of using a single class classifier in an ensemble model is that it significantly increases computation time. Two strategies were investigated to address this issue. A proposal-based technology like Fast R-CNN can be utilized to minimize the calculation time of the underlying classifier. A parallel calculation can be employed to simultaneously calculate the basic classifier. Algorithms such as super resolution of images can be applied for low-resolution images. For segmenting multiple vehicle license plates, a coarse-to-fine methodology may be beneficial. Since OCR has become a commonly used and common tool in recent years, instead of redesigning the entire ANPR, the ANPR developers are focusing on increasing OCR accuracy. Even the Tesseract model (open sources) can be modified to improve the accuracy.

REFERENCES

[1] Koba, S. (2022, January 12). *IOT-based smart parking system development.* MobiDev. Retrieved January 22, 2022, from https://mobidev.biz/blog/iot-based-smart-parking-system

[2] Digitised automated parking: The future of smart cities. (n.d.). Retrieved from www.financialexpress.com/express-mobility/digitised-automated-parking-the-future-of-smart-cities/2381238/.

[3] Yamin Siddiqui, S., Adnan Khan, M., Abbas, S., & Khan, F. (2020). Smart occupancy detection for road traffic parking using deep extreme learning machine. *Journal of King Saud University—Computer and Information Sciences*, 34, 727–733.

[4] Adarsh, P., Rathi, P., & Kumar, M. (2020, March). YOLO v3-tiny: Object detection and recognition using one stage improved model. In *2020 6th international conference on advanced computing and communication systems (ICACCS)* (pp. 687–694). IEEE.

[5] Henry, C., Ahn, S. Y., & Lee, S. W. (2020). Multinational license plate recognition using generalized character sequence detection. *IEEE Access*, 8, 35185–35199.

[6] Zain Masood, S., Shu, G., Dehghan, A., & Ortiz, E. G. (2017). License plate detection and recognition using deeply learned convolutional neural networks. *arXiv:1703.07330*.

[7] Kurpiel, F. D., Minetto, R., & Nassu, B. T. (2017, September). Convolutional neural networks for license plate detection in images. In *2017 IEEE international conference on image processing (ICIP)* (pp. 3395–3399). IEEE.

[8] Pustokhina, I. V., Pustokhin, D. A., Rodrigues, J. J., Gupta, D., Khanna, A., Shankar, K., ... Joshi, G. P. (2020). Automatic vehicle license plate recognition using optimal K-means with convolutional neural network for intelligent transportation systems. *IEEE Access*, 8, 92907–92917.

[9] Laroca, R., Severo, E., Zanlorensi, L. A., Oliveira, L. S., Gonçalves, G. R., Schwartz, W. R., & Menotti, D. (2018, July). A robust real-time automatic license plate recognition based on the YOLO detector. In *2018 international joint conference on neural networks (IJCNN)* (pp. 1–10). IEEE.

[10] Babu, R. N., Sowmya, V., & Soman, K. P. (2019, July). Indian car number plate recognition using deep learning. In *2019 2nd international conference on intelligent computing, instrumentation and control technologies (ICICICT)* (Vol. 1, pp. 1269–1272). IEEE.

[11] Bura, H., Lin, N., Kumar, N., Malekar, S., Nagaraj, S., & Liu, K. (2018, July). An edge based smart parking solution using camera networks and deep learning. In *2018 IEEE international conference on cognitive computing (ICCC)* (pp. 17–24). IEEE.

[12] Redmon, J., & Farhadi, A. (2018). Yolov3: An incremental improvement. *arXiv preprint arXiv:1804.02767*.

[13] Chen, R. C. (2019). Automatic license plate recognition via sliding-window darknet-YOLO deep learning. *Image and Vision Computing*, 87, 47–56.

[14] Yépez, J., Castro-Zunti, R. D., & Ko, S. B. (2019). Deep learning-based embedded license plate localisation system. *IET Intelligent Transport Systems*, 13(10), 1569–1578.

[15] Ghazali, M. N. B. (2018). *Development of car plate number recognition using image processing and database system for domestic car park application.* https://www.semanticscholar.org/paper/Development-of-Car-Plate-Number-Recognition-using-Ghazali/5c7f7d3c324c671508d859a183d625d2181a627f

[16] Kashyap, A., Suresh, B., Patil, A., Sharma, S., & Jaiswal, A. (2018, October). Automatic number plate recognition. In *2018 international conference on advances in computing, communication control and networking (ICACCCN)* (pp. 838–843). IEEE.

[17] Karakaya, M., & Akıncı, F. C. (2018, May). Parking space occupancy detection using deep learning methods. In *2018 26th Signal Processing and Communications Applications Conference (SIU)* (pp. 1–4). IEEE.

[18] Redmon, J., Divvala, S., Girshick, R., & Farhadi, A. (2016, June). You only look once: Unified real-time object detection. *Proceedings of CVPR*, 779–788.

[19] Zhai, X., Bensaali, F., & Sotudeh, R. (2012, July). OCR-based neural network for ANPR. In *2012 IEEE international conference on imaging systems and techniques proceedings* (pp. 393–397). IEEE.

[20] Zhang, Z., & Wang, C. (2012). The research of vehicle plate recognition technical based on BP neural network. *Aasri Procedia*, 1, 74–81.

[21] Öztürk, F., & Özen, F. (2012). A new license plate recognition system based on probabilistic neural networks. *Procedia Technology*, 1, 124–128.

[22] Reddy, C. A., & Bindu, C. S. (2015). Multi-level genetic algorithm for recognizing multiple license plates in a single image. *Journal of Innovation in Computer Science and Engineering*, 4(2), 14–20.

[23] Hima Deepthi, V., Balvinder Singh, B., & Srinivasa Rao, V. (2014). Automatic vehicle number plate localization using symmetric wavelets. In *ICT and Critical Infrastructure: Proceedings of the 48th Annual Convention of Computer Society of India* (Vol. I, pp. 69–76). Springer.

[24] Car license plate detection. (2022). Retrieved January 22, 2022, from www.kaggle.com/andrewmvd/car-plate-detection

5 Enhancing Speaker Diarization for Audio-Only Systems Using Deep Learning

Aishwarya Gupta and Archana Purwar

CONTENTS

5.1 INTRODUCTION

Sustainable development is nothing but consuming the natural resources wisely by keeping in mind the needs of future generations as well. Artificial Intelligence (AI) and its domains have impacted sustainable development goals in one way or another directly or indirectly. All the goals of sustainable development are related to our daily lives, and it's our duty to keep in mind those goals for the betterment of the society, country, environment, and universe as a whole. Artificial intelligent-enabled technology and devices have the capability to achieve the 2030 agenda of sustainable development. Analyzing

DOI: 10.1201/9781003245469-5

deep and vast data with the help of machine learning (ML) and predicting the future consequences is a boon for everyone in all forms of life. There are various requirements of everyone that can be fulfilled and eased out using intelligence and technology.

Various domains of machine learning like deep learning and pattern recognition (stochastic/face/speech/speaker recognition) are the ones to work for in the future [1]. All these variants of ML are helping the individuals in fulfilling the needs of humans or easing them up, which indirectly is accomplishing the goal of sustainable development. This will be opening up various paths to work on, such as recognition and identification of an individual based on their voice and without having any audio-visual data.

Thus, it will be exploring the variants deeply for better performance and reducing noise or error rate. Accuracy will improve after focusing on the specific drawbacks and challenges faced by that problem.

Automatic speech recognition is nothing but the basic process of deriving the word sequences of utterances from a speech waveform, and speaker recognition is something in which we recognize the person who is speaking with the help of speech waveform and the audio inputs [2, 3, 4]. These technologies are domains that are relatable to each other and go hand in hand. The former focuses on identifying the words or speech spoken by any individual whereas the latter deals with recognizing the individual who is speaking.

Speech and speaker recognition are the domains that can be extended further because, nowadays, they can address many issues like automatic responses for flight or telephonic queries, telephone directory assistance, telebanking, automatic speech recognition in real-world meetings, speaker diarization, speaker recognition, speaker verification, law enforcement agencies, and forensics, too [5, 6]. Thus, recognizing the speech of a single speaker in a normal environment is an easy task whereas identifying the speech of a single speaker in a multi-speaker noisy environment with perfect accuracy and clean separation is still a challenging task for the audio-only datasets. This task of achieving a better error rate for an audio dataset is difficult.

Speaker diarization is a process of determining "who speaks when" during any audio or video conversation. It is a task of labelling an audio or video recording corresponding to the speaker identity. Earlier it was just a mere step in the process of automatic speech recognition (ASR). Over the years there has been an improvement in accuracy and robustness found for various application of speech recognition also in the field of speaker diarization. Recently speaker diarization came out as a stand-alone domain in itself with its own challenges and approach to deal with it. It has gained popularity over numerous applications in many areas dealing with day-to-day issues and solutions as well. Diarization is all about knowing the speaker count and the part of speech where they have spoken during the complete conversation.

5.2 LITERATURE SURVEY

5.2.1 Speaker Diarization System

AI-empowered technology added a lot of convenience to the lives of common people. One of the major growing domains of machine learning has multiple applications. One can order pizza sitting at home just by saying it to Amazon Echo or Echo dot.

This advancement has been possible only because of speech recognition. Earlier it was not that hype, but recently with the help of deep learning it was possible to move out of a controlled environment.

By keeping in mind sustainable development, growth in this field is appreciable, as AI is focusing on affecting global productivity, healthcare amenities, environmental concerns, and outcomes. It might affect 17 sustainable development goals as well as 169 as promised for the 2030 agenda for sustainable development. According to the review done by R. Vinuesa [14], 134 targets will be benefitted and have a positive impact whereas 59 targets will be facing the negative impact of AI.

Speech recognition is being divided into two categories as follows:

- **Speaker dependent**: In this type of recognition an individual has to first train the software with his voice so that system can recognize it with its features and characteristics afterward. Voice recognition software recognizes unique characteristics of the individual voice due to which they are called "speaker dependent."
- **Speaker independent**: It requires no prior training at all for any recognition. These are less accurate, but real-time voice recognition is a real challenge to work on. An application like an interactive voice response system and telephonic ones are ideal for this case. As the system is unaware of the voice and speech of an individual prior, it is called "speaker independent."

Speech recognition is just a sub-category of machine learning and a further subsidiary where deep learning models are being implemented. Recurrent Neural Network and Convolution Neural Network have significantly combined with speaker diarization. It has improved the performance and DER (diarization error rate) as well. Deep learning models like LSTM, TDNN, etc. are working well in recent speaker diarization systems.

Speaker recognition is a subdomain of speech recognition. Dealing with the separation of speech of each speaker in a multi-speaker noisy environment is still difficult [7]. Speaker recognition has been an issue for many years. The need now to keep proof of communication in speech is becoming essential. It may be an important office meeting, conference, conversation with a prisoner, multimedia information retrieval, speaker diarization, speaker recognition, speaker identification and verification, speaker turn analysis, audio processing, etc.

Speaker diarization is similar to speaker recognition but with an added feature. It is simply the art of knowing "who speaks when" in any audio or video in a multi-speaker environment with background noises. It enhances the accountability of transcription by specifying which speaker spoke and also their true identity at times.

In a speaker diarization system, there are five basic stages that on combining give the final result. As per the requirement and modifications, these modules can be joined for better performance and to apply different algorithms. One has to provide input of conversation of multiple speakers in either signal form or raw wave file form. Following these stages, a final output with specific slots of the speaker with "what they spoke" is achieved.

Figure 5.1 shows five stages of the speaker diarization system:

- Speech detection
- Segmentation
- Embedding extraction
- Clustering
- Transcription

These modules apply pre-processing and post processing modules as well. But to achieve an enhanced audio data file, pre-processing is a necessary step that should be followed. So, here we have targeted it and improved the performance by removing background noise from an audio file.

In the first stage, the speech detection process takes place. In this step, the human speech is separated from various background noises and disturbances using various noise removal techniques. Here, speech and non-speech parts are being differentiated, and speech enhancement is a second stage that is being followed under this for better results. Many times, there are instances when these background disturbances like clapping, murmuring, overlapped speech segments or reverberation, etc. make audio noisy and unclear. In this case, speech enhancement techniques work by pulling up the speech clearly and ignoring all other noises in the file.

Second one is segmentation, in which an input file is divided into small audio segments. There is a specific homogeneous segment according to the individual speaker identities. It also obtains the speaker changes points within the audio to know "who speaks when".

Third is embedding extraction, in which segment-based features are being extracted such as MFCCs (Mel-frequency Cepstral Coefficient), i-vectors [15],

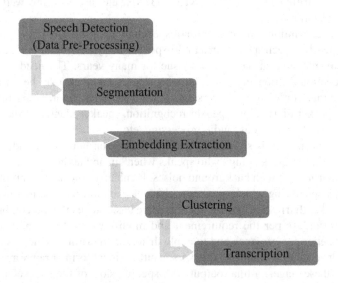

FIGURE 5.1   Modules of speaker diarization system.

x-vectors [16,17,18], and d-vectors [19]. It is an important step in the diarization pipeline. Nowadays audio embeddings are based on neural networks.

Now, the clustering step is followed for combining the same speaker clusters and arranging them in a group. The extracted embeddings from the previous procedure are clustered into various speakers. The number of clusters denotes the speaker count, and labelling of those speakers is done based on that count.

At last, transcription of the received segmented file is done after labelling of the clusters. With the help of a speech text application or using a speech recognition system, one can get the inside content of the conversation between the speakers.

Most of the papers mentioned neural networks and their variants like Deep Neural Network, Recurrent Neural Network, Deep Recurrent Neural Network, Long Short-Term Memory Network, and Convolution Neural Network as the relatable solution in some way or another for speaker recognition [8]. Also, models like the Gaussian mixture model (GMM) and hidden Markov model (HMM) are used for enhancing statistical frameworks in the field [9].

In a paper, "A Memory Augmented Architecture for Continuous Speaker Identification in Meetings," Flemotomos, Nikolaos, and Dimitrios Dimitriadis [10] discussed an extra element of memory augmentation added that makes use of the LSTM network along with RNN. It has a scope of improvements in storing speaker profiles for future scope. Learning about the past experience somehow can be used for the future.

Zhang, Aonan, et al. in "Fully supervised speaker diarization" [11] suggested a zero-clustering technique to resolve the speaker diarization. For future work, we have several options for modification in the field of:

- Speaker change process: In this, the coin-flipping process is used, which can be replaced with the recurrent neural network.
- Speaker assignment: This can be made through a distribution process, whichever will be applicable.
- Training: It can be done with the help of unlabeled data rather than in the space of labeled data.
- Decoding: Instead of online decoding one can use an offline decoding process to make it more feasible and efficient.

In the paper "Speaker diarization with LSTM," Wang, Quan, et al. [12] have mentioned that their model will be useful only for two-speaker environments, but it has a scope for modifying and updating the model for the multi-speaker environment. They achieved a better diarization error rate (DER) as compared to the one used in Feedforward neural network, but the scope remains in other features also like audio embeddings, clustering, and re-segmentation.

In "Unsupervised methods for speaker diarization: An integrated and iterative approach" Shum, Stephen H., et al., [13] limited diarization to telephonic conversation only, which can be tried over-application of broadcast news and meetings. Also, in a multi-speaker environment like crowded parties and outside environments, it can be extended. While applying I-vector embedding, d-vector and x-vector in the latest context can be considered and be used for further references.

5.2.2 CHALLENGES

Every domain has its challenges that can be targeted and overcome slowly over time. Similarly, speaker diarization faces various drawbacks, which are as follows [24]:

- Overlapped speech
- Background noise
- Short conversation
- Multiple speaker interaction
- Different acoustic settings
- Working with real-time data.
- Processing online datasets

5.3 APPLICATIONS

Speaker Diarization can be applied in various real-world scenarios. It is being used for many types of businesses purposes and professional roles [20,21]. Some of the applications are as follows:

- Meetings, news, and broadcasting
- Seminars, workshops, and conferences
- Legal and forensics
- Healthcare and medical services
- Software development

These are some primary application areas of speaker diarization that generally differ in recording quality (background noise, bandwidth, microphone); types and sources of non-speech content, style and structure of speech, and the number of speakers in various domains can vary at maximum range [22].

In call centers conversation [23] as well speaker diarization are difficult tasks to deal with, but the struggle is with channel quality and other non-speech elements that lie within it. In [23] L. Itshak et al. discussed improvements for these issues.

Speaker diarization has a broad area of applications, and all of them are useful in one or another manner for mankind as a whole, as it is easing the complexity faced by an individual while identifying the speaker in any recording.

5.3.1 APPLICATION IN NEWS AND BROADCASTING

Speaker diarization is one of the approaches that helped telecommunication and the news community overall long back. Generally, it is used for recording purposes and video captioning.

In broadcasting, data includes various qualities or types of signals like background noise in the studio, music, speech over music, etc. The system needs to be robust in order to deal with any sort of disturbance in a nearby environment.

One point in this case of news and broadcasting that is different from other applications is that speakers are already known in this, so all that remains is to verify "who spoke when and what."

In 2005, M. Sylvain et al. [25] described two approaches to work on. In the first one speech of an individual speaker is being detected, followed by the clustering process. But in the second one both the methods are processed simultaneously. Both strategies help in testing with various combinations of results of speaker diarization.

In 2006, B. Claude et al. made improvements in the system to achieve high cluster purity, but it tends to split data from speakers with a large quantity of data into several segment clusters [26] by replacing an iterative Gaussian mixture model (GMM) clustering with a Bayesian information criterion (BIC) agglomerative clustering. Also, they added the stage of clustering with the help of the GMM-based speaker identification method. Here, a multistage system is developed in addition to the baseline system on NIST Fall 2004 Rich Transcription (RT-04F).

In the same year, Zhu et al. [27] tried the same previously mentioned speaker diarization system by B. Claude for the lecture dataset i.e. multiple distant microphone (MDM). It used smooth log-likelihood ratio-based speech activity detection (SAD) with acoustic models whereas previously authors used the Viterbi (SAD) system. Thus, it performed well with the MDM lecture data.

5.3.2 Application in Healthcare

Speech recognition has provided solutions in healthcare for various problems. One is medical prescription, and another is patient record. At times it happens that, due to wrong medication or because of misunderstanding of a prescription by a chemist, the wrong medicine is taken by the patient, which can have severe unknown.

So, recently Choudhary K. et al. [28] have countered the issue and suggested the voice e-prescription method. The system records a call and then creates a pdf that will be forwarded to a patient with a secret key and whose password is given only to the concerned patient. This application is a good step for solving the issue as well as in the future a digital ecosystem that is more sustainable for everyone to communicate with doctors and get consultations with ease without any disruption.

Nowadays, with the invention of IoT smart devices, AI technology can be used to ease our daily lives. So, Ismail A et al. [29] suggested an effective solution that is designed with the help of IoT and speech technology. The system defines a flow in which both patient and hospital are connected through an online application. Using dynamic time warping, the algorithm of support vector machine speech recognition process was improved.

In 2020 Dharma Singh Jat [30] described a speech-based automation system for orthopedic patients using voice commands.

During this corona phase, thousands of online virtual solutions have been suggested, but only some of them got implemented and came into real use. Thus, artificial intelligence has been constantly supporting and improving the lifestyle by providing immense facilities by reforming the system. This will also support fulfilling one of the sustainable development goals of good health and well-being.

5.3.3 Application in Forensics

Speaker forensics is one of the important applications of speaker recognition. To use speaker forensics as evidence in the courtroom the diarization system has to achieve maximum accuracy with minimum possible error. Diarization can help the government to deal with a critical matter where it is difficult to understand the audio recording clearly due to a noisy environment.

R J Rouf et al. [31], used joint factor analysis and i-vector. The former one is used for compensating the variability and ignoring the channel factors [32, 33].

5.3.4   APPLICATION IN LEGAL PROCEEDINGS

Speaker diarization is one of the approaches that helped numerous fields to use audio recordings for many years. It added more life to recordings by getting the speaker's identity and also its transcription. This applies to the academic, research, and legal industries. It allows easy access to clean court proceeding text and contributes to new methods to improve access to affordable legal services. Also, it facilitates academic research into the judiciary. In recent years this application seems to be growing at a good pace, and many researchers are looking forward to it. Diarized transcripts should be made publicly available to ease public records.

In the past few years, diarization remains a challenge as various commercial APIs could not provide satisfactory output because audio contains background noise and overlapped speech from the environment. Some authors have worked on this application of speaker diarization where they have found the solution leading to transcription of recorded court proceedings automatically [34].

Recently, T. Jeffery et al., 2021 [35] mentioned the issue being faced by individuals in this legal industry. Here, getting the speaker's annotation from courtroom recordings is a difficult task that is generally not available publicly for any reference if needed.

Last, if one could provide better solutions using speaker diarization or speech processing for resolving the cases with more ease and more smoothly, then this would be completing one of the tasks of sustainable development goals, i.e. peace justice and strong institutions.

5.3.5   APPLICATION IN EDUCATION

The Speaker diarization process has helped students and teachers with the peer-led team learning strategy. In this learning outcomes have improved metrics like course grades overall.

In [36] the author suggested a new speaker diarization system with different and new change detection and clustering algorithm [37, 38]. The DER was improved significantly from the previously used LIUM diarization pipeline. The system is implemented by marking the difference between primary and secondary speakers. But the main disadvantage of this system is that it works only for two speakers at a time for every channel.

This looks forward to resolving more education-related issues like communications problems between students and teachers in normal or even in virtual learning. Also, the interaction between parents and teachers can be improved using this speaker diarization system. Finally, all this will help accomplish one of the sustainable development goals of quality education.

5.4 SYSTEM METHODOLOGY

A speaker diarization system is composed of five modules in which the speech detection phase is focused. In this phase data pre-processing takes place before analyzing it. So, removing the unwanted disturbance, background noise, etc. in this stage is essential. First, different types of noises present in an audio signal like the following are taken care of:

- Broadband noise (hiss or static fall)
- Narrowband noise (incorrect grounding and poorly shielded cables)
- Impulse noise (clicks and pops)
- Irregular noise (background conversation, traffic, and rain).

The main focus is on the technique of noise reduction, which is nothing but the process of adding various filters at each stage to get a better and clearer outcome out of the raw file. In simple words, it can be said that it is a series of filters that allow the removal of specific frequencies of audio where noise occurs.

There are many filtering techniques like adaptive filter, Wiener filtering, Kalman filter, sub-band coding, discrete wavelet-transform, etc. Other than these there are spectral subtraction for broadband noise, high pass filters for narrow-band noise, de-clicking for impulse noise, and various gating techniques are also available. So, the latest out of all noise reduction techniques is the spectral gating technique, in which an unwanted signal is masked and not allowed to propagate forward. The gating method tries to provide a specific window for the particular signal to move forward.

5.4.1 PROPOSED SYSTEM

Figure 5.2 shows our proposed system, in which the first three blocks represent the pre-processing steps (it takes a noisy audio input, processes it through spectral gating technique, and produces a clean audio file). Also, in the case of a clear audio file, any external noise (Gaussian noise) can be added, and differences can be seen. Using the base architecture of the speaker diarization system, only the data pre-processing step has been modified to analyze the changes and improvements in the overall diarization error rate [39].

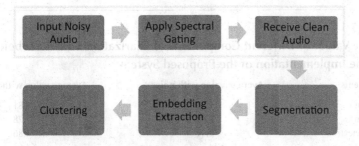

FIGURE 5.2 Block diagram of the proposed system.

5.4.2 Dataset and Implementation Details

For implementing the VoxConverse dataset [40], a large-scale diarization dataset was introduced by authors of Voxceleb [41] and Voxceleb2 [42]. It consists of 50 hours of multi-speaker clips of human conversations in various forms of television interviews and shows, telephonic calls, and other conversations.

Using plan note, metrics, and sci-py, the proposed system is implemented effectively. Diarization error rate (DER) and its three components are:

- False alarm (FA) refers to duration when a non-speech part is considered as speech.
- Missed detection (MD) is the duration when the speech part is missed mistakenly.
- Confusion refers to speaker confusion duration.

All three components have been calculated for the comparison in both scenarios. The comparison can be seen in Table 5.1.

The system adheres to all the modules effectively and smoothly for the VoxConverse dataset and audio clip as well.

5.4.3 Results

The proposed system is analyzed based on the DER performance metric.

Table 5.1 shows that, other than missed detection, both false alarm and confusion were improved. Also, the overall error rate was reduced by a significant amount.

Figure 5.3 clearly depicts the improvement between both the cases after imposing the proposed system on the dataset. The false alarm got reduced, which means the system put up the wrong alarm fewer times than before. Reduction in confusion also suggested improved accuracy in the system. But only missed detection increased, which also affected the overall error rate. The DER was reduced by a large difference, and the system's performance got improved as well. Accuracy will be enhanced when the noise is removed.

TABLE 5.1

Average Values of Different Components of Diarization Error Rate before and after the Implementation of the Proposed System

Components	Average value with noise	Average value without noise
False alarm	11.25	8.81
Confusion	9.41	7.78
Missed detection	2.41	4.86
DER%	74	54.6

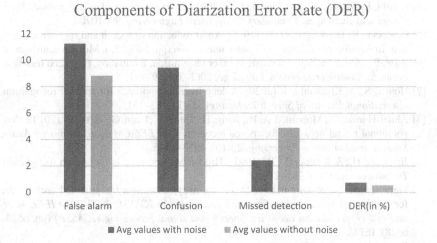

FIGURE 5.3   Comparison of various components of DER on the proposed system before and after the removal of noise.

5.5   CONCLUSION

As the proposed system suggested, the enhancement in the pre-processing step helps in improving the performance and reducing the overall error rate in the system. The diarization error rate was reduced from 74 percent to 54.6 percent, which is significantly less on this VoxConverse dataset. This dataset was recently formed by makers of the Voxceleb dataset for processing audio-only systems like ours for better accuracy and results. The audio-only datasets are less publicly available, so telephonic conversations, broadcast and recorded meetings, seminars, and workshops are the only sources of evaluation for these types of concerns.

So, enhancement at the initial stage is very effective and helpful for further modules as well as removal of waste and unused components at the starting leads to running the system smoothly without any hindrance.

5.6   FUTURE SCOPE

For improving the system's capabilities in parallel it can be tested for various neural networks types in this enhancement technique. Also, the system can be enhanced with a machine learning algorithm at a different stage for different applications.

REFERENCES

[1] Sarker, I.H. (2021). Machine learning: Algorithms, real-world applications and research directions. *SN Computer Science*, 2, 160.
[2] Benzeghiba, M., De Mori, R., Deroo, O., Dupont, S., Erbes, T., Jouvet, D., . . . Wellekens, C. (2007). Automatic speech recognition and speech variability: A review. *Speech Communication*, 49(10–11), 763–786.
[3] Yu, D., & Deng, L. (2016). *Automatic speech recognition*. Springer.

[4] Hanifa, R. M., Isa, K., & Mohamad, S. (2021). A review on speaker recognition: Technology and challenges. *Computers & Electrical Engineering*, 90, 107005.

[5] Peacocke, R. D., & Graf, D. H. (1995). An introduction to speech and speaker recognition. In *Readings in human—computer interaction* (pp. 546–553). Morgan Kaufmann.

[6] Rose, P. (2006). Technical forensic speaker recognition: Evaluation, types and testing of evidence. *Computer Speech & Language*, 20(2–3), 159–191.

[7] Hourri, S., & Kharroubi, J. (2020). A deep learning approach for speaker recognition. *International Journal of Speech Technology*, 23(1), 123–131.

[8] Abdel-Hamid, O., Mohamed, A. R., Jiang, H., Deng, L., Penn, G., & Yu, D. (2014). Convolutional neural networks for speech recognition. *IEEE/ACM Transactions on Audio, Speech, and Language Processing*, 22(10), 1533–1545.

[9] Juang, B. H., & Rabiner, L. R. (1991). Hidden Markov models for speech recognition. *Technometrics*, 33(3), 251–272.

[10] Flemotomos, N., & Dimitriadis, D. (2020, May). A memory augmented architecture for continuous speaker identification in meetings. In *ICASSP 2020–2020 IEEE international conference on acoustics, speech and signal processing (ICASSP)* (pp. 6524–6528). IEEE.

[11] Zhang, Aonan et al. (2019). Fully supervised speaker diarization. In *IEEE international conference on acoustics, speech and signal processing (ICASSP)*. IEEE.

[12] Wang, Q., Downey, C., Wan, L., Mansfield, P. A., & Moreno, I. L. (2018, April). Speaker diarization with LSTM. In *2018 IEEE International Conference on Acoustics, Speech and Signal Processing (ICASSP)* (pp. 5239–5243). IEEE.

[13] Shum, Stephen H. et al. (2013). Unsupervised methods for speaker diarization: An integrated and iterative approach. *IEEE Transactions on Audio, Speech, and Language Processing*, 21(10), 2015–2028.

[14] Vinuesa, R., Azizpour, H., Leite, I. et al. (2020). The role of artificial intelligence in achieving the sustainable development goals. *Nature Communication*, 11, 233.

[15] Sell, G., & Garcia-Romero, D. (2014). Speaker diarization with PLDA i-vector scoring and unsupervised calibration. In *2014 IEEE spoken language technology workshop (SLT)*. IEEE.

[16] Landini, F. et al. (2021). Bayesian HMM clustering of x-vector sequences (VBx) in speaker diarization: Theory, implementation, and analysis on standard tasks. *Computer Speech & Language*, 101254.

[17] Snyder, David, Garcia-Romero, Daniel, Sell, Gregory, Povey, Daniel, Khudanpur, Sanjeev et al. (2018, April 15–20). *X-vectors: Robust DNN embeddings for speaker recognition*. 43rd International Conference on Acoustics, Speech, and Signal Processing ICASSP'18, Calgary TELUS Convention Centre.

[18] Landini, Federico et al. (2021). Analysis of the BUT diarization system for voxconverse challenge. In *ICASSP 2021–2021 IEEE international conference on acoustics, speech, and signal processing (ICASSP)*. IEEE.

[19] Kang, Wonjune, Roy, Brandon C., & Chow, Wesley. (2020). Multimodal speaker diarization of real-world meetings using d-vectors with spatial features. In *ICASSP 2020–2020 IEEE international conference on acoustics, speech, and signal processing (ICASSP)*. IEEE.

[20] Tranter, S. E., & Reynolds, D. A. (2004, May). Speaker diarization for broadcast news. In *2004: A Speaker Odyssey. The Speaker Recognition Workshop (ISCA, Odyssey 2004)*. ISCA.

[21] Cettolo, M., Vescovi, M., & Rizzi, R. (2005). Evaluation of BIC-based algorithms for audio segmentation. *Computer Speech and Language*, 19, 147–170.

[22] Reynolds, D., & Torres-Carrasquillo, P. (2005, March). Approaches and applications of audio diarization. In *Proceedings of international conference on acoustics speech and signal processing (IEEE, ICASSP 2005)*. IEEE.

[23] Mirghafori, N., & Wooters, C. (2006, May). Nuts and flakes: A study of data characteristics in speaker diarization. In *IEEE ICASSP 2006*, pp. 1017–1020, IEEE.

[24] Hernawan, S. (2012, September). Speaker diarization: Its developments, applications, and challenges. *Proceedings International Conference Information System Business Competitiveness*. http://eprints.undip.ac.id/36153/1/Hernawan_Sulity.pdf

[25] Meignier, Sylvain et al. (2006). Step-by-step and integrated approaches in broadcast news speaker diarization. *Computer Speech & Language*, 20(2–3), 303–330.

[26] Barras, Claude et al. (2006). Multistage speaker diarization of broadcast news. *IEEE Transactions on Audio, Speech, and Language Processing*, 14(5), 1505–1512.

[27] Zhu, Xuan et al. (2006). Speaker diarization: From broadcast news to lectures. In *International workshop on machine learning for multimodal interaction*. Springer.

[28] Choudhary, K., Agrawal, T., Dama, R., & Rathod, M. (2021). *Voice-based e-prescription*. Available at SSRN 3867317.

[29] Ismail, Ahmed, Abdlerazek, Samir, and El-Henawy, Ibrahim M. (2020). Development of smart healthcare system based on speech recognition using support vector machine and dynamic time warping. *Sustainability*, 12(6), 2403.

[30] Jat, Dharm Singh, Limbo, Anton S., & Singh, Charu. (2020). Speech-based automation system for the patient in the orthopedic trauma ward. In *Smart biosensors in medical care*, pp. 201–214. Academic Press.

[31] Rouf, R. J., & Arifianto, D. (2021, April). Speaker forensic identification using joint factor analysis and i-vector. In *Journal of physics: Conference series*. IOP Publishing, Vol. 1896, No. 1, p. 012026.

[32] Kenny, P., Dehak, R., & Dumouchel, P. (2011). Front-end factor analysis for speaker verification. *IEEE Trans. Audio, Speech, and Language Processing*, 16(4), 788–798.

[33] Mandasari, M. (2011). Evaluation of i-vector speaker recognition systems for forensic application. *INTERSPEECH'11*, 21–24.

[34] Rusko, Milan, Darjaa, Sakhia, & Pálfy, Juraj. *Automatic speaker diarization in simulated courtroom conditions* Conference Forum Acusticum, Sep 2014.

[35] Tumminia, J., Kuznecov, A., Tsilerides, S., Weinstein, I., McFee, B., Picheny, M., & Kaufman, A. R. (2021). Diarization of legal proceedings. identifying and transcribing judicial speech from recorded court audio. *arXiv preprint arXiv:2104.01304*.

[36] Dubey, H., Kaushik, L., Sangwan, A., & Hansen, J. H. (2016). A speaker diarization system for studying peer-led team learning groups. *arXiv preprint arXiv: 1606.07136*.

[37] Huttenlocher, D. P., Klanderman, G. A., & Rucklidge, W. J. (1993). Comparing images using the hausdorff distance. *IEEE Transactions on Pattern Analysis and Machine Intelligence*, 15(9), 850–863.

[38] Basalto, N., Bellotti, R., De Carlo, F., Facchi, P., Pantaleo, E., & Pascazio, S. (2008). Hausdorff clustering. *Physical Review E*, 78(4), 046112.

[39] Park, T. J., Kanda, N., Dimitriadis, D., Han, K. J., Watanabe, S., & Narayanan, S. (2021). A review of speaker diarization: Recent advances with deep learning. *arXiv preprint arXiv:2101.09624*.

[40] Chung, Joon Son et al. (2020). Spot the conversation: Speaker diarisation in the wild. *arXiv preprint arXiv:2007.01216 (2020)*. Interspeech.

[41] Nagrani, Arsha, Chung, Joon Son, & Zisserman, Andrew. (2017). Voxceleb: A large-scale speaker identification dataset. *arXiv preprint arXiv:1706.08612*.

[42] Chung, Joon Son, Nagrani, Arsha, & Zisserman, Andrew. (2018). Voxceleb2: Deep speaker recognition. *arXiv preprint arXiv:1806.05622*.

[43] Aronowitz, H., Zhu, W., Suzuki, M., Kurata, G., & Hoory, R. (2020). New advances in speaker diarization. *Interspeech*, 279–283.

[44] Siegler, M., Jain, U., Raj, B., & Stern, R. (1997). *Automatic segmentation and clustering of broadcast news audio*. The DARPA Speech Recognition Workshop.

[45] Siu, M.-H., Rohlicek, R., & Gish, H. (1992). An unsupervised, sequential learning algorithm for segmentation of speech waveforms with multi-speakers. In *Proceedings of international conference on acoustics speech and signal processing (ICASSP 92)*, ICASSP, vol. 2, pp. 189–192.

[46] Hain, T., & Woodland, P. (1998). Segmentation and classification of broadcast news audio. In *Proceedings of international conference on spoken language processing (ICSLP 98)*, ICASSP.

[47] Kim, D.Y., Evermann, G., Hain, T., Mrva, D., Tranter, S., Wang, L., & Woodland, P.C. (2003). Recent advances in broadcast news transcription. In *Automatic speech recognition and understanding*, IEEE, ASRU 2003, pp. 105–110.

[48] Chen, S., & Gopalakrishnan, P. (1998). Speaker, environment, and channel change detection and clustering via the Bayesian information criterion. In *DARPA broadcast news transcription and understanding workshop*, DARPA.

[49] Adami, A., Kajarekar, S.S., & Hermansky, H. (2002). A new speaker change detection method for two-speaker segmentation. In *Proceedings of international conference on acoustics speech and signal processing (ICASSP 2002)*, ICASSP, vol. IV, pp. 3908–3911.

[50] Ajmera, J., & Wooters, C. (2003). A robust speaker clustering algorithm. In *Automatic speech recognition and understanding*, IEEE, ASRU 2003, pp. 411–416.

[51] Moh, Y., Nguyen, P., & Junqua, J.-C. (2003, April). Towards domain in-dependent speaker clustering. In *Proceedings of international conference on acoustics speech and signal processing (IEEE, ICASSP 2003)*, IEEE.

[52] Delacourt, P.,& Wellekens, C. (2000). DISTBIC: A speaker-based segmentation for audio data indexing. *Speech Communication*, 32, 111–126.

[53] Zhu, X., Barras, C., Meignier, S., & Gauvain, J. -L. (2005, September). Combining speaker identification and BIC for speaker diarization. *Proceedings of the 9th European conference on speech communication and technology (ISCA Interspeech'05)*, ISCA, pp. 2441–2444.

[54] Anguera, X., Wooters, C., Peskin, B., & Aguilo, M. (2005, July). *Robust speaker segmentation for meeting: The ICSI-SRI Spring 2005 diarization system*. MLMI 2005 Meeting Recognition Workshop.

[55] Tranter, S. E., & Reynolds, D. A. (2004, May). *Speaker diarisation for broadcast news*. Proceedings ISCA Speaker Recognition Workshop Odyssey.

[56] Lapidot, I., Aminov, L., Furmanov, T., & Moyal, A. (2014). *Speaker diarization in commercial calls*.

[57] Ben-Harush, O., Lapidot, I., & Guterman, H. (2012, February). Initialization of iterative-based speaker diarization for telephone conversations. *IEEE Trans. on Audio, Speech, and Language Processing*, 20(2), 414–425.

[58] Lapidot, I., Bonastre, J. -F., & Bengio, S. (2014, June 16–19). *Telephone conversation speaker diarization using mealy-HMMs*. Speaker Odyssey.

[59] Miro, A. X., Bozonnet, S., Evans, N., Fredouille, C., Friedland, G., & Vinyals, O. (2012, February). Speaker diarization: A review of recent research. *IEEE Transactions on Audio, Speech, and Language*, 20(2), 356–370.

[60] Tranter, E., Reynolds, D. et al. (2006). An overview of automatic speaker diarization systems. *IEEE Transactions on Audio, Speech, and Language Processing*, 14(5), 1557–1565.

[61] Gallardo-Antolın, X. A., & Wooters, C. (2006, September). Multi-stream speaker diarization systems for the meetings domain. In *Proceedings of international conference spoken language processing (ICSLP06)*, ICSLP.

[62] Reynolds, D. A., & Torres-Carrasquillo, P. (2005, March). Approaches and applications of audio diarization. In *Proceedings (ICASSP'05): IEEE international conference on acoustics, speech, and signal processing, 2005*, ICASSP, vol. 5, pp. v–953.

[63] https://www2.cs.uic.edu/~i101/SoundFiles/preamble.wav.

[64] Karpagavalli, S., & Chandra, E. (2016). A review on automatic speech recognition architecture and approaches. *International Journal of Signal Processing, Image Processing and Pattern Recognition*, 9, 393–404.

[65] Wyse, L. (2017, May 18–19). Audio spectrogram representations for processing with convolutional neural networks. In *Proceedings of the IEEE international conference on deep learning and music*, IEEE, pp. 37–41.

[66] Pedregosa, F. et al. (2011). Scikit-learn: Machine learning in Python. *Journal of Machine Learning Research*, 12, 2825–2830.

[67] Ning, M. Liu, Tang, H., & Huang, T. S. (2006). *A spectral clustering approach to speaker diarization*. Ninth International Conference on Spoken Language Processing, 2006.

[68] Von Luxburg, U. (2007). A tutorial on spectral clustering. *Statistics and Computing*, 17(4), 395–416.

6 Food Image Recognition Using CNN, Faster R-CNN and YOLO

Megha Chopra and Archana Purwar

CONTENTS

6.1 INTRODUCTION

The quest for a balanced dietary intake is a desire for most of the people in the world today. Everyone wants to eat healthy these days. Dietary habits directly influence the quality of life an individual possesses. Obesity is one of the primary reasons for most chronic diseases. So, to control obesity and to be in shape, one needs to eat healthy and also control the portion size. In this digital era, everyone is dependent on technology for most of their needs, including dietary intake. Technology as a medium could be used to increase awareness amongst people for moving to a healthier life. People are dependent on various mobile food tracking applications and websites for controlling their food intake. Dietary assessment is one crucial thing that needs to be managed accurately and efficiently. Conventional methods are really difficult to use. Nowadays hundreds of applications and softwares are present to monitor diet. While some of them let users manually enter the food type that they are eating, some let the users click a picture to process it. Methods to monitor the diet don't matter much; what matters is the classification accuracy of food. The system should be able to recognize the food accurately and efficiently.

DOI: 10.1201/9781003245469-6

Food image classification is a predominant step of dietary assessment. It is of immense importance as, once the correct food is detected by the system, it becomes easier to process further. This work is going to provide a comparative analysis to recognize food images. For the recognition of any image, segmentation plays a pivotal role. It is necessary to segment the food image properly to get the correct recognition. The food images are really grueling to work upon. It is difficult to categorize food, as many food classes look indistinguishable. It is of utmost importance to categorize it accurately, as if a particular class is detected as another, it changes the whole picture. The calorie intake and nutritional value of each class are completely divergent from each other [1].

Dietary assessment is an essential component for good health. Machine learning has emerged with various applications in diverse fields, and this work has addressed the capability of machine learning in the domain of food computing. In this digital era, we have more accurate methods for calculating food intake and type of food. Many algorithms have been proposed for classification and calorie estimation, but each one has a shortcoming of different kind of pre-set conditions and other issues. In food computing the first and foremost task is classifying the food image. This study initiates with using Convolutional Neural Network (CNN), Faster R-CNN, and YOLO as classifiers for food image recognition and to provide a comparative analysis among them [2]. CNN as a classifier is used for classification of food images. Faster R-CNN has also been used for food image recognition. In addition to this, YOLO as a classifier has also been implemented on the food database for improving the classification accuracy. The purpose of applying the CNN, Faster R-CNN, and YOLO is to classify as efficiently as possible because this can change the whole picture as it is directly dependent on the overall dietary assessment results. This work has been implemented on UEC FOOD-100, a dataset that contains pictures of 100 kinds of food. Results show that YOLO has outperformed all other combinations for classification accuracy. Better accuracy will improve the food dietary assessment process drastically.

This paper has proposed CNN, Faster R-CNN, and YOLO as classifiers for food image classification/recognition. The experimental results have been formulated on real-world dataset UEC-100.

This research work is arranged in following sections. Section 1 describes the Introduction. Section 2 depicts the related work. Section 3 depicts the proposed methodology. Experimental results along with graphs are given in section 4. Section 5 covers the conclusion. Section 6 covers the future work.

6.2   RELATED WORK

Food recognition is an important aspect in monitoring dietary intake. A lot of methods have been proposed in the literature. L. Bossard et al. presented a system for food recognition using random forests [3]. Kagaya, H. et al. [4] tuned the CNN model to achieve better accuracy for food image recognition. In the work Dietcam [5], multiview clustering using multikerel SVM has been used to improve the overall accuracy. Kawano et al. [6] used linear SVM as a classifier along with Grab Cut segmentation. Liu, C. et al. [7] proposed a new dimension in CNN by introducing an inception module in the work to reduce the dimensions and increase the depth. The output received from the module has been used for pooling of the next layer. Liu, C. et al. [8] proposed a CNN-based system where images taken from smartphones have been classified. They used CNN models Inception V3 and Inception V4 for Food-101 dataset, and Yunus et al. [9] built their own dataset and used the Inception V3 model to build a mobile application for food image recognition. Simon Mezgec et al. [10] in their research use GoogleNet and AlexNet as algorithms in contrast with semantic segmentation for the food recognition. Pouladzadeh, P. et al. [11] in their research work used region mining selection as the segmentation technique for food image recognition along with SVM as a classifier on deep belief network (DBN).

Different segmentation techniques yield different results in terms of accuracy and time [12]. It is indeed imperative to opt for a technique that helps in finding the optimized segmented image.

Distinct techniques are available in the literature for recognizing the food images.

Table 6.1 shows a comparative analysis of the standard methods used in the food image recognition domain along with their accuracies. This work has also incorporated some state-of-the art research works as well methods in Table 6.1.

6.3 PROPOSED METHODOLOGY

6.3.1 CNN-Based Technique for Food Image Recognition

This work has proposed a CNN-based food image recognition system. When an input image is given, the CNN manages all the classification steps incorporating extraction of features from the image. Labelling of an image is done as the final output [7].

TABLE 6.1
State-of-the-Art Techniques for Food Image Classification

Method/Research Work	Accuracy	Dataset Used
[3] (RFDC Based Approach)	50.76%	UEC-256
[3] (Food Image Recognition with Convolutional Neural Networks CNN based approach)	56.4%	UEC-256
[13]	72.26%	UEC-100
Extended HOG: Patch-FV + Colour Patch-FV (flip)	59.6%	UEC-100
SURF- BoF+ Color Histogram	42.0%	UEC-100
MKL	51.6%	UEC-100
[14]	70.2%	UEC-100

In comparison to traditional methods, CNN averts the cumbersome procedure of extracting features and reconstructing data as here the image is sent straight away as an input to the network. Eventually we get the class of the object as an output.

As shown in Figure 6.1, the feature set of food images has been mapped, and using the convolution method, a pixel wise matrix has been formed. Further, Rectified Linear Unit functions have been applied to filter the matrix with negative convolution values. Since the feature map matrix is large, for computing reduction of the matrix max pooling has been applied to keep the best feature. Similarly all feature maps have been arranged in a 3D feature map structure. This 3D feature map has been flattened for classification and passed to the classifier where all features have been classified using classification layers.

6.3.2 FASTER R-CNN-BASED TECHNIQUE FOR FOOD IMAGE RECOGNITION

Faster R-CNN is the most extensively used version in the family of R-CNN networks. R-CNN and Fast R-CNN [15] are based on the selective search algorithm. R-CNN and Fast R-CNN begin with generating 'Region—Proposals'. Basically, R-CNN [16] and Fast R-CNN apply the Selective Search algorithm to subsegment an image. It works on four characteristics of the images based on color, size, shape, and textures. Now the similar regions based on previously mentioned characteristics are combined to form a lager region. The selective search algorithm takes a longer computational time, hence it affects the network performance.

Here in Faster R-CNN, the selective search algorithm is eradicated by a separate network to calculate 'Region—Proposals'. Proposed regions are reconfigured by utilizing a ROI pooling layer that is further operated for the classification of an object.

Figure 6.2 shows food image classification using Faster R-CNN. A feature set of food images are mapped and, using the convolution method, a pixel wise matrix has

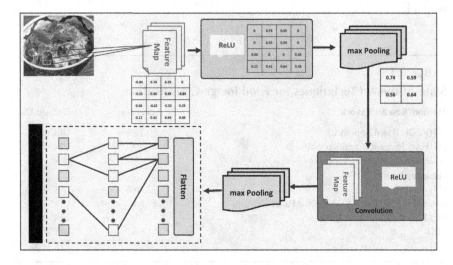

FIGURE 6.1 Food image classification using CNN.

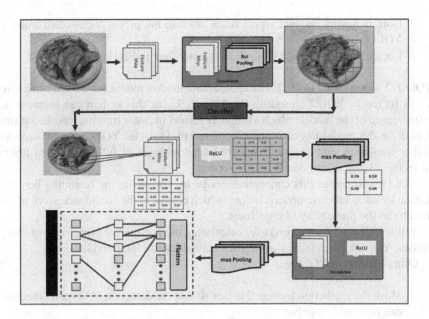

FIGURE 6.2 Food image classification using Faster R-CNN.

been formed. Further, using region of interest pooling (RoI), various regions have been identified and marked via bounding box. These bounding boxes then pass to CNN for classification.

6.3.3 YOLO-Based Technique for Food Image Recognition

This work has also proposed YOLO Based Technique for Food Image Recognition. YOLO stands for You Only Look Once. YOLO considers the complete image in one instance and partitions the image into a grid (vXv). Image localization is performed for all the grids, and then bounding boxes are envisioned [15]. Class probabilities are determined for all the respective bounding boxes. YOLO has a faster computational speed as compared to many other food recognition techniques for calorie estimation [2, 17–19].

The conceptual design of YOLO is faster in comparison with CNN and F-CNN; initially dividing the image and computing five most necessary parameters i.e. x, y, w, h, and a confidence score results in very fast detection of the object.

YOLO has three versions; initially in 2016, YOLOv1 was released, and YOLOv1 has the loss function to reduce the computation time in predicting bounding boxes with no object. A bounding box with no object has a zero confidence score. To handle these, YOLO reduced the weight for those bounding boxes. YOLOv1 has the following limitations

1. Small object detections that appear in groups are very difficult using YOLO V1 because of its reduced weight loss design.

2. Objects having unusual aspect ratios are also not properly classified using YOLOv1.
3. Localization errors in YOLOv1 are greater than in Fast R-CNN.

YOLOv2 was released in 2017. The architecture made a number of iterative improvements on top of YOLO, including BatchNorm. Using this system can increase the convergence of the model, which ultimately results in faster training; resolution limitation for the model has also been increased to 448x448. YOLOv2 now supports higher resolution, and anchor boxes are added in place of fully connected layers. Objectness score is introduced in this version.

YOLOv3 improves this conceptual design by improving the bounding box prediction by more intuitive understanding, which considers the confidence score of box directly as the probability of objectness.

While YOLOv2 uses softmax for calculating the conditional probability of image classes, YOLOv3 uses individual logistics classification for each class.

Other benefits of YOLOv3:

1. Move the prediction from cell to box so that the detection of several objects can be done in one box.
2. Use logistic activation/classification for each class to keep the range.
3. Use logistic classification to enable multi-labelling.

Figure 6.3 shows food image classification using YOLO in which a vector for the food image has been prepared with a class probability map and used as training data. Similarly, predicting vectors with class probability map have been matched.

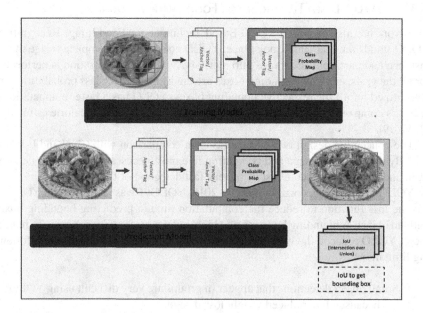

FIGURE 6.3   Food image classification using YOLO.

6.4 EXPERIMENTAL WORK

To implement the previously proposed methodology, dataset UEC-100 was used and all three methodologies discussed earlier were implemented on it. UEC-100 is an openly available standard dataset used in the food recognition domain.

Table 6.2 depicts the results after applying CNN, F-CNN, and YOLO on UEC Food-100 data set.

Figure 6.4 shows the comparative analysis of classification accuracy that shows an improvement using YOLO.

Figure 6.5 depicts the comparison of estimated accuracy by CNN, F-CNN, and YOLO for various food items of the UEC Dataset. YOLO has obtained higher accuracy in most images as compared with CNN and F-CNN.

TABLE 6.2
Experimental Result

Details		Model Testing		
Model optimization	Dataset	Classification accuracy	Precision	Recall
CNN	UEC FOOD-100	0.728	92.05%	90.98%
F-CNN	UEC FOOD-100	0.732	94.5%	90.62%
YOLO	UEC FOOD-100	0.741	95.82%	91.23%

Figure 6.4 depicts the comparative analysis of classification accuracy. It shows the improvement in classification accuracy using YOLO.

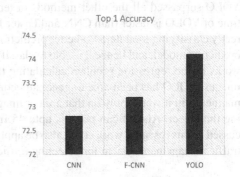

FIGURE 6.4 Accuracy comparison.

Figure 6.5 depicts the comparison of estimated accuracy by CNN, F-CNN, and YOLO for various food items of the UEC dataset. YOLO has obtained the higher accuracy in most images as compared with CNN and F-CNN.

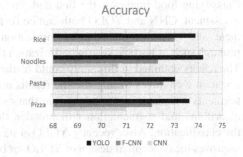

FIGURE 6.5 Food item wise accuracy comparison.

Figure 6.6 depicts the comparison of estimated accuracy and precision by CNN, F-CNN, and YOLO for the UEC Dataset. YOLO has obtained the higher accuracy and precision in most images as compared with CNN and F-CNN.

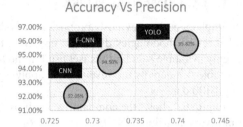

FIGURE 6.6 Accuracy vs precision.

Figure 6.6 depicts the comparison of estimated accuracy and precision by CNN, F-CNN, and YOLO for the UEC dataset.

6.5 CONCLUSION

It has been observed lately that Machine Learning and Artificial Intelligence have assisted in building computer aided dietary assessment systems. In this research work, three methodologies were presented for food image recognition. Experimental work was done on all three of them using the dataset UEC-100. Results show that YOLO surpassed all the other methods in terms of accuracy. Also, the processing time of YOLO is faster than CNN and Faster R-CNN. To broaden the scope for correctly classifying food items, a better prediction model is needed. YOLO has a better prediction model, and hence it is able to classify food images more efficiently. YOLO works as the regression problem calculating the class probabilities of the identified images. YOLO has been able to generate better results as it works in a probabilistic manner as in it works only on that area of image that has a higher chance of containing the object. Also, it can process upto 45 frames per second. The classifiers discussed in this research work could also be applied in medical imaging, improving the traffic management system for detecting traffic violations using image processing.

6.6 FUTURE WORK

Classifying food images is the first and foremost step that can improve the dietary assessment. CNN and YOLO both can be further optimized for much better assessment and classification. Many optimization techniques can be applied to this research such as particle swarm optimization (PSO) [19] and genetic algorithms [20]. The results obtained in this study could further be optimized using the various optimization techniques to improve the results and to extend this research work. Also, segmentation plays an important role that can be applied for improvement. These algorithms and segmentation will broaden this research and can further improve the classification. Also, recently YOLOv4 has also been introduced, which further improves the conceptual design of YOLO for better classification accuracy and faster training along with improved results. YOLOv4 could also be incorporated in the future to improvise the classification accuracy.

REFERENCES

[1] Yu, Qing, et al. *Food image recognition by personalized classifier.* 2018 25th IEEE International Conference on Image Processing (ICIP), IEEE, 2018.

[2] Shen, Zhidong, et al. "Machine learning based approach on food recognition and nutrition estimation." *Procedia Computer Science* 174 (2020): 448–453.

[3] Bossard, Lukas, Matthieu Guillaumin, and Luc Van Gool. *Food-101—mining discriminative, components, with random forests.* European Conference on Computer Vision, Springer, 2014.

[4] Kagaya, Hokuto, and Kiyoharu Aizawa. *Highly accurate food/non-food image classification based on a deep convolutional neural network.* International Conference on Image Analysis and Processing, Springer, 2015.

[5] He, H., F. Kong, and J. Tan. "DietCam: Multiview food recognition using a multikernel SVM." *IEEE Journal of Biomedical and Health Informatics* 20.3 (2015): 848–855.

[6] Kawano, Y., and K. Yanai. "Foodcam: A real-time food recognition system on a smartphone." *Multimedia Tools and Applications* 74.14 (2015): 5263–5287.

[7] Liu, Chang, et al. *Deepfood: Deep learning-based food image recognition for computer-aided dietary assessment.* International Conference on Smart Homes and Health Telematics, Springer, 2016.

[8] Liu, Chang, et al. "A new deep learning-based food recognition system for dietary assessment on an edge computing service infrastructure." *IEEE Transactions on Services Computing* 11.2 (2017): 249–261.

[9] Yunus, Raza, et al. "A framework to estimate the nutritional value of food in real time using deep learning techniques." *IEEE Access* 7 (2018): 2643–2652.

[10] Mezgec, Simon, and Barbara Koroušic Seljak. *NutriNet: A deep learning food and drink image recognition system for dietary assessment.* Information and Communication Technologies, Jožef Stefan International Postgraduate School, June 2017.

[11] Pouladzadeh, P., and S. Shirmohammadi. "Mobile multi-food recognition using deep learning." *ACM Transactions on Multimedia Computing, Communications, and Applications (TOMM)*, 13.3s (2017): 36.

[12] Chopra, M., and A. Purwar. "Recent studies on segmentation techniques for food recognition: A survey." *Archives of Computational Methods in Engineering* (2021): 1–14.

[13] Kawano, Yoshiyuki, and Keiji Yanai. *Food image recognition with deep convolutional features.* Proceedings of the 2014 ACM—International Joint Conference on Pervasive and Ubiquitous Computing, Adjunct Publication, 2014.

[14] Zhang, Weishan, et al. *Food image recognition with convolutional neural networks.* 2015 IEEE 12th International Conference on Ubiquitous Intelligence and Computing and 2015 IEEE 12th International Conference on Autonomic and Trusted Computing and 2015 IEEE 15th International Conference on Scalable Computing and Communications and Its Associated Workshops (UIC-ATC-ScalCom), IEEE, 2015.

[15] Ren, Shaoqing, et al. "Faster R-CNN: Towards real-time object detection with region proposal networks." *arXiv preprint arXiv:1506.01497* (2015).

[16] Xu, Hang, et al. *Reasoning-RCNN: Unifying adaptive global reasoning into large-scale object detection.* Proceedings of the IEEE/CVF Conference on Computer Vision and Pattern Recognition, 2019.

[17] Ahmad, Tanvir, et al. *Object detection through modified YOLO neural network.* Scientific Programming, 2020.

[18] Ege, Takumi, and Keiji Yanai. "Image-based food calorie estimation using recipe information." *IEICE Transactions on Information and Systems.*

[19] Sengupta, S., S. Basak, and R. A. Peters. Particle swarm optimization: A survey of historical and recent developments with hybridization perspectives. *Machine Learning and Knowledge Extraction* 1.1 (2019): 157–191.

[20] Katoch, S., S. S. Chauhan, and V. Kumar. "A review on genetic algorithm: Past, present, and future." *Multimedia Tools and Applications* (2020): 1–36.

7 Machine Learning/ Deep Learning for Natural Disasters

Tripti Sharma, Anjali Singhal,
Kumud Kundu and Nidhi Agarwal

CONTENTS

7.1 INTRODUCTION

Natural disasters (e.g. landslides, floods, earthquakes, tropical cyclones, etc.) are complex phenomena that affect not only the environment of the area but also assets

DOI: 10.1201/9781003245469-7

of that area like agriculture, infrastructure, and economic assets. The destruction of homes, industries, public offices, and other properties due to floods results in huge loss of economy along with the loss of human lives. This loss is unrecoverable, but by endorsing appropriate structural and non-structural measures the damage by floods is often minimized as tried in Mosavi, A. et al. (2018). Similarly, an earthquake's occurrence disturbs not only the living beings of that area but the complete ecosystem of that area is disturbed. Seismological analysis involves magnitudes estimation, hypo centric location detection, and various other parameters that are important to construct a seismic system or logical catalogue. The manual methods using handpicked data are not very reliable because of humongous seismic databases, the frequency of earthquakes with small and low magnitude, and the likely presence of some number of the backgrounds' noisy disturbances. Therefore, automated methods and algorithms are more powerful to be applied for the identification and detection of earthquakes. One algorithm that is used most frequently for the detection and identification of earthquakes is short timing average and long timing average. However, this method also faces some of the shortcomings like inaccuracy in initializing the parameters—which may lead to some falsification in alarm ringing, making this method unsuitable in the presence of life background noise. So, there is an urge to have efficient earthquake detection and prediction algorithms using machine learning and other deep learning algorithms (Gao, Z. et al. 2019; Allen, R.V. 1978). Seismic activity, changes to vegetation affecting soil composition, heavy rain, and floods can trigger the destabilization of ground. Deforested mountains also increase the risk of landslides. In India, most landslides are triggered by intense rainstorms. This destabilization of ground due to heavy rainfall causes high speed downward movement of mud, rocks, boulders, etc. In India, more than 12% of the total landmass, 420,000 sq km approximately, is susceptible to landslides. As per the Geological Survey of India (GSI), the landmass of seven eastern states of India, Mizoram, Sikkim, Tripura, Assam, Meghalaya, Arunachal Pradesh, and Nagaland is quite prone to landslides. From the western Ghats, the Nilgiri hills of Tamil Nadu are prone to landslides. In Northern India, from Himachal Pradesh, the upper catchment areas of the Beas River like Chamba, Lahaul-Spiti, and Kinnaur districts are prone to landslides. In Jammu and Kashmir, Leh, Poonch, Udhampur, and Ramban districts are landslide susceptible areas while in Uttrakhand the Tehri-Garhwal region is prone to landslide hazards.

Another natural disaster—Tropical Cyclones (TC)—also creates havoc in the life of the people living near coastal lines. TCs originate due to the disturbances in the oceans. They are basically high-speed circulatory wind systems. They vary from region to region and year to year and cause huge loss of life and properties in the tropical coastal areas. The main elements of the prediction are intensity, speed of the wind, and the eye of the TC. Analysis on these elements is required in the accurate prediction of intensity, track, and cyclogenesis of the TC. This is quite complex and there is a huge amount of data on climate, damage, and exposure. So, in recent years, many researchers have summarized and analyzed the challenges of TC. These reviews have helped them in coming up with different models using ML to predict TC. The role of disaster management is to reduce the effect of disasters on its area residents. In a developing country like India, where economic development is the major concern, allocating a sufficient budget for disaster management is very difficult. But past records of disaster management show that spending money and manpower on the steps taken for recovering from disaster is less efficient than spending on steps taken to prevent from that disaster. These disasters can be managed if there are some early warning systems; also their susceptibility of occurrence mapping may help in damage management. ML has sped up the potential of understanding natural hazards effectively. Various natural disasters like floods, earthquakes, wildfires, landslides, etc. cause damage to life, flora, and fauna of that area. This chapter delves into the application of ML/DL techniques for the prediction of future landslides (landslide susceptibility maps/landslide hazard maps/landslide risk maps) and the estimation of risks associated with induced landslides at a site or locale. Four types of natural disasters discussed in this chapter include floods, earthquakes, landslides, and tropical cyclones. Primary challenges faced in their applications and the potential solutions to the challenges are also discussed at length. We explicitly bewilder the pertinent challenges in the utilization of prevailing deep learning approaches for disaster prediction, which indirectly aids in stable economic growth and sustainable development. Data used in this chapter for study comprises publicly available datasets of Google Earth Engine. In the subsequent sections we aim at potential for upcoming research in this emerging area by tracing feasible contemporary paths based on the existing results. Finally, this chapter presents some hybrid approaches that can be explored and expected future trends of applications of ML and DL in natural disaster management.

7.2 IMPACT OF NATURAL DISASTERS

Natural disasters are the geo-induced threats arising from heavy rainfall, cloudbursts, volcanic eruptions, Tsunamis, mining, or earthquake shocks. This section sheds light on some of the broad impacts of natural disasters. Following are the broader impacts of the natural disasters:

7.2.1 DESTRUCTION OF INFRASTRUCTURE

Floods occur frequently both nationally and globally and have become increasingly common and severe in recent decades as a result of land use and climate change. A major flood incident that happened in early 2021, across the Chamoli region of Uttarakhand (India), completely destroyed the Dhauliganga Dam at the confluence of the Rishiganga and Dhauliganga rivers by the flood waters. The repercussions rising from the earthquakes depend mainly on the area of the earth where they have arisen. The effect of the impact depends to a great extent on where the quake is located—if it is a highly urbanized and populated area, a land area with less population, forest, industrial area, developed area, an underdeveloped area, or the area covering the sea or ocean beds. Landslide threat happens mainly in hilly areas. The Geological Survey of India estimates that the economic loss due to landslides is approximately between 1–2% of the Gross

National Product (GNP) of India. Occurrence of landslides destroys communication, electricity infrastructure, national highways, residential and commercial
building complexes, sewage disposal systems, water supplies, etc. In India, the
most tragic Kedarnath landslide that occurred in Uttarakhand in 2013 witnessed a
complete wash of infrastructures of many nearby villages. The most recent series
of landslides that occurred in India—landslides that occurred in the suburb areas
of Mumbai (Vikhroli and Chembur (July 2021)), in Kinnaur district of Himachal
Pradesh (August 2021) and in Nainital region of Uttara Khand (October 2021)—
are some among many landslides that caused infrastructure losses worth crores.
India is a peninsula; the Bay of Bengal, the Indian Ocean, and the Arabian Sea
surround India from three sides. Out of the 7,516 km coastline of India, 5,700 km
are prone to cyclones. Almost 6% of the world's cyclones hit the Indian subcon-
tinent. This results in lots of destruction within 100 km from the center of the
cyclones. Infrastructure worth crores is destroyed in the repeated occurrence of
storm surge and gale cyclones in India.

7.2.2 Loss of Natural Resources and Life

Floods have the potential to kill a lot of people and create a lot of damage. The flood
incident that happened in Chamoli in 2021 resulted in the loss of more than 200
lives. Recently, in November 2021, floods in four districts of Andhra Pradesh (India)
impacted about 70,000 people. Many times when the impact of an earthquake is
severe, it can break the electrical wires and fuel pipelines. As per the Geological
Survey of India (GSI), 21 landslides have been witnessed already by India in the
year of 2021. Landslides lead to wiping off the top layer of fertile soils from the
slopes of hilly areas, badlyimpacting the organic carbon content of soils and leading
to destruction of the wildlife habitat and forest areas, thereby impacting the carbon
cycle. Landslide debris can impact the water quality of rivers of the affected areas
and hence the fish habitat of that area. More than 56,000 people have been killed by
landslides worldwide during this time, most of them affected by one slope, according
to a study based on the Global Fatal Landslide Database (GFLD). Also, high tidal
waves cause 90% of deaths during the tropical cyclone. It also causes lots of destruction to human-made structures and can even stir up the waters of coastal estuaries
affecting fish breeding locales, resulting in damage to marine life in coastal estuaries
and destruction of sanitation facilities.

7.2.3 Reconstruction of the Lost Infrastructure

Approximately Rs 13,800 crore is required for flood relief and reconstruction for
the victims of the 2013 Kedarnath flood tragedy. The impact of a landslide can be
extensive, including loss of life, destruction of infrastructure, damage to land, and
loss of natural resources. Approximately Rs 100 crore to Rs 150 crore per annum at
2011 prices was lost due to the landslides occurring in India (Disaster Management
in India, MHA, Govt. of India, 2011 [https://reliefweb.int]). A huge amount of effort
is required to clear off the disruption caused on national highways due to landslide

debris. Building up the lost infrastructure requires both money and effort. Recently, several redevelopment projects worth several hundred crores have been initiated to restore the lost infrastructure in Keday valley after tragic Kedarnath flash floods that devastated the area completely in 2013. Because tropical cyclones (TC) are a multihazard weather phenomenon, reconstruction of sanitation, drinking water, electricity supplies, and transportation facilities in the impacted area are required (www.downtoearth.org.in).

7.2.4 REHABILITATION OF IMPACTED MASSES

Rehabilitation involves a large amount of financing in the reconstruction of settlements. There may be the urgent need to provide relief and care to the impacted people. It may entail relocating displaced people, reconstructing schools, hospitals, water and gas infrastructure, and restoring power, transportation, and communication. Disease surveillance is also very important after any natural disaster to maintain the good state of public health. Robust surveillance is also important after any disaster occurrence to guide local and regional health service delivery and also to prevent the outbreak of infectious diseases. Floods have the potential to enhance the spread of water-borne infections (typhoid fever, leptospirosis, cholera, and hepatitis) and vector-borne diseases, (dengue and dengue hemorrhagic fever, malaria and Kalaazar, West Nile Fever). Landslide occurrence increases the burden of building hospitals, relief, and care centers on an urgent basis. After tropical cyclones comes setting up relief centers, providing a safe shelter, and providing regular supplies of food to the affected families.

7.3 LITERATURE REVIEW

This section comprehensively surveys approaches of ML and DL utilized for the detection of natural disasters, early prediction of them, and the assessment of the risks associated with their occurrence. The existing studies of ML/DL application to the natural disasters can be broadly classified on the basis of their application areas: natural disaster occurrence prediction based on the past data or the assessment analysis of the risks/hazards associated with its occurrence. Figure 7.1 depicts use of ML/DL techniques in natural disaster studies.

7.3.1 APPROACHES OF MACHINE LEARNING AND DEEP LEARNING

Prediction of natural disasters is a growing field that has a lot of voluminous data wrapped with csv-file types or other excel files, which can be dealt with quite efficiently by adopting the help of ML and DL techniques. Till now disaster monitoring has been done through signal detection and phase picking, which are the major domains for the early detection and prediction of disasters. To produce better results for sensitivities, specificities, and various prediction parameters, "machine learning (ML)" and "deep learning (DL)" techniques/models have emerged as preferred

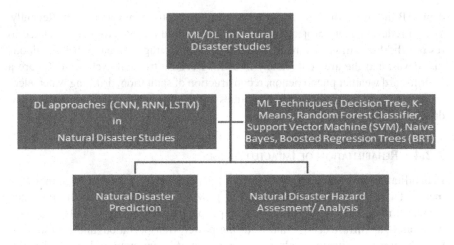

FIGURE 7.1   ML/DL application in natural disaster studies.

choices. They are able to produce better results for sensitivities, specificities, and various prediction parameters. It has been observed that the accuracy is improved and the error rate is reduced if ML/DL techniques are adopted. This section delves into the standard machine learning and deep learning approaches employed in the studies of flood occurrence, earthquake magnitude prediction or forecasting, landslide occurrence prediction, or tropical cyclone prediction.

Decision Tree: This is one of the supervised machine learning techniques, in which data is continuously split according to certain parameters. It can be used for both regression as well as classification problems. There are two nodes, decision node and leaf node. For making decisions, decision nodes are used. This has many branches. Outputs of the decision node are shown using leaf nodes. As the leaf node gives the output, it does not have further branches.

Random Forest: Random Forest is a supervised learning technique, which is used in classification and regression problems. It contains a number of decision trees for different subsets of datasets, and the average of all the results of the decision tree is taken into consideration for the final result (Breiman, L. 2001). This improves the prediction accuracy. The greater the number of decision trees, the higher the accuracy.

Support Vector Machine (SVM): SVMs are supervised machine learning algorithms used for classification and regression. They are used to find a hyper-plane in an N-dimensional space that distinctly classifies the data points. The main objective is to find a plane with a maximum margin (Cortes, C. et al. 1995). This provides better classification in future data points.

K-Means: KNN is the simplest classification algorithm in machine learning that is used for regression as well as classification. A k-nearest neighbor is a classification algorithm that attempts to determine how near the group of data points are around it.

Naïve Bayes: A Naïve Bayes is a supervised learning algorithm. This is basically the implementation of Bayes' theorem in mathematical statistics. Conditional probability is used for classification of data. This algorithm takes an assumption that data points are strong or naïve independent. These classifiers are broadly used in text categorization-based problems because they are easy to carry out.

Support Vector Selection Adaptation (SVSA): SVSA is a machine learning algorithm in which support vectors of the linear SVM is used for classification. This classification is done in two steps, selection and adaptation. In the first step, support vectors are obtained from linear SVM and then the selected support vectors are iteratively adapted in the training algorithm.

Boosted Regression Trees (BRT): This is a combination of two other techniques i.e. decision tree and boosting methods. Many decision trees are used to improve the accuracy of the model, and the boosting method is used to apply the weights on that input data, which was poorly modelled by previous trees and now has the higher probability of being selected in the new tree.

Bayesian Networks: Bayesian networks are probabilistic graphical models consisting of a structure and parameters. In this the structure part is the directed acyclic graph used to express conditional independencies and dependencies among random variables, which are associated with nodes. The parameters are the conditional probability distributions, which are associated with each node. The Bayesian network is a flexible, interpretable, and compact representation of joint probability distribution.

Naïve Bayes classifier: This is a supervised learning machine algorithm, based on Bayes' theorem. It is mainly used to solve text classification problems, which includes a high dimensional training dataset. It is most effective in building fast machine learning models.

Convolutional Neural Network (CNN): This is a deep learning algorithm used for image recognition and their processing. It performs generative as well as descriptive tasks like recognizing image and video along with NLP and recommender systems. In CNN, layers of neurons are arranged in a way so that the entire visual field should be covered and piecemeal image procession problems can be avoided (Perol, T. et al. 2018).

Latent Dirichlet Allocation (LDA): LDA is a topic modelling method for unsupervised classification of documents. This helps in organizing, searching and summarizing large electronic archives.

7.3.1.1 Related Work for Flood Warning Systems Using ML

Chang, L.-C. et al. (2010) developed a system for flood prediction that was based on a back propagation neural network and K-means clustering analysis method. A real time flood forecast system using a neural network was also developed. A clustering based hybrid inundation model was proposed using linear regression models and artificial neural networks. The first stage included data preprocessing; in this K-means clustering was used to classify the data points of different characteristics of flooding and identify the control points from the flooding clusters. In the second stage, a model was developed, which included the flood depth forecasting model from three classes, applying a back propagation neural model for each control point and linear regression models for finding linear correlation with the control points. Koem et al. (2021) proposed a classification-based real-time model for predicting floods. This model was basically based on 2D flood analysis and the Environmental Protection Agency stormwater management model. The working of this model was done in two steps: first zone wise prediction for cumulative volume and second predicting flood maps through spatial expansion of that predicted cumulative volume. Researchers tried to combine a numerical analysis model with a machine learning model. The damage caused by floods was estimated using a flood damage curve showing the depth of flood, its duration, frequency, and the damage caused. These results were then summarized and sampled using a Latin Hypercube sampling technique. Then that data was classified under five classes using a probabilistic neural network classification method taking average flood depth criteria into consideration. Each class has flood depth information by duration. Total flood depth calculated from all five classes provides the information of flood volume. Then the linear relationship was examined by making a graph between hourly volume with cumulative volume in each class. From this linear relationship an average coefficient of determination and regression constants were calculated. Based on that, the spatial distribution of the flood impact was calculated with the help of global information systems and weight arithmetic mean. The weights of the different disaster impacts were assigned based on the Jenks Natural Break classification. The model was applied to some actual basins for forecasting the chances of flood.

Ho jun keum Keum, H.J. et al. (2020) tried to propose a real-time flood prediction model. This was based on the relationship between the cumulative rainfall and its volume using a neuro fuzzy model. Cumulative volume was predicted for each zone, and the depth of flood for each grid was calculated and a linear relationship was analyzed. A database for floods was created according to the scenarios of rainfall. This included the data having average flood depth, max flood depth, the frequency of flood, and the average velocity of rainfall. This data was summarized and was sampled using Latin hypercube sampling. The probabilistic neural network classification method was used to classify data into five classes based on the flood depth. The total flood, depth calculated was then multiplied by the area to calculate the volume using Gamma Test. Gamma Test was used to analyze the uncertainty in advance. The flood depth was predicted using the calculated representative cumulative volumes. The predicted results were compared with the simulated result to assess the validity of the prediction results.

Khalaf, M. et al. (2018) proposed to apply machine learning models for prediction of flood severity and tried to improve the intelligent system. The dataset for this was collected from different cities from various countries and was divided under three levels (normal, abnormal, and dangerous). Random Forest classifier, Support Vector machine, and linear combiner network classifier were used in this algorithm. Various types of features were integrated with the dataset. These features include duration of the day, number of humans dead, number of people displaced, total area affected, magnitude of the flood, centroid, number of floods per year, torrential rain, heavy rain, tropical storm, snowmelt, and monsoon rain. The dataset was then visualized using Principal Component Analysis and evaluated using Holdout technique. This helped in performing linear mapping with lower dimensional space of the dataset so that dataset representation could be maximized. They used the t-Distributed Stochastic Neighbor Embedding technique for visualization of the datasets with high dimensionality. After that, holdout technique was used for evaluating the result. They found a Random Forest classifier, more accurate for analyzing the datasets of flood for prediction. The reason behind it was the use of a different method instead of a cross validation method. This in turn increases the stability of results.

Panda, R.K. et al. (2010) have designed a data-driven model to predict flood, which is based on artificial neural network design. The aim was to compare the performance of various artificial neural network techniques with manual-based hydrodynamic models. The values of root mean square error and Nash-Sutcliffe index were more accurate using the ANN model as compared to the values of these using hydrodynamic models. The difference between the observed peak water level and the simulated peak level calculated using the ANN model was much better than that of the hydrodynamic model.

Snehil et al. (2020) have used Gaussian Naive Bayes, tree-based approaches, and K Nearest Neighbor-like machine learning algorithms for implementing the heavy rain damage prediction model. The performance of different models was analyzed on the basis of their mean square error and the coefficient of determination. R^2 value for machine learning technique in regard to damage to crops and loss of human lives were separately calculated. Similarly, mean squared error of human lives lost and damage to crops were calculated with different ML techniques. They concluded that K Nearest Neighbor and random forest algorithm gave better results as compared to Gaussian Naive Bayesian.

Table 7.1 depicts a concrete comparative analysis of most of the approaches discussed earlier for flood warning systems using ML.

7.3.1.2 Related Works for Earthquake Warning Systems Using ML

An important algorithm, Akaike information concept (AIC), is used widely among the research community for earthquake detection. One of the major drawbacks of this algorithm is that it is unable to estimate the minimal globally incorrect values when the backdrop of the noisy disturbance increases (Leonard, M. et al. 1999). In order to overcome these disadvantages a group of researchers suggested automatic detection of earthquakes using various machine learning methods like K-means algorithm (Chai, X. et.al 2020, Chen, Y. 2018), templates match method (Beroza, G.C. 2019), support vector machine (Chen, Y. 2020; Aad, O.M. et al. 2020a), fuzzification algorithms (Chen, Y. 2020), wavlytic transformational changes (Hafez, A.G. et al. 2013), and various threshold controlling algorithms. As discussed earlier in Jozinovic, D. (2019) and Harirchian, E.ALAMR et al. (2020), various deep learning algorithms are gaining more importance for early detection and prediction of earthquakes. Deep learning-based algorithms are used like CNN, RNN, and auto-encoder (Zhu, W. and Saad, O.M. et al. 2018; Beroza, G.C. et al., 2019, Evries, P.M. 2018). The most important aspect of using CNN is that it is capable of extracting the most significant attributes from the input data. But it suffers from various shortcomings also. The Mousavi, S. M. et al. (2019) various layers of CNN architecture do not completely contribute to enhancing the dimensional value of the network. There is one layer present in the CNN architectural framework that is called the pool layer, which reduces the dimensional value of the whole structure. The prediction model (due to pool layer presence) does not have the scope to lose important data. This problem reduces the predictive performance of the CNN algorithm (Sabour, G.E. et al. 2017; Jia et al. 2020). Capsule neural network, which was suggested for the first time by

TABLE 7.1

Comparative Analysis of Flood Warning System Approaches

Earlier Studies in Flood Prediction Using ML/DL

Reference	Addressed issue	ML/DL Approach used	Datasets	Accuracy/prediction metrics used
M. Khalaf et al. (2018)	To investigate the potential of an empirical dataset for the classification of flood severity, using various machine learning algorithms to get better accuracy	Random Forest Classifier, Support Vector Machines, Levenberg- Marquardt training algorithm	"Flood Data and resources" [Online]. Available at: https://public.tableau. com/s/sites/default/fil es/media/Resources/FlooddataMasterListrev. xlsx. [Accessed: 10-Oct- 2016].	Water level, the duration of the day, and magnitude
Snehil et al. (2020)	To assess the implications of flood risk analysis	Gaussian naive bayes, Decision Tree, Support Vector Machine (SVM), K Nearest Neighbor (KNN), Random Forest	Open available dataset for states Bihar, Uttar Pradesh and Kerala	Mean square error (MSE) and the coefficient of determination (R2)
RabindrA.K. Panda (2010)	Performance of an ANN model was compared with MIKE 11 hydrodynamic model to predict hourly water level at the Nimapara gauging site of the river branch Kushabhadra	Artificial neural networks, the hydrodynamic model	Hourly water level data of the period June–September 2006 at the Nimapara gauging site of the river branch Kushabhadra, Central Water Commission (CWC)	Coefficient of determination (R2), Nash–Sutcliffe Coefficient (E), root mean square error (RMSE), index of agreement (IOA) and difference in peak (DP) values
Tabbussum R., Dar A. Q. (2021)	Three different AI techniques are evaluated for simulation of stream flow	ANN, fuzzy and ANFIS algorithms	The data was collected from the different agencies accredited by the government of India for a total period of 28 years, from 1990 to 2018. The stage and runoff data of river Jhelum and its tributaries was obtained from the Irrigation and the Flood Control Department, Srinagar (I&FC 2018). The daily rainfall data was procured from the Indian Meteorological Department (IMD 2018).	Mean squared error (MSE), root mean squared error (RMSE), coefficient of determination (R2), Nash-Sutcliffe model efficiency (NSE), mean absolute error (MAE), combined accuracy (CA)
Li-Chiu Chang et al. (2010)	To build the regional flood inundation forecasting model	Back- propagation neural network (BPNN)	Water Resources Agency (WRA), Taiwan,	Mean absolute error (MAE), root mean square error (RMSE) and total inundated volume percentage error (TIVPE)

Sabour et al. (2017), was helpful in removing the drawbacks of the CNN algorithm. Capsule neural network is the improved technology for the deep learners' architecture, which provides the ability to the framework to adapt and enhance learning even with the absence of a pool layer. In such a case it proceeds without the loss of any data and provides training to the model with even a lesser number of samples of data to finally achieve generalized enhanced capability. It is able to achieve remarkable improvements as it uses a vectored output for the simple network framework. This technology is being used efficiently by many researchers inclined toward the earthquake seismic data analysis from various areas like South California (Ross, Z.E. et al. 2018), Asia and Europe. The end results are obtained with better predictive performance accuracy than the various other existing algorithms used in previous studies (Ross, Z.E. et al. 2018).

Automatic arrival picking of certain seismic or micro seismic phases has been studied for decades. However, automatic detection of continuous signal waveforms has been seldom addressed. Researchers have proposed many novel approaches for automatically detecting the waveforms in the micro seismic data (Chen, Y. 2018; Mousavi, S. M. et al. 2020). The waveform detection can be formulated into a classification-based machine learning (ML) problem, i.e., each data point in the micro seismic record needs to be classified as either waveform or non-waveform (El Zini et al. 2020). The classic K-means clustering-based unsupervised machine learning algorithm can be used to solve this problem. Some other researchers have used mean, power, and spectral centroid as the three features to help the machine to characterize each data point. Both synthetic and real micro seismic data examples are used to demonstrate the feasibility of the proposed algorithm (Chen, Y., 2018; Zhu, W. et al. 2018; He, Z. et al. 2020).

Table 7.2 shows a concrete comparative analysis of most of the approaches discussed earlier for earthquake warning systems using ML.

7.3.1.3 Related Work for Land Susceptibility Mapping Systems Using ML

Accurate estimation of location of susceptible areas helps to an extent to manage its hazards. Prediction of landslide susceptibility can be done through mathematical evaluation models, GIS/RS image analysis, and ML techniques. Recently, there has been a surge in interest in the utilization of ML/DL models for landslide susceptibility predictions and damage assessment due to landslide disasters. The key aim in the computer-assisted prediction of landslide prone areas is to prepare landslide inventory maps using either conventional mathematical models or ML techniques.

Pourghasemi, H.R. et al. (2013) utilized a GIS-based support vector machine (SVM) with different kernel functions (Linear, Polynomial (of degree 2,3 and 4), sigmoid function, and Radial Basis Function (RBF)) for the mapping of landslide susceptibility in Kalaleh Township area of Golestan province, Iran. Their studies found prediction accuracy of 85% with RBF kernel and 83% with polynomial kernel with a degree of 3 for the SVM Model. They also pointed out that the frequency ratio between landslide occurrence and altitude is consistent. Jena, R. et al. (2021) and Jebur (2017) explored the efficiency of simple ML techniques K-Nearest Neighbor (K-NN) and Logistic Regression (LR) in mapping of landslide susceptibility across Ulu Klang, Malaysia. They found prediction accuracies of 82.64 and 72.18%, for KNN and LR respectfully.

TABLE 7.2

Comparative Analysis of Earthquake Warning System Approaches

Earlier Studies in Earthquake Magnitude Prediction/Earthquake Susceptibility Mapping Using ML/DL

Reference	Addressed Issue	ML/DL Approach Used	Datasets	Accuracy/Prediction Metrics Used
Yoon, C. E. (2015)	Detecting, identifying the tremors	Short-tenure mean/long-tenure mean (STA/LTA), ANN procedure	Stanford's tremordataset (STEAD) /training, validation, and test: https://github.com/sm ousavi05/STEAD.	Accuracy, precision, recall, f1 score, PPV, NPV
Leonard, M. et al. (1999)	Detecting, identifying the tremors	Akaike information criterion (AIC)	http://service.iris.edu/fdsnws/dataselect/1/	Accuracy, precision, recall, f1 score
Chen, Y. (2020), Chen, Y. (2018)	Detecting, identifying the tremors	K-means procedure	Arkansa's dataset	Accuracy, precision, recall, ROC
Lomax, A. (2019) Mousavi, S. M et al. (2020)	Detecting, identifying the tremors	Template matching	(www.jma.go.jp/jma	Accuracy, precision, recall, f1 score
Beroza, G. C. (2019)	Detecting, identifying the tremors	Support vector machine	http://service.iris.edu/fdsnws/dataselect/1/	Accuracy, Precision, recall, f1 score, PPV, NPV
McClellan, J et al. (2018)	Detecting, identifying the tremors	Fuzzy algorithm	(www.jma.go.jp/jma	Accuracy, AUC, sensitivity, specificity
Jozinović, D. (2019)	Detecting, identifying the tremors	Wavelet transformation	The continuous data for the Totorri region www.hinet.bos ai go.jp/about_data/?LANG=en	Accuracy, precision, recall, f1 score, PPV, NPV
Beroza, G. C. et al. (2020)	Detecting, identifying the tremors	Thresholding algorithm	Southern California seismic open-source data, "Stanford Earthquake Dataset" (STEAD)	Accuracy, precision, recall, f1 score, PPV, NPV

Reference	Task	Method	Dataset	Metrics
Zhu, W. and Beroza, G. C. (2018)	Detecting, identifying the tremors	Convolutional neural network (CNN), deep recurrent neural network (RNN), and autoencoder	Stanford earthquake dataset (STEAD) used for the training, validation, and test: https://github.com/sm ousavi05/STEAD	Accuracy, precision, recall, f1 score, PPV, NPV
Sabour, S. et al. (2017) Huang, Q. (2020)	Detection and identification of the earthquakes	CNN	(www.jma.go.jp/jma)	Accuracy, precision, recall, f1 score, AUC, ROC
Aad.O.M. et al. (2020b)	Detection and identification of the earthquakes	Variant of CNN	Arkansas datasets are open-source dataset	Accuracy, precision, recall, f1 score, PPV, NPV
Chen,Y. (2008), Zhu, W. et al. (2018), He, Z. et al. (2020)	Detection and identification of the earthquakes	CapsNet	http://service.iris.edu/fdsnws/dataselect/1/	Accuracy, recall, f1 score, sensitivity, specificity

Kadavi, P. R. et al. (2018) investigated 20 landslide condition factors using AdaBoost, Bagging models, LogitBoost, and Multiclass Classifier for the landslide susceptibility analysis in Sacheon-myeon area of South Korea. Landslide susceptibility maps were validated using the area under the curve (AUC) method. The Multiclass Classifier offered of prediction accuracy 85.9%, Bagging classifier offered 85.4% accuracy, and LogitBoost offered an accuracy of 84.8%. All the ensemble models offered better prediction accuracy than the traditional frequency ratio model for the 20 conditioning factor maps and the landslide inventory of the selected region. Due to the involvement of dynamic nature of various factors on landslide susceptibility analysis, Lee, S. et al. (2020) in their work investigated Naïve Bayes and Bayesian network to predict the influence of forest, topographical, and soil characteristics in Umyeonsan (region of Korea) landslide susceptibility. Accuracies for susceptibility analysis for Naïve Bayes and Bayesian network models accounted for 79.2% and 83.5%, respectively. Table 7.3 highlights a comparative analysis of most of the approaches discussed earlier for Land Susceptibility mapping systems using ML.

7.3.1.4 Related Work for Tropical Cyclones Warning Systems Using ML

India has a long coastline, and almost every year India faces one or more tropical cyclones. The people living in that area are badly affected. Challenges across NDRF are to make effective plans for disaster management at the root level. The main stress is on how tropical cyclones can be forecasted and early warning can be sent at the local level and effective measures can be taken far in advance to reduce the effect. Many researchers are trying to explore in-situ, radar, and satellite data using ML for predicting TC so that more accuracy can be achieved. They divided their study into three categories: selection, clustering, and classification. Irrelevant attributes are eliminated in feature selection. They used K-means algorithm and Finite mixed model for clustering and SVM for classification. Basically, stress is on using predictive algorithms so that the central location and intensity of TC can be predicted.

To implement ML in prediction of TC, they divide their data-driven model into following five aspects:

1) Genesis forecasts, which includes short term forecasting and long-term forecasting.
2) Track forecasts, which includes long term path prediction, predictors mining, and similarity search.
3) Intensity forecast, which includes intensity estimation, intensity prediction, and intensity change prediction.
4) TC weather and the disastrous impacts forecasts, which include TC wind field forecast, TC rainfall forecast, and storm surge forecast.
5) Improving the numerical model, which includes improvement in data processing, parameterization, and application of model outputs.

A different strategy has been developed by Giffard-Roisin et al. (2020), to apply the input of data similar to an image as training data in a deep learning network

TABLE 7.3
Comparative Analysis of Land Susceptibility Mapping Systems Approaches
Earlier Studies in Landslide Susceptibility Mapping Using ML/DL

Reference	Addressed Issue	ML/DL Approach Used	Datasets	Accuracy/Prediction Metrics Used
Van Dao, D. et al. (2020)	Determination of landslide prone areas along national highways by finding the persuading factors that contribute toward the mapping of landslide susceptibility locations	Bagging Logistic Model Tree (BLMT), Dagging Logistic Model Tree (DLMT), Adaptive Boost Logistic Model Tree (ABLMT), MultiBoostABLMT (MBLMT) Cascade GeneralizationLogistic Model Tree (CGLMT)	Open-source data collected from Landsat, Google Earth, ASTER GDEM along National Highway 6, passing through Hoa Binh province, Vietnam	Statistical indices like specificity, sensitivity, area under ROC curve (AUC), accuracy (ACC), k-index and RMSE
Mandal, K. et al. (2021)	Quantification of the roles of 20 land conditioning factors for landslide susceptibility analysis, in the Rorachu river basin of the Sikkim state, India	Three ML models (Bagging, Random Forest, ANN) and one Deep Learning Model (CNN)	Satellite Images (Landsat 8), Rainfall data (www.worldclim.org/), Land Conditioning Factors (elevation, slope aspect, plan curvature, slope and profile curvature) from ALOS PALSARDEM of 12.5 m ground resolution. ALOS PALSAR Digital Elevation Maps (DEM) from Alaska Satellite Facility	Information gain ratio (IGR) for the role quantification of land conditioning factors, tolerance and variance inflation factor for multicollinearity analysis among various LCF. Area under the curve (AUC) of the receiver operating characteristics (ROC) curve and statistical indices like root mean square error (RMSE) and mean absolute error (MAE)

(Continued)

TABLE 7.3   (Continued)

Earlier Studies in Landslide Susceptibility Mapping Using ML/DL

Reference	Addressed Issue	ML/DL Approach Used	Datasets	Accuracy/Prediction Metrics Used
Sachdeva, S. et al. (2020)	Landslide susceptibility assessment for the Brahmaputra valley region (Assam and Nagaland), India	Ensemble technique of majority-based voting among Gradient Boosted Decision Trees (GBDT), Logistic Regression (LR), and Voting Feature Interval (VFI)	System for Automated Geoscientific Analyses (SAGA GIS) resource, advanced spaceborne thermal emission and reflection radiometer Global digital elevation model (ASTER GDEM) for the Brahmaputra valley region	Gini index for weightage of 16 causative land conditioning factors (profile curvature, plan curvature, slope, elevation, surface roughness, topographic wetness index, slope length, normalized difference vegetation index, stream power index, distance from rivers, railways, roads) area under receiver operating characteristic curve (AUC of ROC)
Rachel, N., & Lakshmi, M. (2016)	Prediction of landslide occurrence on the basis of rainfall received	Kernel-SVM with Radial Basis as kernel function	Dataset comprises seven years of daily recorded data for rainfall from 2009 to 2015 for Cherrapunjee region of Meghalaya (www.cherrapunjee.com/weather-info/daily- weather-data-2009–2015)	Prediction accuracy
Bera. S. et al. (2021)	Investigation of landslide behavior for susceptibility analysis and sustainable landslide mitigation	Bayesian Network (BN), Random Forest (RF), Backpropagation Neural Network (BPNN)	Secondary data source (reports prepared by Geological Survey of India; www.gsi.gov.in) for hilly regions located on the banks of the river Bhagirathithe of Uttarkashi region of Uttarakhand, India QGIS software for generation of LSM for the selected land conditioning factors of the selected locale	Sensitivity, specificity, precision, recall, accuracy, and area under the curve (AUC)

Reference	Application	Model	Data used	Metrics / Link
Li, Y. et al. (2020)	Human activity (road construction, urban expansion, commercial cultivation) impact analysis on landslide susceptibility development	CNN	Geological map, a relief map, and remote sensing images from Google earth, Sentinel-2A, multitemporal monitoring data (in the three years of 2010, 2015 and 2019) in the Zigui–Badong section of the Three Gorges area	Sensitivity, specificity, precision, recall, accuracy, and area under the curve (AUC)
Yunus, A.P. et al. (2021)	Impact analysis of major topographic factors and anthropogenic disturbances on the landslides in the Western Ghat region (Kerala, India)	Random Forest classifier and generalized feature selection algorithm	Climate, digital elevation model (DEM), soil, and land-use data of Three Gorges Reservoir	http://download.geofabrik.de/asia/india.html

for a 24-hour forecast tracking task in the storm center. This provides the estimation of the displacement vector between the locations of current and future purpose. Researchers also proposed the reanalysis of data by cropping images centered on the locations of storms. Due to the reduced computation time, the information of storms coming from a large number of tropical cyclone basins can be inferred.

Table 7.4 shows a concrete comparative analysis of most of the approaches discussed earlier for the Tropical Cyclone Prediction system using ML.

Though DL/ML techniques application in natural disaster studies has given better prediction accuracies, still a lot more efficient DL methods need to be explored to address the problem of fully accurate disaster detection. These methods can also be deployed along with other real-time analysis techniques like fast Fourier transformation to improve the prediction accuracy. The various challenges faced by ML and DL methods also need to be addressed in the future for enhancing the predictions and thus making improvements in this direction.

TABLE 7.4

Comparative Analysis of Tropical Cyclone Prediction System Approaches

Earlier Studies in Tropical Cyclone Prediction Using ML/DL

Reference	Addressed Issue	ML/DL Approach Used	Datasets	Accuracy/ Prediction Metrics Used
Rui Chen, Weimin Zhang and Xiang Wang (2020)	Forecasting Tropical cyclone	Support Vector Machine (SVM), Decision Tree (DT), K-mean	satellite, radar, in-situ data	accuracy of (F1- score), prediction accuracy
Michael B. Richmana, Lance M. Lesliea (2012)	Seasonal predictions of TC frequency and intensity	Support Vector Regression (SVR) models, multiple linear regression	www.bom.gov.au; www.esrl.noaa.gov/ psd/data/gridded/	R2, accuracy of prediction
Wenwei Xu et al. (2021)	Forecasting TC intensity	Deep learning-based multilayer perceptron	http://rammb.cira. colostate.edu/ research/ tropical_cyclones/ ships/index.asp. https://ftp.nhc. noaa.gov/atcf/ archive/	R2, accuracy of prediction
Giffard-Roisin S, Yang M, Charpiat G, Kumler Bonfanti C, Kégl B and MonteleoniC (2020)	To estimate the longitude and latitude displacement of tropical cyclones	CNN, RNN	NOAA database IBTrACS (International Best Track Archive for Climate Stewardship, Knapp, K. R. and Kruk, M. C., 2010)	root mean square error (RMSE)

7.4 SYSTEM DESIGN/METHODOLOGY

7.4.1 DESIGN METHODOLOGY FOR FLOOD WARNING SYSTEMS

Researchers in Chang, L.C. et al. (2010) proposed a two-stage approach to build the regional flood inundation forecasting model that includes data preprocessing and model building. The whole study area is partitioned into three subsets. The first subset is training, the second is validation, and the third is a testing subset for crossvalidation to assess the proposed approach. The K-means clustering analysis is done, which provides help in identifying various groups that both minimizes within the group variation for data in a cluster and also maximizes between-group variation to identify potential difference between the clusters. In Khalaf, M. et al. (2017), the researchers establish an optimized and reproducible standard of intelligent systems by proposing a model that consist of the key procedures such as data collection, preprocessing data, splitting data into three major parts, building the model based on the training data, and evaluating the model. Dependent on the testing set, this will lead to selecting the appropriate model.

7.4.2 DESIGN METHODOLOGY FOR EARTHQUAKE WARNING SYSTEMS

Autoregressive methods proved to be quite helpful in characterizing the seismic record. They calculate the power spectral for the seismic record to determine the first and subsequent arrivals at various time intervals divided into classes. If the time series is represented with an autoregressive record followed by magnitude then it can be applied to both multiple components and single components for broadband short and long graphic records. If fast Fourier transformation methods are applied for the autoregressive methods then it proves to be more effective for the incoming data of seismic record, as the waves obtained through fast Fourier transformation methods are more uniform and less prone to background noises too. In another work the researchers have described digital implementation for a novel transformation called the second-generation transformation for both two and three dimensions. The first digital transformation is based on an equally spaced fast Fourier transformation, and the second one is based on the wrapping of specially selected Fourier samples. However, for both the implementations, all details are different while selecting the spatial grades used for the translation of colors for every scale and angle. But for both of them the digital transformation returns the digital curvelet coefficients in tabular form, which has an index of the scaling parameters and orientation parameters.

An important algorithm, Akaike information concept (AIC), is used widely in the researcher community for earthquake detection. One of the major drawbacks of this algorithm is that it is unable to estimate the minimal globally incorrect values when the backdrop of the noisy disturbance increases (Leonard, M. et al. 1999). In order to overcome these disadvantages, a group of researchers suggested automatic detection of earthquakes using various machine learning methods like K-means algorithm (Chai, X. et.al 2020, Chen,Y. 2018), templates match method (Bergen, K. J. et al. 2019), support vector machine (Chen, Y. 2020; Aad, O.M. et al. 2020a), fuzzification algorithms (Chen, Y. 2020), wavlytic transformational changes (Hafez, A.G. et al. 2013), and various threshold controlling algorithms (Nakata, N. & Beroza, G. C. 2015). As discussed earlier in Jozinovic, D. (2019) and Harirchian, E.ALAMR et al. (2020), nowadays various deep learning algorithms are gaining more importance for early detection and prediction of earthquakes. Deep learning-based algorithms such as CNN, RNN, and auto-encoder are used (Li, Z. et. Al 2018, & Beroza, G.C. 2018; Beroza, G. C. et al. 2019; Evries, P.M. 2018). The most important aspect of using CNN is that it is capable of extracting the most significant attributes from the input data. But it suffers from various short-comings also. The Beroza, G. C. et al. (2019) various layers of CNN architecture do not completely contribute to enhancing the dimensional value of the network.

7.4.3 Design Methodology for Landslide Susceptibility Map

Recently, various machine learning models like LR, SVM, FLDA, Bayesian Networks, and Naïve Bayes have been applied for landslide susceptibility assessment (Pham, B. T. et al. 2016; Nguyen, V. V. et al. 2019; Wang, Z. et al. 2020; Juyal, A. and Sharma, S. 2021). Conventional machine learning classifiers were also integrated with deep learning models (Xiao, L. et al. 2018; Fang, Z. et al. 2020; Bera, S. et al. 2021; Ngo, P. T. T. et al. 2021). Building landslide susceptibility models is inherently a complex task due to the dynamic nature of landslides and the various land condi-tioning factors on which it is built. In India, a national program "National Landslide Susceptibility Mapping" (NLSM) has been used by GSI to map 0.42 million sq. km of landslide-prone areas of the country. System design for LSM for the given locale is generally a series of steps starting from building landslide inventory maps.

Following are the steps in building landslide susceptibility models:

1) The first step in building landslide susceptibility models is the building of landslide inventory maps either through the multiple field studies or aerial photo interpretation.
2) The second step consists of the selection of landslide conditioning factors like topological information, such as altitude, slope, aspect, and curvature, and hydrological parameters, such as stream power index (SPI) and topo-graphic wetness index (TWI). Other conditioning factors are soil, geology, land use/land cover (LULC), distance from rivers, and distance from roads.
3) The third step consists of quantitative analysis of land conditioning factors (LCF) for finding the conditioning factors with which the training layer was utilized for analysis to measure the probability map of landslides.
4) The fourth step is Selection of machine learning/deep learning model for training/cross validation/testing. While training a model all the landslide conditioning factors are exploited to establish the spatial relationship between the landslides and the factor to detect the most susceptible areas.

7.4.4 Design Methodology for Tropical Cyclone Warning Systems

For forecasting TC, Rui Chen et al. (2020) first try to predict the central loca-tion of the TCs along with their intensity and effect on the shore. The different techniques mainly used for forecasting are numerical forecasting, statistical forecasting, and empirical forecasting. They also used the path prediction for getting more accurate results. For applying ML for forecasting TC, they divide this forecast into five aspects like genesis forecasts, track forecasts, intensity forecasts, TC weather, and the disastrous impacts forecast and improving numerical model. In the genesis forecast, the aim is better monitoring of the tropical ocean with respect to place and time by generating a measurable forecast of cyclone genesis and probabilistic forecasts of fixed regions in real time. For TC genesis prediction, they use algorithms like Random Forest, SVM, AdaBoost, and logistic regression. For track forecasting of TC, nonlinear mapping is used. For predicting the position of TCs, MLP and RNN algorithms are used. To generate predicted TC cloud images, generative adversarial networks are used. For intensity forecasts, CNN is used to detect TC from satellite cloud images. For predicting the intensity, the full sequence and track from origin to the end of TC is observed so that the intensity of the next moment can be predicted using MLP or RNN. At present these methods are helpful for short term forecasting. In terms of TC weather and disastrous impact forecast, MLP and SVR are used. They use the relevant precipitation, tide information, and characteristics of TCs to predict the rainfall or height of storm. For forecasting wind fields near the ocean surface, the CNN model is used.

Figure 7.2 shows system design for landslide susceptibility maps.

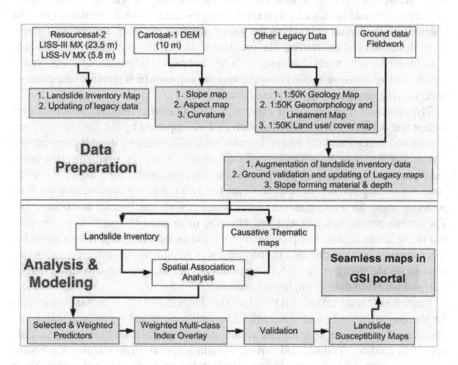

FIGURE 7.2   System design for landslide susceptibility maps (source: gsi.gov.in).

7.5   FINDINGS/RESULTS WITH DISCUSSIONS

We study results of work done up to now in the natural disaster assessment and management and then draft the prospective open issues for the effective use of deep learning in natural disaster prediction and management.

7.5.1   FINDINGS FOR FLOODS WARNING SYSTEM

RFC is the strongest model to analyze flood data sets that has been proven in the data science methodology that combines various features from the vast records for flood protection. RFC preserves a number of features that are very important for the decision trees with respect to the training module. This model is a faster model in comparison with other models considered in the research. Random forest and KNN perform better than normal regression algorithms. EV bass performs very low because of various reasons like discretization of the data and sensitivity to low-quality data.

7.5.2   FINDINGS FOR EARTHQUAKE WARNING SYSTEM

The researchers were successfully able to explain that, if the methodology is implemented using autoregressive format, it proves to be successful to the three-component-based systemic record to estimate the spectral and characterized records at the arrival of the timing seismic data. This method coupled with fast Fourier transformation has more advantages than taking only the autoregressive method or the fast Fourier transformation method, as the waves obtained are more uniform and give a better description of the results. The problem is dealt with very nicely with the help of an autoregressive method coupled with fast Fourier transformation, as the system becomes more sensitive to the incoming specific waves, which increases the chances of registering changes in the incoming data quality.

The improved working for the duo coupled with autoregressive mode for the prediction and detection of earthquakes with FFT is supported by the fact that the foreground and background voices are detected more effectively by the AR components. The model performs best for the onset of seismic waves by predicting the noisy disturbances of the signal. Predictions for various components for both seismic components and diagonal components that even have a lot of signal to noise ratio are done perfectly using an autoregressive model coupled with fast Fourier transformation. The double parameters followed one after the other used by researchers are quite fast in the terms of time complexity and for rapid evolutionary algorithms about the space complexity. The digitally transformed waves improve the implementation that is based on the first generation for corporates as they are conceptually quite simple, fast, and nonredundant to implement initially.

Empirical outputs reveal that the algorithm helps detect the dominant waveforms for the data effectively and efficiently. The automatic found waveforms accelerate obtaining the microseismic images results using an amplitude-based reverse time migration method. As discussed earlier, nowadays various deep learning algorithms are gaining more importance for early detection and prediction of earthquakes. Deep learning-based algorithms are used like CNN, RNN, and Auto-encoder (McClellan, J. et al. 2018).

7.5.3 Findings for Landslide Warning System

Researchers were successfully able to design landslide susceptibility models using machine learning techniques that predict the probability of its occurrence on the basis of past landslide data and the various land conditioning factors of that area.

Van Dao, D. et al. (2020) in their work attempted to couple various logistic model tree models and random subspace. They utilized the OneRAttributeEvalation method to quantify the contribution of 14 landslide-conditioning factors (curvature, elevation, slope, sediment transport index (STI), Topographic Wetness Index (TPI), geomorphology, spectstream power index (SPI), weathering crust, lithology, river density, structural zone, rainfall, and fault density) in the identification of landslide-prone areas along National Highway-6, passing through Hoa Binh province, Vietnam.

Mandal, K. et al. (2021) in their work studied the correlation between selected land conditioning factors. Further, they measure the impact of a factor by computing the information gain ratio (IGR) for every factor. They found that rainfall, elevation, and soil texture are the dominant factors having a high impact on landslide occurrence. The landslide susceptibility index (LSI) was constructed from the prediction values computed via Random Forest, CNN, Bagging, and ANN models. The GIS environments to construct the landslide susceptibility index (LSI). LSI values were further categorized into very high (VH) landslide, high (H), moderate (M), low (L), and very low (VL) landslide susceptibility zones.

Sachdeva, S. et al. (2020) explored ensemble technique of majority-based voting, combining few of the most popular machine learning models, such as Logistic Regression and a variant of Decision Tree called Gradient Boosted Decision Trees, for landslide susceptibility assessment for the Brahmaputra valley region (Assam and Nagaland). They mapped 436 landslide locations of the selected landslide inventory. Their proposed ensemble model's accuracy is 3% higher than their standalone classifier counterparts.

In another study, different types of landslides (debris slides and rockslides–rockfalls) in the Kalimpong region of northeastern Himalayan region of India were studied. Bera, S. et al. (2021) in their investigation found that slope between 34° and 42°, and aspect (south, southeast, and southwest), the concavity of slope, high TWI, denudation hill, and deciduous forest are the most favorable conditions to initiate debris slides, while for rockslides–rockfall, steep slope (42°–75°), low TWI, soil depth less than 3 m, nearest distance to drainage, highly dissected terrain, and barren lands are the favorable land conditioning factors. It was found that the steep slope of the mountains, types of rocks, overburden depth, rainfall, and human activity influence the frequency of occurrence of landslides in that area (Ghosh, S. et al., 2012; Bera, S. et al., 2021). Different geo-environmental conditions account for the spatial distribution of debris slides and rockslides. Kainthura, P. and Sharma, N. (2021) in their work found that Random Forest performed better than Bayesian network and Backpropagation neural network during their investigation into prediction of landslide susceptibility in the Uttarkashi region. Li, Y. et al. (2020) in the study utilized multiresolution segmentation and CNN to point out that landslide susceptibility has a general consistency with the amount of human activity (like excavation for road construction, commercial planting) and rainfall received in the selected study area. Multiresolution segmentation and CNN are employed to improve the accuracy of LSE. Yunus, A. P. et al. (2021) investigated the drivers of intensified landslide regimes in Western Ghats, India and found that road cutting and slopes modified for plantations are the strongest environmental variable (~72%) related to the landslide patterns, whereas short-duration intense precipitation in the elevated terrain, profile concavity, and stream power contributed to the initiation of landslides.

7.5.4 Findings for Tropical Cyclones Warning System

There were two stages in TC genesis; the first was transformation of tropical depression from tropical disturbance, and the second was transformation of tropical storm from tropical depression. They used different algorithms discussed earlier, but they found AdaBoost to be the most effective algorithm, with maximum accuracy. Robustness was also increased due to extension of lead time to 48 hours. In TC, track factors like sea surface temperature, atmosphere temperature, weather pattern, TC structure, TC intensity, and features of land topography were used. The normal path of TC was predicted well, but there was a problem in predicting abnormal path, which happens when there is sudden movement in the speed, stagnation, and curvature. For TC path prediction, researchers used satellite images and achieved 86% of correctness in segmentation rate and 97% of correctness in classification. The prediction error of the RBF network improved by 30%. In the model proposed by Michael B. Richmana and Lance M. Lesliea (2012), Nonlinear SVR gives better accuracy, measured by R2, by~ 40% as compared to MLR. On adding a new variable in SVR, its prediction accuracy improved by approximately 121%.

7.6 CHALLENGES AND OPEN-ENDED ISSUES

7.6.1 Challenges

The various challenges that are faced by the research community with respect to the earthquake detection are first unavailability of fully accurate methods to detect the seismic waves that are in the form of incoming data as true earthquake waves, to determine whether the waves are from micro earthquakes or greater intensity earthquakes. To detect the amount of destruction that the earthquake waves may create (depending upon their intensity), depending upon the duration of the wave we need to decide to apply some better methods like fast Fourier transformation to do the real-time analysis of the seismic waveforms. According to the duration of the earthquake, it may be classified as a micro-earthquake or a macro earthquake. The various previously discussed challenges need to be looked after as, ultimately, we are working upon the intensity of destruction that is being caused to the living beings present. The history of the type of landform of the area where the earthquake is taking place also matters, as some areas are more prone to quakes where the soil is softer. So, the hardness and softness of the soil in the upper layers and the lower layers of the areas where earthquakes are taking place is also to be considered as an important challenge in detecting tremors. Proper utilization of data including data from satellites, observation, and analysis is still required. Use of machine learning on detection of various problems in remote sensing is still challenging. More exploration in prediction of TC and the intense precipitation caused by TC is required.

7.6.2 OPEN-ENDED ISSUES

When the ML-DL methods are applied on datasets, then there is a big challenge to clean the dataset, as most of the time the available dataset has certain noise. Now the extent of cleaning depends upon the quality of the dataset that we get. The cleaning of the dataset can also be done only to some extent.

7.6.2.1 Availability of Relevant Clean Datasets

The various methods that can be applied to clean the dataset are checking for the missing values by replacing them with their median values. Many times, the obtained data has a lot of gaps in the range of the values, meaning some of the data values belong to a very low range and some of them belong to a very high range, i.e., the data is not evenly distributed. So, in such a case the prediction accuracy may be affected. In such a scenario we need to just normalize the data to bring it to a uniform scale. All these challenges need to be issued well before going for accurate predictions. Many times, there are many disparities in the number of values belonging to various classes especially when multiclass classification is concerned. In such a scenario we need to sample the data based on either the oversampling or the undersampling. Sometimes we need to apply the fast Fourier transformation method to process the data if there are many outliers present in the data. The best way to accomplish this is to first do the descriptive analysis of the data based on various parameters like maximum value for a particular column value, minimum value for a particular column, 25% quartile value, 50% quartile value, and finally 75% quartile value. The descriptive statistics helps us in understanding our data to a good extent. So, it is always best to first clean the data by pre-processing it and then go for application of any machine learning for the deep learning prediction model.

7.7 CONCLUSION

It has been concluded that the increase in data will reflect the improvement in disaster prediction techniques. Adding more variables will not affect the performance in the prediction technique. More data, even if it comes from a different source with low quality, will improve flood damage modelling compared to a test set, as many models become susceptible to data mistakes if the dataset is small. Regression trees are outperformed by Random Forests and KNN. The Gaussian Naive Bayesian algorithm is not very effective as compared to other ways. As far as the detection of earthquakes' signals and wave catching is concerned, it has been observed from the previous data that the machine learning models are proving slightly less efficient as compared to the deep learning methods. So, there has been a challenge to involve more and more deep learning methods that can help in the aforementioned work. The deep learning models help in improving the performance of catching the systemic waves and then converting them to the complete waveform for the detection and interpretation of seismic waves. It has been observed that the various deep learning methods are being applied, but still many more methods need to be explored and implemented by the researchers to further improve the accuracy of the prediction models. High efficiency and prediction power models based on deep learning algorithmic concepts that are further based on various statistical parameters will help in detecting and characterizing the seismic data in a better way. Earthquake monitoring has been an urgent need among the population, and various new and efficient improved methodologies need to be developed that can cater to a large amount of input seismic data. A lot of work still needs to be done especially in exploring deep learning methods, preferably in hybrid or ensemble or in a coupled way with FFT or other methods. This will certainly help the research community and the inhabitants of the areas that are quite prone to the frequent tremors.

As far as the prediction of landslides is concerned, landslide susceptibility evaluation (LSE) is inherently a challenging problem as the factors that influence its estimation (rainfall, soil moisture, rock weathering, human activities in the locale, etc.) are dynamic in nature and keep on changing. Occurrence of landslides depends on the complex relationship among geological, topographic, and hydrological factors of the selected locale. Owing to complex relationships, selection of the influential land conditional factors becomes subjective as well as challenging.

As far as the prediction of tropical cyclones is concerned, this has been a challenging task for meteorologists for over ten decades. Many researchers have done deep studies on various issues related to TC. Many machine learning algorithms have been used on different datasets for detecting, analysing, and predicting TC. This has led to a new way of solving many issues regarding TC prediction. Numerical models using different techniques of machine learning algorithms and data-driven approaches have improved the forecasting of TC to a great extent. By integrating machine learning, researchers have done lots of progress in various aspects like TC weather prediction, intensity prediction, path prediction, genesis forecasts, and improving numerical models. But some aspects still remain to be studied and improved. Use of machine learning in large scale data of TC is still under study. There are several challenges in accurate forecasting of TC that require a deeper understanding of TC dynamics. In this regard India has established a dedicated authority, the National Disaster Management Force (NDRF). Its role is to take all necessary steps pre disaster as well as post disaster starting from prediction of disaster, doing risk analysis, preparing locals of disaster-prone areas, and taking countermeasures so that disaster does not affect India socially as well as financially to a large extent.

7.8 FUTURE RESEARCH DIRECTIONS

Machine learning techniques have previously been shown to be successful for preprocessing environmental time-series datasets as a preliminary to the classification of flood data in previous research. We will continue to improve in the future. Other forms of machine learning approaches should be considered. For example, the application of global optimization techniques like the genetic algorithm can broaden the reach and the scope of the research. In the context of a natural disaster, "earthquake" detection and prediction using deep learning methods is proven to be more effective in terms of predicting various parameters. Earthquake monitoring has been an urgent need among the population, and various new and efficient improved methodologies need to be developed that can cater to a large amount of input seismic data. One such work has been done that has proven quite successful in which the signals are coupled with fast Fourier transformation to give a better real-time effect and also enhance the accuracy of predictions. But still we have been able to achieve only 95% of the accuracy that further needs to be improved. Still there are many waves that are generated as fake waves for the earthquake detection that should be reduced if not completely eliminated. A lot of work is going on in the research community especially for the areas that are quite prone to the earthquake detection.

In the context of the natural disaster "landslides," landslides are a very complex geo-environmental phenomenon. Management of risk and the loss associated with the landslide occurrence can be minimized by enhancing the reliability of landslide prediction models. However, there is subjectivity associated with the consideration of mapping units and the resolution of the remote sensing data sources. This subjectivity of different mapping units affects the landslide risk maps, which is one of the essential tools while studying landslide susceptibility. In order to achieve precision in predictions, the future scope can be development of standard units and standard quantification of various landslide causative factors. The predictions models can be further improved by taking into account the dynamism involved in important land conditioning factors like green vegetation area or land cover area, stream erosion, etc. Another scope is to consider the hidden region-specific factors associated with the climate change of those regions. Incorporation of event-based landslide inventories may help in overcoming overfitting problems of the deep learning models. DL/ML models can be utilized to study less-explored types of landslides like rotational landslides and translational landslides. These models can be utilized for the analytical analysis of the triggering factors, simulation of the nonlinear relationships between these factors, and their influence on landslide susceptibility of the locale as well as the surrounding areas that are likely to be affected by the debris flows. Further studies are also required for debris flow susceptibility assessment triggered by either natural landslides or human engineering activity-induced landslides. TC has been a major concern for more than ten decades. Numerous researchers have done in-depth research in finding out key issues regarding TC along with forecasting techniques. Studies show that better prediction accuracy is obtained after applying machine learning methods in predicting TC. This increases the efficiency of traditional approaches like calibration of damage function. Cross validation can become an intuitive tool to estimate uncertainty linked with forecasting models. Machine learning is challenging as well as promising in its techniques in predicting TC. It requires deep knowledge of machine learning and the key problems so that they can be solved by building suitable ML models. Improvement in performance of numerical models in TC forecasts is required. Numerical models are very expensive to use. In numerical model solving, error propagation becomes a cause of inaccurate forecasting of TC. So a pure data-driven TC prediction system is required to ensure low cost and good efficiency.

REFERENCES

Aad, O. M., & Chen, Y. (2020a). Deep denoising autoencoder for seismic random noise attenuation. *Geophysics*, 85(4), 67–V376.

Aad, O. M., & Chen, Y. (2020b). Automatic waveform-based source-location imaging using deep learning extracted microseismic signals. *Geophysics*, 85(6), Art. no. 6. doi:10.1190/geo2020-0288.1

Allen, R.V. (1978). Automatic earthquake recognition and timing from single traces. *Bull. Seismological Soc. Amer.*, 68(5), 1521–1532.

Bera, S., Upadhyay, V.K., Guru, B., & Oommen, T. (2021). Landslide inventory and susceptibility models considering the landslide typology using deep learning: Himalayas, India. *Natural Hazards*, 1–33.

Bergen, K. J., & Beroza, G. C. (2019). Earthquake fingerprints: Extracting waveform features for similarity-based earthquake detection. *Pure Appl. Geophys.*, 176(3), 1037–1059.

Breiman, L. (2001). Random forests. *Machine Learning*, 45(1), 5–32.

Chai, X., Tang, G., Wang, S., Peng, R., Chen, W., & Li, J. (2020). Deep learning for regularly missing data reconstruction. *IEEE Trans. Geosci. Remote Sens.*, 58(6), 4406–4423.

Chang, L. C. et al. (2010). Clustering-based hybrid inundation model for forecasting flood inundation depths. *Journal of Hydrology*, 85(1–4), 257–268.

Chen, R., Zhang, W., & Wang, X. (2020). Machine learning in tropical cyclone forecast modeling: A review. *Atmosphere*, 11(7), 676.

Chen, Y. (2018). Fast waveform detection for microseismic imaging using unsupervised machine learning. *Geophys. J. Int.*, 215(2), 1185–1199.

Chen, Y. (2020). Automatic microseismic event picking via unsupervised machine learning. *Geophys. J. Int.*, 222(1), 1750–1764.

Cortes, C. et al. (1995). Support-vector networks. *Machine Learning*, 20(3), 273–297.

El Zini, J., Rizk, Y., & Awad, M. (2020). A deep transfer learning framework for seismic data analysis: A case study on bright spot detection. *IEEE Trans. Geosci. Remote Sens.*, 58(5), 3202–3212.

Evries, P. M., Viégas, F., Wattenberg, M., & Meade, B. J. (2018). Deep learning of aftershock patterns following large earthquakes. *Nature*, 560(7720), 632–634.

Fang, Z., Wang, Y., Peng, L., & Hong, H. (2020). Integration of convolutional neural network and conventional machine learning classifiers for landslide susceptibility mapping. *Computers & Geosciences*, 139, 104470.

Gao, Z., Pan, Z., Zuo, C., Gao, J., & Xu, Z. (2019). An optimized deep network representation of multi mutation differential evolution and its application in seismic inversion. *IEEE Trans. Geosci. Remote Sens.*, 57(7), 4720–4734.

Ghosh, S., van Westen, C. J., Carranza, E. J. M., & Jetten, V. G. (2012). Integrating spatial, temporal, and magnitude probabilities for medium-scale landslide risk analysis in Darjeeling Himalayas, India. *Landslides*, 9(3), 371–384.

Giffard-Roisin, S., Yang, M., Charpiat, G., Kumler Bonfanti, C., Kégl, B., & Monteleoni, C. (2020) Tropical cyclone track forecasting using fused deep learning from aligned reanalysis data. *Front. Big Data*, 3(1). doi:10.3389/fdata.2020.00001

Hafez, A. G., Rabie, M., & Kohda, T. (2013). Seismic noise study for accurate P-wave arrival detection via MODWT. *Comput. Geosci.*, 54, 148–159.

Harirchian, E., Lahmer, T., & Rasulzade, S. (2020). Earthquake hazard safety assessment of existing buildings using optimized multi-layer perceptron neural network. *Energies*, 13(8), 2060.

He, Z., Peng, P., Wang, L., & Jiang, Y. (2020). PickCapsNet: Capsule network for automatic P-wave arrival picking. *IEEE Geosci. Remote Sens. Lett., Early Access.* doi:10.1109/LGRS.2020.2983196.

Jena, R., Pradhan, B., Naik, S. P., & Alamri, A. M. (2021). Earthquake risk assessment in NE India using deep learning and geospatial analysis. *Geosci. Front.*, 12(3), 101110.

Jia, B., & Huang, Q. (2020). De-capsnet: A diverse enhanced capsule network with disperse dynamic routing. *Appl. Sci.*, 10(3), 884.

Juyal, A., & Sharma, S. (2021, February). A study of landslide susceptibility mapping using machine learning approach. In *2021 third international conference on intelligent communication technologies and virtual mobile networks (ICICV)* (pp.1523–1528). IEEE.

Kadavi, P. R., Lee, C. W., & Lee, S. (2018). Application of ensemble-based machine learning models to landslide susceptibility mapping. *Remote Sensing*, 10(8), 1252.

Kainthura, P., & Sharma, N. (2021). Machine learning driven landslide susceptibility prediction for the Uttarkashi region of Uttarakhand in India. *Georisk: Assessment and Management of Risk for Engineered Systems and Geohazards*, 1–14.

KeumKun, H. J. et al. (2020). Real-time flood disaster prediction system by applying machine learning technique. *KSCE Journal of Civil Engineering*, 24(9), 2835–2848.

Khalaf, M. et al. (2017). *A performance evaluation of systematic analysis for combining multiclass models for sickle cell disorder data sets*, 115–121. Cham: Springer International Publishing.

Khalaf, M. et al. (2018). A data science methodology based on machine learning algorithms for flood severity prediction. *IEEE Congress on Evolutionary Computation (CEC)*, 1–8. doi:10.1109/CEC.2018.8477904.

Knapp, K. R., & Kruk, M. C. (2010). Quantifying interagency differences in tropical cyclone best-track wind speed estimates. *Mon. Weather Rev.*, 138(4), 1459–1473.

Koem, C. et al. (2021). Flood disaster studies: A review of remote sensing perspective in Cambodia. *Geographia Technica*, 16(1), 13–24.

Lee, S., Lee, M. J., Jung, H. S., & Lee, S. (2020). Landslide susceptibility mapping using naïve bayes and bayesian network models in Umyeonsan, Korea. *Geocarto International*, 35(15), 1665–1679.

Leonard, M., & Kennett B. L. N. (1999). Multi-component autoregressive techniques for the analysis of seismograms. *Phys. Earth Planet. Interiors*, 113(1–4), 247–263.

Li, Y., Wang, X., & Mao, H. (2020). Influence of human activity on landslide susceptibility development in the Three Gorges area. *Natural Hazards*, 104(3), 2115–2151.

Li, Z., Peng, Z., Hollis, D., Zhu, L., & McClellan, J. (2018). High-resolution seismic event detection using local similarity for large-N arrays. *Sci. Rep.*, 8(1), 1646.

Lomax, A., Michelini, A., & Jozinovic, D. (2019). An investigation of rapid earthquake characterization using single- station waveforms and a convolutional neural network. *Seismological Res. Lett.*, 90(2A), 517–529.

Mandal, K., Saha, S., & Mandal, S. (2021). Applying deep learning and benchmark machine learning algorithms for landslide susceptibility modelling in Rorachu river basin of Sikkim Himalaya, India. *Geosci. Fronti.*, 12(5), 101203.

Michael, B. R., & Lesliea, L. M. (2012). Adaptive machine learning approaches to seasonal prediction of tropical cyclones. *Procedia Comp. Sci.*, 12, 276–281.

Mousavi, S. M., Ellsworth, W. L., Zhu, W., Chuang, L. Y., & Beroza, G. C. (2020). Earthquake transformer—an attentive deep-learning model for simultaneous earthquake detection and phase picking. *Nature Commun.*, 11(1), 1–12.

Mousavi, S. M., Zhu, W., Sheng, Y., & Beroza, G. C. (2019). CRED: A deep residual network of convolutional and recurrent units for earthquake signal detection. *Sci. Rep.*, 9(1),1–14.

Mosavi, A. et al. (2018). Flood prediction using machine learning models: Literature review. *Water*, 10(11), 1536. doi:10.3390/w10111536

Nakata, N., & Beroza, G. C. Stochastic characterization of mesoscale seismic velocity heterogeneity in Long Beach, California. *Geophys. J. Int.*, 203, 2049–2054.

Ngo, P. T. T., Panahi, M., Khosravi, K., Ghorbanzadeh, O., Kariminejad, N., Cerda, A., & Lee, S. (2021). Evaluation of deep learning algorithms for national scale landslide susceptibility mapping of Iran. *Geosci. Front.*, 12(2), 505–519.

Nguyen, V. V., Pham, B. T., Vu, B. T., Prakash, I., Jha, S., Shahabi, H., & Tien Bui, D. (2019). Hybrid machine learning approaches for landslide susceptibility modeling. *Forests*, 10(2), 157.

Panda, R.K. et al. (2010). Simulation of river stage using artificial neural network and MIKE 11 hydrodynamic mode. *Computers & Geosciences*, 36(6),735–745.

Perol, T., Gharbi, M., & Denolle, M. (2018). Convolutional neural network for earthquake detection and location. *Sci. Adv.*, 4(2), Art. no. e1700578.

Pham, B. T., Pradhan, B., Bui, D. T., Prakash, I., & Dholakia, M. B. (2016). A comparative study of different machine learning methods for landslide susceptibility assessment: A case study of uttarakhand area (India). *Environmental Modelling & Software*, 84, 240–250.

Pourghasemi, H. R., Jirandeh, A. G., Pradhan, B., Xu, C., & Gokceoglu, C. (2013). Landslide susceptibility mapping using support vector machine and GIS at the Golestan Province, Iran. *Journal of Earth System Science*, 122(2), 349–369.

Pradhan, B., & Jebur, M. N. (2017). Spatial prediction of landslide-prone areas through k-nearest neighbor algorithm and logistic regression model using high resolution airborne laser scanning data. In *Laser scanning applications in landslide assessment* (pp. 151–165). Springer.

Rachel, N., & Lakshmi, M. (2016). Landslide prediction with rainfall analysis using a support vector machine. *Indian Journal of Science and Technology*, 9(21).

Ross, Z. E., Yue, Y., Meier, M., Hauksson, E., & Heaton, T. H. (2019). PhaseLink: A deep learning approach to seismic phase association. *J. Geophys. Res. Solid Earth*, 124(1), 856–869.

Ross, Z. E., Meier, M., Hauksson, E., & Heaton, T. H. (2018). Generalized seismic phase detection with deep learning. *Bull. Seismological Soc. Amer.*, 108(5A), 2894–2901.

Saad, O. M., Inoue, K., Shalaby, A., Samy, L., & Sayed, M. S. (2018). Automatic arrival time detection for earthquakes based on stacked denoising autoencoders. *IEEE Geosci. Remote Sens. Lett.*, 15(11), 1687–1691.

Sabour, S., Frosst, N., & Hinton, G. E. (2017). Dynamic routing between capsules. *Proc. Adv. Neural Inf. Process. Syst*, 3856–3866.

Sachdeva, S., Bhatia, T., & Verma, A. K. (2020). A novel voting ensemble model for spatial prediction of landslides using GIS. *International Journal of Remote Sensing*, 41(3), 929–952.

Snehil, C. et al. (2020). *Flood damage analysis machine learning techniques*. International Conference on Smart Sustainable Intelligent Computing and Applications under ICITETM2020.

Tabbussum, R., & Dar, A. Q. (2021, May). Performance evaluation of artificial intelligence paradigms-artificial neural networks, fuzzy logic, and adaptive neuro-fuzzy inference system for flood prediction. *Environ Sci Pollut Res Int.*, 28(20), 25265–25282. doi:10.1007/s11356-021-12410-1. Epub 2021 Jan 16. PMID: 33453033.

U, S., Guan, Z., Verschuur, E., Chen, Y. (2020). Automatic high-resolution microseismic event detection via supervised machine learning. *Geophys. J. Int.*, 222(3), 1881–1895.

Van Dao, D., Jaafari, A., Bayat, M., Mafi-Gholami, D., Qi, C., Moayedi, H., & Pham, B. T. (2020). A spatially explicit deep learning neural network model for the prediction of landslide susceptibility. *Catena*, 188, 104451.

Wang, Z., Liu, Q., & Liu, Y. (2020). Mapping landslide susceptibility using machine learning algorithms and GIS: A case study in Shexian County, Anhui Province, China. *Symmetry*, 12(12), 1954.

Xiao, L., Zhang, Y., & Peng, G. (2018). Landslide susceptibility assessment using integrated deep learning algorithm along the China-Nepal highway. *Sensors*, 18(12), 4436.

Xu, W. et al. (2021). Deep learning experiments for tropical cyclone intensity forecasts. *Weather and Forecasting*, 36(4).

Yunus, A. P., Fan, X., Subramanian, S. S., Jie, D., & Xu, Q. (2021). Unraveling the drivers of intensified landslide regimes in Western Ghats, India. *Science of the Total Environment*, 770, 145357.

Zhu, W., & Beroza, G. C. (2018). PhaseNset: A deep-neural-network-based seismic arrival-time picking method. *Geophys. J. Int.*, 216(1), 261–273.

8 A Recent Survey on LSTM Techniques for Time-Series Data Forecasting

Present State and Future Directions

Sardar M N Islam, Narina Thakur, Kanishka Garg and Akash Gupta

CONTENTS

8.1 INTRODUCTION

The forecasting of an outcome based on time-dependent inputs is known as time series forecasting. Stock market data, which illustrates whether stock prices change over time, is an illustration of time series data. Stock market data is an example of time series data since it shows how stock prices change over time. Similarly, the temperature of a certain region might be called time series data because it fluctuates hourly. These time dependent problems are often referred to as sequence problems since they involve data acquired progressively through time. Since all inputs in a feed-forward neural network are independent of one another or are usually referred to as independent identical distributed (IID), sequential data processing is not acceptable. Only a short-term memory can benefit from a simple recurrent neural network. The vanishing gradient problem affects recurrent neural network (RNN) with a larger time dependency. Long short term memory network (LSTM), a version of RNN, is now being utilized to solve sequence problems in various disciplines. It was designed to work around the constraints of an RNN. The typical extensive datasets

DOI: 10.1201/9781003245469-8

of time series can be a time-consuming process, retarding the RNN architecture training. The data volume can be reduced; however, this may lead to information loss. In any time-series data set, it is essential to understand previous trends and the periodicity of data to make accurate forecasts. This demonstrates work done recently in LSTM in time series analysis. There were many ensembles and hybrids on LSTM in time series forecasting. It was mainly used in power consumption or electricity forecasting because, as the world is moving toward the smart home and smart grid systems, accurate power consumption is necessary. Other hybrid LSTM techniques were explored as well, as they could also be applied in our main problem statement.

8.2   RELATED WORK

Researchers are presently focused on recently created deep learning-based time series data forecasting algorithms, such as LSTM, which outperform previous algorithms. The technique of making scientific predictions based on prior time series data is known as time series forecasting. It entails emerging models based on previous data that are used to make judgments and guide future strategic decisions. The fact that the future outcome is fully unknown at the time of the work is a key distinction in forecasting. LSTM researchers have recently emphasized time series forecasting in an attempt to solve the disadvantages of traditional forecasting approaches, which are inefficient and complex. With the increased availability of large volumes of past data and the necessity for precise forecasting, a sophisticated forecasting technique that can infer the stochastic dependency between previous values has become essential for forecasting future values.

Several techniques have been used in forecasting the future values, including univariate autoregressive (AR), univariate moving average (MA), simple exponential smoothing (SES), autoregressive integrated moving average (ARIMA), and its variations to effectively predict the next lag of time series data. The ARIMA model, in particular, has exceeded in regard to precision and accuracy in anticipating future lags. Algorithms for analyzing and forecasting time series data are being developed as a result of recent advances in computers' computational capacity and, more crucially, the advancement of more advanced machine learning algorithms and methodologies such as deep learning. Researchers emphasize new deep learning-based methodologies for forecasting time series data, such as LSTM, that outperform traditional algorithms.

Figure 8.1 depicts how researchers' interest in LSTM time series forecasting internet search has grown from January 2012 to July 2021. Google Trends and the search term "Time Series Forecasting with LSTM" were used to collect the data. Since the numbers are z-normalized using the highest interest rate, the highest interest rate has a value of 100. The time series forecasting with LSTM had already recently attracted a lot of attention from academia and industry, particularly in the last few years. Furthermore, the related topics and linked terms long-term memory (LTM), forecasting for time series data and ARIMA models are also on the rising trend.

According to Runge et al. [1] deep learning algorithms are useful in the field of building-level demand forecasting. Previous research, on the other hand, lacked comparative assessments of deep learning methods. The usefulness of deep

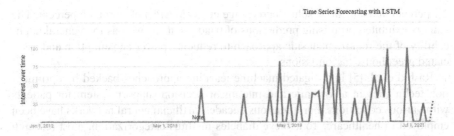

FIGURE 8.1 Google Trends for time series forecasting with LSTM.

learning-related load forecasting is investigated in this study utilizing three-layer deep-learning architectures: LSTM sequence to sequence (LSTM S2S), unidirectional LSTM, and CNN. The data collection comprises electrical usage data for a residential building over four years, taking into account individual resident consumers; the data collection includes electrical consumption data. According to our findings, the LSTM is best adapted to forecasting the forecast's trend. The LSTM is unable to work for such alterations, according to this article. In comparison to the CNN classifier, the S2S models can also predict data reliably.

Manic et al. [2] highlighted that the forecast for electricity demand has important consequences for the cost and security of energy supply. For safe and reliable energy system activity, accurate forecasting models are needed. In this chapter, the author presents and proposes two novel time-series methodologies for load forecasting in the short term for the Singapore power system data. The authors concluded that multiplicative decomposition model performance is better than the ARIMA model, and both the models showed favorable predictive performance. In addition, the author noted that the multiplicative decomposition models mean absolute percentage error was much lower than the ARIMA model, particularly on Tuesday and Wednesday. In today's culture, accurate power load demand forecasting is crucial to the successful implementation of energy policies. Algorithms utilizing deep learning for time series datasets in several industries have recently presented ensemble techniques.

Qiu et al. [3] have introduced an ensemble deep learning approach by combining deep belief network (DBN) and support vector regression (SVR). The authors have tested the regression methodology for the Mackey-Glass' time series data set and three electricity-load-demand. The four benchmark approaches have been compared with the proposed technique. For the root mean square error (RMSE) and mean absolute scaled error (MASE) evaluation metrics, the proposed deep learning hybrid strategy outperformed the four benchmark approaches. The proposed method can handle huge and complex datasets.

Jiménez et al. [4] in 2020 proposed an algorithm for the detection and predicting occasions of the high demand of power, controlled in demand-side solution programs at the national scale, for United Kingdom triads. The technique has two stages: LSTM and filtering of the likely highest requirement peaks for electricity by the exponential-moving average. The United Kingdom real-time data was considered as a use case/sample study. The approach has been validated with an RMSE of about

2.2 percent and a mean absolute percentage error (MAPE) of about 1.6 percent. The authors exhibited promising predictions of triad events as well as the general applicability of the requirement side approach to reducing power consumption and associated greenhouse gas emissions.

Rashid et al. [5] investigated machine-learning approaches backed by optimization techniques in a dynamic and intelligent decision-support system for patients with various conditions. For numerous decades, artificial neural networks have been employed in healthcare. To achieve diabetes mellitus categorization, most research works use a multilayer layer perceptron trained with a backpropagation learning technique. Instead of using the backpropagation learning process, this chapter shows how to train an adaptive LSTM with two optimization algorithms. Biogeography-based optimization and genetic algorithm are the optimization algorithms used. Finally, for classification purposes, the datasets were input into adaptable models such as LSTM with biogeography-based optimization (BBO) and LSTM with genetic algorithm (GA).

Siami-Namini et al. [6] have compared the overall accuracy of ARIMA and LSTM algorithms. The experimental results obtained for the finance data highlight that the ARIMA is way better than LSTM. The LSTM algorithm precisely enhanced the accuracy by about 85% to the ARIMA algorithm. In addition, the number of epochs is altered, and no change was reported. The proposed research work discussed and highlighted the advantages of using deep learning algorithms and techniques for economic and financial data. There are many prediction problems in finance and economics, which can be solved efficiently using deep learning methods.

Chung et al. [7] in 2018 proposed an LSTM-based share price forecasting model, which is based on the RNN basic deep learning algorithm. To look into the time attributes of the share market, the authors have combined the GA and LSTM network and used the model's architectural factors. The LSTM network selected in this work consists of two hidden layers, which use deep neural network architecture to more efficiently communicate complicated features of the stock market that are not linear. For the time window width and the number of LSTM nodes in the neural network, GA was used to search for the optimal or near-optimal value. The experiments have been performed on the Korean stock market dataset to predict the ending price and daily price. The work findings showed that there is a lower mean square error (MSE), mean absolute error (MAE), and MAPE for the proposed method, and the changes are statistically important. These final results show that the genetic algorithm with a long short term memory strategy can be an efficient tool for forecasting share markets to reflect time trends.

Kuremoto et al. [8] in 2012 have proposed a unique neural network model for forecasting time-series data. Here, the type of machine learning model is a DBN made up of many limited restricted Boltzmann machines (RBMs). The model's structure was z-optimized by an algorithm for particle swarm optimization (PSO). The pre-processing of the original time series data was also used to obtain time-series differential data for neural-net models such as the RBM and multi-layer perceptron data (MLP). Using the CATS benchmark, the proposed model was validated as a priority for traditional neural-net models such as MLP and the mathematical ARIMA model. Although the prediction accuracy did not reach the maximum level of the IJCNN'04 competition system of participant prediction, this research first illustrates the availability of RBM that can be used to predict time-series data.

Pan et al. in 2016 [9] highlighted that forecasting the number of epidemics is critical for the Canter for Disease Control and Prevention (CDC). This paper has proposed an ARIMA-based model to improve forecast and the real data from the CDC between January and August 2014 have been used. To begin, autocorrelation (AC) and partial autocorrelation (PAC) analysis have been used to generate a stationary time series and then estimate the autocorrelation order, moving average order, and different order. Also, the Least-Squares (LS) method has been used to estimate the prediction model's parameters. The proposed model obtained 92.1% forecast accuracy and significantly outperforms the simple moving average method currently used by the CDC.

The issues faced in forecasting the time series environmental dataset were emphasized by Alhirmizy et al. in 2019 [10]. The authors proposed an LSTM-based forecasting method for the air quality and levels of pollution in Madrid, Spain, for every hour for two years from 2015 to 2016. The data includes the date and time as well as pollution concentrations, principally SO2, NO2, NO, and CO.

Siami et al. [11] present a behavioral review and proposed Bi-LSTM and LSTM benchmarking forecasting models. The proposed model aims to tune the parameters that are involved in successive levels of data training. It has been highlighted that Bi-LSTM-based modeling produces more accurate forecasting results than regular LSTM-based models, and Bi-LSTM models, as compared to other models, have been found better predictions. The experimental results finding indicated that Bi-LSTM can capture some additional data-related features that unidirectional LSTM models can't since training is only one way. As a result, instead of employing LSTM to foresee the problem in time series analysis, this research recommends using Bi-LSTM.

Masum et al. [12] emphasized the intricacies of multiple-step forecasting and its methodologies in 2018. The authors propose a two-step forecasting algorithm for three non-linear electric load datasets extracted from a publicly available power system dataset in this paper. The efficiency of the ARIMA, LSTM, and multiple-step forecasting models has been compared, and the results indicated that the LSTM significantly outperforms the ARIMA model for multi-step electric load forecasting.

Yadav et al. [13] addressed the research gap that there are no predefined rules for configuring LSTM because it is a relatively new model. Because of the presence of a long-term trend, seasonal and cyclical fluctuations, and random noise, LSTM has been applied to time-series forecasting, which is a particularly difficult problem to solve. The performance of LSTM varies on the selection of various hyperparameters, which must be done with care to achieve good results. An LSTM model was designed for a dataset derived from the Indian stock market for four companies. It was then tuned for the number of hidden layers by comparing stateless and stateful models. The authors concluded that a stateless LSTM model is recommendable for time series forecasting challenges due to its higher stability. Increasing the number of hidden layers seems to have the benefit of ensuring the LSTM is very much stable, as demonstrated by the reducing standard deviation values and variance in the box and whisker plot diagram in the experimental results.

Zhang et al. [14] proposed an LSTM cyclic neural network (NN) for gas concentration forecasting used for coal mineral mine production. Efficient gas concentration forecasting leads to the rational development of corresponding safety measures, which are critical for a leading role in the improvement of coal mine safety management. To implement the forecasting model, the authors considered various features such as gas production monitoring data model, structural design, model training, model prediction, and model optimization. The proposed model is implemented with both objective functions as the minimization goal. Adam optimization technique is employed to update the neural network's weight, and the network layer and batch size are changed to select the best one. The coal mine gas concentration prediction model uses the number of layers and batch size as parameters. The experiment suggested that the LSTM gas concentration forecasting model has better accuracy than the bidirectional recurrent neural network (Bi-RNN) model as well as the gated recurrent unitmodel using large data volume sample prediction. The prediction model's average means square error can be minimized up to 0.003, and the predicted mean square error can be minimized up to 0.015, resulting in better reliability in gas concentration time series forecasting. The gas concentration at the time inflection point can be forecasted more accurately, and the mean square error at the inflection point can be lessened to 0.014, boosting the applicability of the proposed model.

Sagheer et al. [15] proposed a deep LSTM (DLSTM) architecture approach as an extension of the traditional RNN. The genetic algorithm has been used to optimally configure DLSTM's architecture. Two different case researches from the petroleum industry field have been carried out as well as performance evaluation, using production data from two actual oilfields. To ensure a fair evaluation, the proposed approach's performance is compared to that of several standard methods. The empirical results, using multiple evaluation criteria, show that the proposed DLSTM model outperforms other standard approaches.

Sahoo et al. [16] evaluated and illustrated the efficiency of LSTM-RNN with RNN for the hydrological time series dataset and proposed LSTM-RNN forecasting of low flow time-series data. The proposed forecasting models used regular output data from the Mahanadi River in India. Some statistical metrics have been used to evaluate results by comparing the outputs of different algorithms. The research authors demonstrated that the LSTM-RNN may be utilized to model the hydrological time series dataset low flow river data, outperforming the RNN and other basic models by a large margin.

Laptev et al. [17] proposed an RNN solution for accurate, complete trip forecasting using Uber data and public monthly data from M in 2017 [17]. During special events, Uber forecasting can lead to more efficient driver allocation and shorter wait times for riders. Existing methods frequently use an amalgamation of multivariate regression forecasting methods and machine learning. Such a system, however, is difficult to tune and scale. The experimental results demonstrate that three key parameters must be high in order to select an appropriate NN model for time series: time-series length, the correlation between time series and the number of time series, time-series correlation, and time-series data duration. The authors of this chapter proposed an accurate prediction for Uber completing trips using a novel RNN architecture that outperforms existing forecasting methods on Uber data and generalizes well to the M3 dataset.

Kasun Bandara et al. [18] demonstrated the LSTM multi-seasonal net (LSTM-MSNet) methodology and proposed multi-seasonal decomposition techniques as a supplement to the LSTM for M4 forecasting competition. This research paper proposes a framework for forecasting time series with multiple seasonal patterns. The proposed algorithm can forecast the number of linked time series with different seasonal cycles.

Zhao et al. [19] elaborated in 2017 that short-term LSTM-based traffic forecasting is an important problem in intelligent transportation systems. This paper examines a short-term traffic prediction model as well as traffic data analysis approaches. In traffic management, an accurate forecast result allows commuters to choose appropriate travel modes, routes, and departure times. This study focuses on traffic volume prediction, but for commuters, a comprehensive road-traffic forecast that includes transit time, traffic speed, and traveler accommodations would be more useful. The availability of large amounts of traffic data and computational capabilities has increased recently, motivating the researchersto use deep learning methods for improving the accuracy of short-term traffic forecasts. When compared to other algorithms, the suggested LSTM method for forecasting road traffic performs better.

Hu, Y. et al. addressed their work on time-series data for safety and proposed an ensemble LSTM-based forecasting in 2019 [20–21]. Trends vastly differing in time length have been further predicted and analyzed from the perspective of safety forecasts. The superiority of the proposed algorithm has been demonstrated experimentally by applying it to the electromagnetic radiation intensity TSD sampled from an actual coal mine and PM2.5 in the UCI repository.

Table 8.1 shows a critical examination of various recent time-series algorithms.

TABLE 8.1
Critical Analysis of the Recent Time Series Algorithms

Reference No.	Dataset type	Time series algorithm used	Advantages
[1]	Individual customer electricity consumption data	LSTM S2S	Outperformed LSTM, CNN and ANN
[2]	Singapore electricity data	Multiplicative decomposition model	Performs better than the seasonal ARIMA model
[3]	Electricity load datasets, California housing dataset	Ensemble Deep Learningapproach—DBN + SVR	DBN + SVR have better results than SVR, feedforward Neural Network, ensemble feedforward NN, and DBN
[4]	UK triads Case study forecasting	LSTM, ANN, SVM, Random Forests, Bayesian Regression	LSTM outperforms other models
[5]	Breast cancer dataset and diabetes dataset	LSTM + BBO LSTM + GA	LSTM + BBO is better than LSTM + GA
[6]	Time series data of financial kind	LSTM model	Outperforms other models

(*Continued*)

TABLE 8.1   (Continued)

Reference No.	Dataset type	Time series algorithm used	Advantages
[7]	Korean stock exchange data	GA z-optimizing the LSTM time window	The new GA improved LSTM outperforms benchmark with a huge margin
[8]	CATS benchmark data	DBN + RBM + PSO	Outperformed MLP and ARIMA
[9]	The everyday number of patients for a pandemic spread between January and August 2014	ARIMA	Better result than a simple moving average algorithm for a time frame of a week
[10]	Raw pollution dataset of Madrid	LSTM model	Good prediction for a dataset less than 2,000 rows and more epochs with more layers improves the result
[11]	Stock market data from Jan 1985 to Aug 2018	Bi-LSTM	Bi-LSTM outperformed LSTM and ARIMA
[12]	Three nonlinear electric load datasets	LSTM	LSTM outperformed ARIMA
[13]	Indian stock market data	Stateless LSTM	Stateless LSTM performed better than stateful LSTM
[14]	Coal mine production monitoring data	LSTM with modified hidden layers and batch size	The proposed algorithm performs better than other basic algorithms
[15]	Petroleum time series data	Various LSTM blocks stacked on top of each other	The proposed algorithm outperforms RNN and GRU
[16]	Hydrological time series data	LSTM-RNN	The proposed algorithm performs better than RNN
[17]	Special event forecasting at Uber	LSTMautoencoder + LSTM forecaster	Better than classical approach only if correlation in your series is of a high extent
[18]	One competitive dataset, two energy used time-series datasets, and one traffic movement dataset	LSTM-MSNet	The proposed method has more efficiency than many forecasting methods that have one variable
[19]	Short term traffic forecast	Cascaded LSTM in both time and spatial domain	Better than all basic algorithms
[20]	Time series data for safety forecasts.	PSO and gradient descent aggregated LSTM	Better than normal LSTM

8.3   DISCUSSION OF FINDINGS

We looked at and summarized the various types of time-series data forecasting methods that are currently available. This section discusses challenges, patterns, and complexities that have evolved as a result of the time series data. In the literature, various time series datasets were analyzed using various time series forecasting methodologies. Recent research indicates that LSTM and LSTM variants (such as Bi-LSTM, DLSTM, ensemble LSTM, LSTM-MSNet, LSTM-RNN) algorithms easily surpass existing machine learning and ARIMA models. When the LSTM method was compared to ARIMA, the average error rate reduction achieved by LSTM was between 84 and 87%, demonstrating its superiority. Furthermore, epochs—or the number of training iterations—have no influence on the trained prediction model's performance, and it demonstrates stochastic behavior. The best algorithm for this was LSTM in the majority of the research papers. The major research gap identified in the study was that the performance of the LSTM could be improved if we knew what weights should be initialized to the weights of the LSTM Neural Network, the number of hidden layers used for your particular dataset. Time-series datasets all individualistic characteristics, and after preprocessing the dataset, LSTM architecture, like the number of its hidden layers and the optimization function it is using, could be modified to provide better solutions. For such modification with infinite possibilities, nature-inspired techniques could be used.

8.4 CONCLUSION AND FUTURE SCOPE FOR RESEARCH

Time-series data prediction is an important field of work. For example, we can predict the usage of electricity for a house or place accurately. There is no extra electricity generated, and we would not need a sophisticated setup to store that. This could be the future of solar electricity smart grids or time-series data like the estimated number of Coronavirus cases, which is of utmost importance in itself. The more accurate the prediction, the more prepared the nation can be with its health facilities and administrative planning. We read upon many other kinds of time-series data in this study. And we need a sophisticated and accurate method to do so. Many algorithms and techniques are discussed in the papers with LSTM and GRU giving considerably good results. For future work, researchers can try to improve the accuracy of these algorithms.

REFERENCES

[1] Runge, J., & Zmeureanu, R. (2021). A review of deep learning techniques for forecasting energy use in buildings. *Energies*, 14(3), 608.
[2] Manic, M., Amarasinghe, K., Rodriguez-Andina, J. J., & Rieger, C. (2016). Intelligent buildings of the future: Cyberaware, deep learning-powered and human interacting. *IEEE Industrial Electronics Magazine*, 10(4), 32–49.
[3] Deng, J., & Jirutitijaroen, P. Short-term load forecasting using time series analysis: A case study for Singapore. In *2010 IEEE conference on cybernetics and intelligent systems*. IEEE, 2010.
[4] Qiu, X., Zhang, L., Ren, Y., Suganthan, P. N., & Amaratunga, G. (2014, December). Ensemble deep learning for regression and time series forecasting. In *2014 IEEE symposium on computational intelligence in ensemble learning (CIEL)* (pp. 1–6). IEEE.
[5] Jiménez, J. M., Stokes, L., Yang, Q., & Livina, V. N. (2020). Modeling energy demand response using long-short term memory neural networks. *Energy Efficiency*, 13(6), 1263–1280. Springer.
[6] Rashid, T. A., Hassan, M. K., Mohammadi, M., & Fraser, K. (2019). Improvement of variant adaptable LSTM trained with metaheuristic algorithms for healthcare analysis. In *Advanced classification techniques for healthcare analysis* (pp. 111–131). IGI Global.

[7] Siami-Namini, S., Tavakoli, N., & Namin, A. S. (2018, December). A comparison of ARIMA and LSTM in forecasting time series. In *2018 17th IEEE international conference on machine learning and applications (ICMLA)* (pp. 1394–1401). IEEE.

[8] Chung, H., & Shin, K. S. (2018). Genetic algorithm-optimized long short-term memory network for stock market prediction. *Sustainability*, 10(10), 3765.

[9] Kuremoto, T., Kimura, S., Kobayashi, K., & Obayashi, M. (2012, July). Time series forecasting using restricted Boltzmann machine. In *International conference on intelligent computing* (pp. 17–22). Springer.

[10] Pan, Y., Zhang, M., Chen, Z., Zhou, M., & Zhang, Z. (2016, June). An ARIMA-based model for forecasting the patient number of epidemic diseases. In *2016 13th international conference on service systems and service management (ICSSSM)* (pp. 1–4). IEEE.

[11] Alhirmizy, S., & Qader, B. (2019, March). Multivariate time series forecasting with LSTM for Madrid, Spain pollution. In *2019 international conference on computing and information science and technology and their applications (ICCISTA)* (pp. 1–5). IEEE.

[12] Siami-Namini, S., Tavakoli, N., & Namin, A. S. (2019). A comparative analysis of forecasting financial time series using Arima, LSTM, and Bi-LSTM. *arXiv preprint arXiv:1911.09512*.

[13] Masum, S., Liu, Y., & Chiverton, J. (2018, June). Multi-step time series forecasting of electric load using machine learning models. In *International conference on artificial intelligence and soft computing* (pp. 148–159). Springer.

[14] Yadav, A., Jha, C. K., & Sharan, A. (2020). Optimizing LSTM for time series prediction in the Indian stock market. *Procedia Computer Science*, 167, 2091–2100.

[15] Zhang, T., Song, S., Li, S., Ma, L., Pan, S., & Han, L. (2019). Research on gas concentration prediction models based on LSTM multidimensional time series. *Energies*, 12(1), 161.

[16] Sagheer, A., & Kotb, M. (2019). Time series forecasting of petroleum production using deep LSTM recurrent networks. *Neurocomputing*, 323, 203–213.

[17] Sahoo, B. B., Jha, R., Singh, A., & Kumar, D. (2019). Long short-term memory (LSTM) recurrent neural network for low-flow hydrological time series forecasting. *ActaGeophysica*, 67(5), 1471–1481.

[18] Laptev, N., Yosinski, J., Li, L. E., & Smyl, S. (2017, August). Time-series extreme event forecasting with neural networks at uber. In *International conference on machine learning* (Vol. 34, pp. 1–5). ICML.

[19] Bandara, K., Bergmeir, C., & Hewamalage, H. (2020). LSTM-MSNet: Leveraging forecasts on sets of related time series with multiple seasonal patterns. *IEEE Transactions on Neural Networks and Learning Systems*, 32(4), 1586–1599.

[20] Zhao, Z., Chen, W., Wu, X., Chen, P. C., & Liu, J. (2017). LSTM network: A deep learning approach for short-term traffic forecast. *IET Intelligent Transport Systems*, 11(2), 68–75.

[21] Hu, Y., Sun, X., Nie, X., Li, Y., & Liu, L. (2019). An enhanced LSTM for trend following of time series. *IEEE Access*, 7, 34020–34030.

9 Conflicting Statements Detection Using Bi-Directional LSTM

Jeena Varghese and Aswathy Wilson

CONTENTS

9.1   INTRODUCTION

Identifying discrepancy, inconsistency and arrogance in the text are the major functions in automatic contradiction detection or conflicting statements detection in texts. For example, if one of the candidates says "I support the new anti-corruption law" and another candidate says "I do not support the new anti-corruption law". Here, the statements indicate negation. Therefore it is contradictory. Another example of statements consisting of numeric mismatch is: "More than 50 people died in the plane crash" and "10 people died in the plane crash" [1]. These are relatively simple examples of conflicting statements, but statements can be more complex and require in-depth inference, understanding and comprehension of text. Different types of contradictory statements mentioned in my work are negation, numeric mismatches and antonyms [2–3].

Conflict detection has many real-world applications. So, the areas of machine learning, information retrieval and natural language processing (NLP) are attractive to many researchers for finding solutions for automatic conflict detection. Conflict

DOI: 10.1201/9781003245469-9

detection has many technical challenges and has a rigorous NLP task. Clashing articulation recognition between wellsprings of texts or two-sentence sets can be outlined as a classification issue [4]. The most well-known way to find contradictions in the text is to train or learn a classifier from hand-deciphered guides to derive linguistic features from the text and perform the classification task. Observing inconsistencies in the text is as yet not a totally tackled issue, and there are numerous limits and exploration gaps in the current work. Deep artificial neural networks have become extremely well known as of late because of their viability in solving many pattern recognition and AI issues. The use of artificial neural networks and deep learning is a somewhat neglected and unused region for the issue of discrepancy in texts.

The main goal is to explore the use of deep learning and artificial neural network to detect contradictions in statements. Likewise, technologies and strategies such as embeddings Global Vectors for Word Representation (GloVe) and Bi-directional long short-term memory (LSTM) have gained prominence in NLP and ML literature. The utilization of these methods has not been explored for the detection of contradictions and the problem of finding conflicting statements. The primary motivation is to inspect the use of GloVe embeddings and bi-directional LSTM for feature extraction from sentences and text portrayal.

9.2   LITERATURE REVIEW

The literature survey includes many papers that cover almost all aspects of conflicting statements detection.

9.2.1   IDENTIFYING CONFLICTING INFORMATION IN TEXTS

Understanding the relationship between the sentences is important for information analysis. Here, focus on the contradiction relationship, and build a system for detecting conflicting statements. Such a system requires more subtle variations than general systems [5]. Also, argue for the centrality of the event conference and therefore include a component based on the topicality. The typology of discrepancies naturally arises in the text, giving the first detailed breakdown of the conflicting statements detection task.

The proposed system for detecting discrepancies is compatible with the Stanford recognizing textual entailment (RTE) system. The multi-stage engineering of the Stanford framework is followed by it. The initial step is to consider the etymological portrayals that contain data about the semantic substance of the sections: the text and derivation are changed over into composed dependency graphs created by Stanford Parser. The second step provides an alignment between the graphs, consisting of a mapping from each node in the hypothesis to a unique node in the text or to null. Subtleties on the scoring arrangement scale and search calculation can be found. In the last step, we extract the conflicting features that logistics regression applies to arrange the pair as contradictory or not. Features are hand-held, directed by phonetic awareness. Inconsistency highlights depend on confusion between the sentences and the speculation. Notwithstanding, sets of verses that don't depict a similar occasion may not thus contradict each other, although they may contain inconsistent information. So it adds an extra step of filtering out non-corporate events before feature extraction [5]. The bungle of data between sentences is regularly a decent sign of non-edification, yet it isn't adequate to identify logical inconsistency and requires a more accurate understanding of the consequences of the sentences.

9.2.2 EARLY CONFLICT DETECTION WITH MINED MODELS

Designers are progressively embracing source code for the executives (SCM) frameworks with broad help for mergers such as branching, parallel development and Git and Mercurial [6]. Procedures for early discovery of contradictions and unexpected connections between changes in different branches will lessen the work needed to manage programming advancement and contemporary improvement by a significant element. The primary thought introduced in this chapter is to run a multi-branch server-side unique investigation on each submitted activity. The experiments accessible in the SCM framework will be dissected to decide the idea of the application, and models that catch how the program chips away at various scales will be automatically obtained. For example, the functional behavior of a program can be addressed by technique pre-post conditions, programming interface utilization conventions and rules of inclination between strategy summons. Models that catch angles, for example, time cutoff points, periodicity and timeline controls of tasks can represent the temporal nature of a program. These models are utilized on the server side to enable automatic conflict recognition.

Models are created for every form in each branch and, consequently, analyzed as each change is presented. Comparison of models permits recognizing conduct inconsistencies paying little heed to the presence of text-based inconsistencies that shouldn't be settled for investigation. This investigation is called behavioral driven continuous integration (BDCI). By elevating the examination to a conduct level, BDCI can significantly work fair and level of contentions that are identified and settled toward the start of the cycle, just like the pace of bugs caused by contemporary updates to the software. Subsequently, the expense and exertion needed to finish the consolidation will be significantly reduced, and the capacity to foster programming over the long haul will be significantly improved.

9.2.3 NORM CONFLICT IDENTIFICATION USING DEEP LEARNING

Agreements will be arrangements with at least two gatherings officially as deionic articulations or models inside their provisions. If not painstakingly planned, such discrepancies can discredit a whole agreement, so human critics take incredible measures to compose tedious and defective struggle-free agreements for perplexing and extended agreements. In this work, a methodology has been created to computerize the distinguishing proof of errors between the norms in agreements. A two-venture approach utilizing conventional AI and deep learning were created to recognize and contrast the standards with distinguishing discrepancies between them. With a bunch of inconsistencies deciphered as a train and a test set, our methodology accomplishes 85% precision and sets up new complexity [7].

Our approach to identifying inconsistencies between the standards in agreements is isolated into two phases. In the main part, we recognize the models inside the contract clauses via a support vector classifier utilizing a physically deciphered dataset. In the subsequent part, we order ordinary sets as clashing or not utilizing a CNN. The first venture toward standard conflict identification is to distinguish which sentences of an agreement contain deontic explanations (norms). For this errand we consider contract sentences to be of two restrictive sorts: norm sentences and non-norm sentences. To isolate normal sentences from the remainder of the agreement message, we train a classifier dependent on support vector machines (SVMs) utilizing a physically interpreted dataset. We made the dataset utilizing the first agreements separated from the site, particularly the agreements of the development area. We physically interpreted the conditions of every agreement, either normal or non-sentence, bringing about 699 non-sentences and 494 non-sentences out of an aggregate of 22 agreements, which we use as train and test sets.

In this work, we fostered a two-venture way to deal with distinguishing clashes between the norms in the agreements. Our significant commitments are: (1) a dataset containing physically interpreted criteria and non-regularizing sentences from the first arrangements; (2) an AI model to order legally binding sentences as standardizing and non-regularizing; (3) a self-interpreted dataset of agreements containing irregularities between standards; (4) a deep learning model for classifying normal sets as contradictory and non-contradictory.

9.2.4 PRECLUDE: CONFLICT DETECTION IN TEXTUAL HEALTH ADVICE

With the fast digitalization of the wellbeing area, individuals frequently go to portable applications and online wellbeing sites for wellbeing counsel. Because they address various parts of wellbeing (e.g., weight reduction, diet, disease) or in light of the fact that they are unconscious of a client's unique circumstance (for example age, gender, physical condition), wellbeing guidance created from various sources might be inconsistent. Inconsistencies might occur because of lexical highlights (like negation, antonyms or numerical mismatch) or might be changed by time and/or physiological status. We figure out the issue of tracking down incongruous wellbeing guidance and foster a thorough classification of contradictions. As a comparable exploration region in the natural language processing domain investigates the issue of technical conflict identification, the discovery of contradictions in health counselling raises its own novel lexical and semantic difficulties. These incorporate the enormous primary distinction between the reading material and theory combined, the revelation of the reasonable cross-over between the pair's recommendation and the guess of the semantics of the counsel (i.e., what to do, why and how). In this way, we foster Preclude, a clever semantic principle-based answer for finding contradictions in wellbeing guidance assembled from an assortment of sources utilizing phonetic laws and outside information bases. Since our answer is interpretable and far reaching, it can likewise lead clients to compromise. We assess the Preclude utilizing 1,156 unique guidance articulations covering 8 significant wellbeing points gathered from cell phone wellbeing applications and famous wellbeing sites. We block results at 90% accuracy and surpass the essential methodology exactness and F1 score by 1.5 multiple times, separately.

Observing inconsistencies in text-based wellbeing directing raises both lexical and semantic difficulties. This undertaking is lexically difficult as the lexical design of the text can shift impressively, depending on the length of the instruction and additionally the tone of the exhortation. The semantic difficulties of conflict recognition are multifold. First of all, we really want to remove the indirect activity of the exhortation from the text and the outcomes. Second, we need to find out if there is any ideological overlap between two or more counseling statements. Observing applied crossover regularly requires progressive connections between various subjects, like diet, medications, and exercise. At long last, clashes are frequently transitory or contingent, i.e., a contention happens when a brief/state of being is correct. Consequently, it requires a far-reaching suspicion of the semantics of counsel, as well as tracking down irregularities with 90% accuracy, Preclude gives exact data on possible reasons for struggle to assist with settling on informed choices for compromise [8].

9.3 METHODOLOGY

One of the most difficult problems to uncover in machine learning has nothing to do with complex calculations: it is the problem of getting the right datasets in the right organization. Obtaining accurate information refers to the aggregation or separation of information related to the results to be anticipated. The collection and development of a training set—a large collection of known information—requires some investment and area-specific study of where and how relevant data should be collected. Test sets that require more time or skill can fill a particular niche in the field of information science and critical thinking. Choosing the right dataset is very important for machine learning.

The extracted data is passed on to the data preprocessing unit. Pre-preparation refers to changes associated with information before being fed into the calculation. Data preprocessing is a method used to convert crude information into a perfect information index. In order to achieve the best results from the model applied in a machine learning project, it is essential that the configuration of information should be in a lawful manner. This unit consists of various processes like Tokenization and Stemming [9]. The outputs from these processes are used for analysis. Then fitting on the text will help us create a vocabulary so that each word is assigned with a unique integer. We then convert the entire sentence of Statement 1 to a series of numbers assigned by the tokenizer. The Tensorflow Machine Learning System is used to train and test the prediction model to detect inconsistencies in the text. Padding the sequence helps in making all the sentences of the same length. Maxlen is the parameter that decides the length we want to assign to all the sentences.

Padding is done by adding 0 on either the end of a sentence or prior to the sentence if the sentence has a length less than max length. This is also a parameter that the user can changeby defaulting its prefix. If the length of the sentence is more than 100,

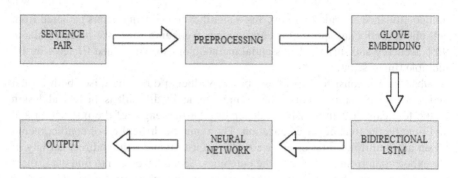

FIGURE 9.1 A simple diagrammatic representation for conflicting statements detection using bi-directional LSTM.

then it is pruned, which brings down the length to 50 (maxlen). Then, create a corpus by combining training and test events for a specific dataset. Then, compute all the unique terms in the corpus. An index ID is given to each of the terms in the corpus. Then convert all the sentences in the dataset to a vector containing the index ID of the word. After that, GloVe embedding performs an operation in which each word in the sentence is changed over into a vector (from word to vector), which is utilized as a NLP task to track down the inconsistency in the text.

Use word embeddings so that words with similar words have similar representation in vector space [10]. It represents every word as a vector. The words that have similar meanings are placed close to each other. There are many words in the training data that are not there in the glove embeddings. So we take the mean of the embeddings and replace the absent words in the embeddings with mean. We can perform functions like vector ("research")—vector ("journal") on the vector of the actual value acquired because of GloVe embedding [11]. Now, we convert each word in our vocabulary into word embeddings. This embedding is a vector of 150 dimension, which represents each word as a vector and places them into a vector space. An embedding matrix is created in which the number assigned to the word by the tokenizer is assigned with the corresponding vector that we get from the glove embeddings. The vector is utilized to make one of the highlights of the textual classification task called contrast or unbiased discovery.

After that, the bi-directional LSTM approach, an augmentation of the conventional LSTM, is applied. A bi-directional LSTM deep neural network is utilized because it has shown encouraging results across various domains and datasets [12]. For issues where all the time phases of the input sequence are accessible, two bi-directional LSTMs are polished rather than one LSTM in the input sequence. The first is the input sequence and the second is the converse duplicate of the input sequence. This will give the network more context and result in a faster and more complete study of the problem. The utilization of bi-directional LSTMs may not bode well for all sequence prediction problems, yet it can offer some advantages as far as better outcomes to those spaces where it is proper. This methodology has been utilized with incredible impact with LSTM recurrent neural networks. For encoding an English sentence into a vector, recurrent neural networks (RNN) with LSTM can be utilized. An RNN with an LSTM-based methodology is applied on the grounds that concentrates on the great portrayal of the sentence pair to be ordered, for the errand of tracking down inconsistencies or neutral sentences. With an end goal to work on the precision of the framework, not many features have been designed. When these features are combined with the features created by the neural network, there is a significant increase in accuracy.

1) Jaccard Coefficient:

Jaccard Coefficient (otherwise called Intersection over Union—IOU) is a measurement broadly utilized in information recuperation applications to gauge the likeness between two texts [1]. For our situation, the quantity of words normal to the two sentences is a negligible portion of the quantity of words in the numerator, and the complete number of words in the two sentences is the denominator. The coefficient catches the connection between the level of likeness between two sentences and the presence of a logical inconsistency between them. It is helpful to gauge the comparability between sentences of a similar subject and when utilizing comparable wording.

2) Negation:

It is a double component that labels the qualities as valid or not. It is actually the case that one of the words in a given pair of sentences doesn't contain any of these words or phrases: *never, not, not much, without, none,* and *nothing* from our predetermined invalidation list. The thought here was to catch the logical inconsistencies that one refrain communicates while a different one communicates negative opinion. Obviously, this component alone can't recognize an inconsistent and a non-problematic assertion. Be that as it may, this component is valuable when breaking down sentence sets (short sentences) on a similar subject and utilizing comparative phrasing.

3) IsAntonym:

IsAntonym is exceptionally instinctive and self-explanatory. In the event that none of the words in a single sentence have antonyms in the other sentence, it will have a value of 0. Else, value 1. After adding 47 antonyms from our ending, we analyze the words from each sentence against a bunch of antonyms we have incorporated in the non-official characterization (NOC). The last set contains 3,714 antonyms. On the off chance that we observe something from any sentence in our antonym list, we take its antonym from the set and check if that word is in the other sentence. In the event that it is available, the worth is 1, else, the worth is 0. The rundown is explicit to our dataset and can be improved as increasingly more datasets are added.

4) Overlap Coefficient:

It estimates the cross-over between two sets, determined as the size of the convergence partitioned by the littlest size of the two sets. Overlap coefficient catches the similitude well when the distinction between the measures of the two sentences is enormous.

The embedding layer is important as this will help us in the training of the sentences with their respective embeddings that we have assigned previously. The first parameter is the size of our vocabulary. The second parameter is the output embeddings length, which is 50 in this case as we used the 50 glove embeddings of each word. The length of each observation that is expected by the network is given by the input_length parameter. We have padded all the observations to 50, hence we set input_length = 50. The weights parameter shows that the embedding that we want to use is embeddings_matrix, and it should not be altered, hence trainable is kept false. If we want to train our own embeddings, we can simply remove the weights and trainable parameters.

Next is building an LSTM model. LSTM networks are valuable in arrangement information since they can retain the previous words that assist to comprehend the significance of a sentence that aids in message order [13–14]. The bidirectional layer is useful as it assists with understanding the sentence from start to finish and from end to start [15–16]. It works in the two ways. This is valuable on the grounds that the reverse order LSTM layer can learn designs that are unrealistic for ordinary LSTM layers, going from the start to the furthest limit of the sentence in the typical request. Bi-directional layers along these lines are valuable in text-based order issues as they can catch various examples from two headings. There are several pooling techniques used. Average Pooling, Global Average Pooling, Max Pooling and Global Max Pooling. Average Pooling and Max Pooling take in kernel size as argument. If the input of the max pooling layer is 0,1,2,2,5,1,2 and the global max pooling output is 5, then the pool size of ordinary max pooling layer is 3 outputs 2,2,5,5,5 (assuming stride = 1). The same is the case with Average Pooling, only instead of taking the maximum values we do the average of all the values.

Dense Layers is a completely associated layer that is each input node in one layer associated with every one of the neurons in the next layer. It is basically helping us out in the classification of the dependent variable with an activation function that tells the neuron when to be activated. Rectified Linear Activation Function (ReLU) activation function is used here. There are many different activation functions that can be used like sigmoid, tanh, ReLU etc. Dropout layers help us to avoid overfitting. In dropout, a few neurons are randomly turned off. This is done in order to remove the dependency of neurons on each other. No neuron is particularly responsible for learning a specific feature. Argument 0.2 specifies that 20% of the neurons in these layers are going to the turned off. This value ideally ranges between 0.1 to 0.5.

The last layer is the dense layer with the quantity of neurons equivalent to the quantity of classes of dependent variable (1 for this situation). The activation used here is softmax. We use sigmoid because it is a multi-label classification problem (one observation can belong to more than one class). This is one of the special cases, but when we are working on a multi-label classification problem where one observation is assigned to a single class then we use the softmax activation function in our final layer. It performs the same as a sigmoid function, that is it provides us with the probabilities of the observation belonging to a particular class. We pick the maximum probability and assign that observation to the class having maximum probability. Sigmoid is used when we are dealing with a binary classification problem. Probability less than 0.5 is assigned to 0 class and more than 0.5 is assigned to

1 class. ReLU is a piecewise direct capacity that will yield straightforwardly to the input in the case that it is positive; any other way, it will yield to nothing. Adam is an advancement calculation that is utilized to refresh network loads iterative dependent on training information.

Compiling the model is the final step to complete building the model. Optimizer is like the cost function (similar to gradient descent). Adam is one of the best performing functions that is used in deep learning models. We can change and test with different optimizers too. Binary cross entropy is the loss function that checks the loss in predictions made by the model by calculating the distance between the predicted value and the actual value. We use categorical_crossentropy when dealing with multiclass classification problems and binary_crossentropy when dealing with binary classification. This is a special case for the multi-label classification problem, hence we are looking for the probability of observation belonging to each class. The top few classes are taken for classification.

Accuracy is being used for checking the performance of the model. Accuracy is one of the evaluations that can be used to check if the model is overfitting or underfitting. If the training accuracy is very high as compared to testing accuracy we can see that the model is overfitting as it is not generalizing well on the unseen data (test data) [17–18]. To avoid overfitting we can reduce the complexity of the model, add more data and add more regularization (dropout, batch normalization). If the training accuracy is much lower than the testing accuracy, the model is inadequate, i.e. the model cannot capture the features, so we need to train the model with more epochs.

Batch size is the parameter that defines how many observations are fed into the model at a time for training. The greater the batch size the more computational resources and memory are required for processing. This is one of the hyperparameters that can be used for tuning the model. Batch size of one will feed one observation at a time, which is really useful, but the training and convergence speed will be very slow, and the model might overfit easily.

The test data is preprocessed so that model can easily make its prediction as it should be in the same format as that of our training data. Note that we are using the same tokenizer in our testing and we are not fitting it again because this might change the numbers assigned to words that are there in the training data. Dropout argument is done for dropping inputs to the LSTM cell, and recurrent dropout is done for dropping the long-term memory of the LSTM connection.

9.4   RESULTS AND ANALYSIS

The results have two phases; one is the claim extraction result and the next step is the contradiction detection result. Initially test the machine learning model is trained using bi-directional LSTM with GloVe word embeddings. The Pheme dataset on figshare is used for training and testing the Bi-LSTM model. The next phase is the logical inconsistency discovery.

Conflicting statements detection results are calculated using three features: accuracy, recall and F1 score [19–20]. Here accuracy is the ratio between the number of accurate predictions and the absolute number of data tests. To evaluate the performance of the algorithms for the conflicting statement detection problem, various evaluation measures were used.

- True positive (TP): When fake articles are predicted it is actually interpreted as fake article.
- True negative (TN): When true articles are predicted it is actually interpreted as true article.
- False negative (FN): When true articles are predicted it is actually interpreted as fake article.
- False positive (FP): When fake articles are predicted it is actually interpreted as true article.

The formulas for calculating f1-score, accuracy and recall are shown below.

$$Precision = True\ positive/(True\ positive + False\ Positive)$$

$$Recall = True\ positive/(True\ positive + FalseNegative)$$

$$F1\ score = (2 * Precision * Recall)/(Precision + Recall)$$

Precision is characterized as the negligible part of the significant occasions over the recovered examples as evident. Recall is characterized as the small part of the pertinent cases over the complete significant cases.

F1-score can be characterized as the harmonic mean of the precision and recall. Based on the confusion matrix, we can understand that bi-directional LSTM is performing better than other algorithms.

The accuracy score of 96% clearly describes how a model works on test data that is not used for training. The exactness score clearly describes how a model chips away at test information that isn't utilized for preparing. We can clearly pretend that conflicting statements detection using bi-directional LSTM is performing best with an accuracy of 96% dependent on the exactness diagram (i.e., accuracy graph).

9.5 CONCLUSION

In this study, a neural network was introduced to learn bi-directional LSTM with GloVe word embeddings for the task of finding contradictions in sentences. A strategy is presented dependent on deep learning, artificial neural networks, long short-term memory and global vectors for word portrayal to identify clashing articulations in the text. The primary objective is to fabricate a framework for recognizing irregularities and inconsistencies in the text. The issue of observing irregularities in the text is outlined as an classification problem, which accepts a couple of sentences as information sources and yields a binary value showing whether the sentence sets are contradictory or not. To do this, first get the linguistic proof and printed highlights from the sentence pair, like the presence of negation, antonyms, intersection and string overlaps. Then, we apply artificial neural networks, a feature based on long short-term memory and GloVe embedding. The contradiction class has 96% accuracy. Accuracy scores are another important measure of a machine learning model.

This score clearly explains how a model works on test data not used for training. The spread and descriptive statistics for features are different; they are unrelated and give different perspectives. Overall, experimental analysis shows that it is possible to accurately detect contradictions in short sentence pairs, including negation, antonym and numeric mismatch using deep learning techniques.

9.6 FUTURE SCOPE

This research proposes a novel approach to tackle the complex, recurring and relevant problem of detecting contradiction of statements. Therefore, new opportunities, different approaches and possible improvements never cease. In this proposed system, we use only a pair of two sentences as input. In future work, we are planning to take a pair of two various documents as input and detect whether those documents are contradictory or not.

Author Contributions: Jeena Varghese conceived, designed and built the model; Mrs Aswathy Wilson supervised her student Jeena Varghese to write the chapter; Together they read and approved the final manuscript.

Conflicts of Interest: The authors declare no conflict of interest.

REFERENCES

[1] Lingam, V., Bhuria, S., Nair, M., Gurpreetsingh, D., Goyal, A. and Sureka, A. "Deep learning for conflicting statements detection in text," *PeerJ Preprints*, vol. 6, 2018.

[2] Harabagiu, Sanda and Hickl, Andrew. "Methods for using textual entailment in open-domain question answering," *Proceedings of the 21st International Conference on Computational Linguistics and the 44th Annual Meeting of the Association for Computational Linguistics*, pp. 905–912. Association for Computational Linguistics, 2006.

[3] Padó, S., de Marneffe, M.C., MacCartney, B., Rafferty, A.N., Yeh, E. and Manning, C.D. "Deciding entailment and contradiction with stochastic and edit distance-based alignment," *Theory and Applications of Categories*, 2008, November.

[4] Dragos, Valentina. "Detection of contradictions by relation matching and uncertainty assessment," *Procedia Computer Science*, vol. 112, pp. 71–80, 2017.

[5] de Marneffe, Marie-Catherine, Anna R. Rafferty and Christopher D. Manning. "Identifying conflicting information in texts," 2009.

[6] Mariani, L., Micucci, D. and Pastore, F. "Early conflict detection with mined models," *IEEE International Symposium on Software Reliability Engineering Workshops*, pp. 126–127, 2014.

[7] Aires, João Paulo, and Felipe Meneguzzi. "Norm conflict identification using deep learning," *International Conference on Autonomous Agents and Multiagent Systems*. Springer, 2017.

[8] Preum, S. M., Mondol, A. S., Ma, Meiyi, Wang, Hongning and Stankovic, J. A. "Preclude: Conflict detection in textual health advice," *IEEE International Conference on Pervasive Computing and Communications (PerCom)*, pp. 286–296, 2017.

[9] Abadi, M., Agarwal, A., Barham, P., Brevdo, E., Chen, Z., Citro, C., Corrado, G., Davis, A., Dean, J., Devin, M., Ghemawat, S., Goodfellow, I., Harp, A., Irving, G., Isard, M., Jia, Y., Jozefowicz, R., Kaiser, L., Kudlur, M., Levenberg, J., Mané, D. and Monga, R. "Tensorflow: Large-scale machine learning on heterogeneous distributed systems," Preliminary White Paper. *arXiv Vanity*, 2015.

[10] Almuzaini, Huda Abdulrahman and Azmi, Aqil M. "Impact of stemming and word embedding on deep learning-based Arabic text categorization," *IEEE Access*, vol. 8, pp. 127913–127928, 2020.

[11] Kamyab, Marjan, Liu, Guohua and Adjeisah, Michael. "Attention-based CNN and bi-LSTM model based on TF-IDF," *Applied Sciences*, vol. 11, no. 23, 2021.

[12] Liu, Gang and Guo, Jiabao. "Bidirectional LSTM with attention mechanism and convolutional layer for text classification," *Neurocomputing*, vol. 337, pp. 325–338, 2019.

[13] Abduljabbar, Hussein Dia, Pei-Wei Tsai, "Unidirectional and Bidirectional LSTM Models for Short-Term Traffic Prediction," *Journal of Advanced Transportation*, p. 16, 2021.

[14] Singh, Rusul L., Kumar, Jitendra, Goomer, Rimsha and Kumar, Ashutosh. "Long short term memory recurrent neural network (LSTM-RNN) based workload forecasting model for cloud datacenters," *Procedia Computer Science*, vol. 125, pp. 676–682, 2018.

[15] Schuster, M. and Paliwal, K.K. "Bidirectional recurrent neural networks," *IEEE Transactions on Signal Processing*, vol. 45, pp. 2673–2681, 1997.

[16] Basaldella, Marco, Antolli, Elisa, Serra, Giuseppe and Tasso, Carlo. "Bidirectional LSTM recurrent neural network for keyphrase extraction," *Italian Research Conference on Digital Libraries*, pp. 180–187. Springer, 2018.

[17] Kobayashi, K. "Learning from conflicting texts: The role of intertextual conflict resolution in between-text integration," *Reading Psychology*, vol. 36, no. 6, pp. 519–544, 2015.

[18] Zhang, H., Yan, Z., Sun, C. and Wei, S. "Based on entities behavior patterns of heterogeneous data semantic conflict detection," *2015 12th Web Information System and Application Conference (WISA)*, pp. 169–174. IEEE, 2015.

[19] Sardinha, R. C., Weston, N., Greenwood, P. and Rashid, A. "Ea-analyzer: Automating conflict detection in aspect-oriented requirements," *IEEE/ACM International Conference on Automated Software Engineering*, pp. 530–534, 2009.

[20] Shih, C., Lee, C., Tsai, R. T. and Hsu, W. "Validating contradiction in texts using online co-mention pattern checking," *ACM Transactions on Asian Language Information Processing (TALIP)*, vol. 11, no. 4, 2012.

10 Big Data and Deep Learning in Healthcare

S Jahangeer Sidiq, Ovass Shafi, Majid Zaman and Tawseef Ahmed Teli

CONTENTS

10.1 INTRODUCTION

Information analysis plays an important role in modern world. Multiple terms are coming out day by day from the concept of data science. Data science is one of the popular topics that floods the news, videos, social media and print media these days. Machine learning is one of the subtopics of artificial intelligence that includes algorithms for enabling machines to automate the problem solving process without any specific computer program. The term *artificial intelligence* refers to a broad area and has a wide range of applications in digital personal assistants, self-driving cars, the ranking of products and much more. The most exciting application of artificial intelligence is in healthcare. As artificial intelligence enables the machines to self-learn continuously, there is a need to temper advanced machine

DOI: 10.1201/9781003245469-10

learning algorithms against the problems of applying such tools in medical practices. In broader terms one has to consider three aspects before the implementation of machine learning in healthcare. The first one is the practical appreciation of machine learning tools along with the limitations, understanding and interpretation of the findings. The second one is regulatory, ethical and legal framework for proper and secure application of these tools in clinical healthcare. The third consideration is the government framework and platform for reasonable use of data. As far as the technical aspect is concerned, machine learning is the application of a specific class of algorithms called deep neural networks of deep learning on big datasets, also known as Big Data. A subset of a dataset is used to train the neural network, which in turn can provide highly accurate results and helps to find complex patterns in the datasets. There are a number of application of deep learning in the healthcare sector. As an enormous amount of data is produced by medical organizations like hospitals, the data is disordered and complex. In order to efficiently apply deep learning on such data, the data needs to be preprocessed and mapped appropriately before applying it to the training of machines. The preprocessing step is the most important step as the success of the model depends on the reliability of the data fed to the machine. If a model is designed and trained using the wrong data, it might recommend the wrong medication and hence lead to bad consequences. As the case is with any engineering tool, the machine learning algorithm needs to be carefully engineered in order to be effective. When machine learning is applied in the healthcare context, the driver and the locus point for machine learning applications must be the medical problem itself. With the main emphasis on the clinical problem, machine learning has the ability to integrate and examine complex and large clinical datasets and this makes it a useful aid for clinicians to make better decisions for the treatment of patients. A second consideration while applying machine learning in healthcare is ethics. There were no clear guidelines about ethical use of clinical datasets until recently; however, a set of guidelines has been issued recently by Singapore's Model Artificial Intelligence Governance Framework that provides proper guidelines to private organizations regarding the use of artificial intelligence in an ethical way. The machines trained with artificial intelligence can help humans to perform routine tasks for which they are well trained and suited, freeing the human resource so that they can devote more effort to tasks that require human judgment. Another issue in the use of a dataset for machine learning application in medical care is data security and privacy. In order to provide the data security and privacy, there is a need for the implementation of sound data anonymization along with strong security measures so that data theft can be avoided.

10.2   MOTIVATION BEHIND RESEARCH

Diabetes mellitus is one of the deadly diseases that not only affects people around the globe but affects people of all age groups irrespective of gender. Diabetes mellitus is the main cause behind some serious ailments like cardiac problems, kidney related diseases, eye related problems, liver ailment etc. With the advancement in technology and with the availability of a huge collection of data related to diabetes mellitus, it is now possible to use artificial intelligence techniques along with big datasets for early diagnosis of the disease so that the disease can be controlled before its onset, which will help people around the globe to live healthy lives.

10.3 LITERATURE REVIEW

A research study (Naz and Ahuja 2020) gives a method for predicting diabetes using a diverse machine learning algorithm applied on a PIMA dataset. The authors use ANN, Naïve Bayes, Decision Tree and Deep Learning algorithms on a dataset and find that deep learning achieves an accuracy of 90–98%. The authors propose that the system gives an opportunity prognostic tool for healthcare personals. They further suggest that the accuracy can be enhanced by including the omics data for prediction of the onset of the diabetes (Reddy et al. 2020). The research work uses principal component analysis grounded on a model of deep neural network by using the Grey Wolf Optimization algorithm for classification of the features extracted from diabetic retinopathy dataset with the help of the standard scaler normalization method. The authors also reduce the dimensionality using PCA after application of ideal hyperparameters, and finally the dataset is trained using a DNN model. The model evaluation was done using performance measures like recall, accuracy, sensitivity and specificity. The authors compare the model with various machine learning algorithms. The results of the evaluation show that the proposed model offers promising performance as compared to other machine learning algorithms. They propose the use of the model on larger datasets with high dimensionality for the purpose of improving performance of the proposed model. Chaki et al. (2020) give a review an analyze the diagnosis and detection of diabetes mellitus from various points like feature extraction, datasets pre-processing, classification and diagnosis of DM, machine learning-based identifica-tion, artificial intelligence-based intelligent DM assistance and performance measures. They further provide understanding about some research issues in the field of diabetes mellitus, namely diagnosis and detection and personalization and self-management. Parte et al. (2020) suggest the method based on breath as the sample for detection of the diabetes in place of a blood or urine sample. They read the acetone level in breath and then use a deep learning algorithm, namely a con-volutional neural network, along with a support vector machine for calculating the automated features from the raw signal and then classify the derived features. Rahman et al. (2020) in a research study developed a novel diabetes detection model using convolutional long short-term memory (Conv-LSTM), but practically the model was never applied. Instead they applied convolutional neural network (CNN), traditional long short-term memory (TLSTM) and CNN-LSTM and then analyzed the performance with their developed model by applying the model on a PIMA dataset. They also use the Boruta algorithm to extract features from the dataset and then apply hyperparameter optimization by the application of the grid search algorithm in order get the optimized parameters for the model. The devel-oped model (Conv-LSTM) gives an accuracy of 91.38%. After the application of the cross validation technique, the accuracy of the model improved to 97.26% and thus outperformed other deep learning models. In a research study, Zhu et al. (2020) provide the review of the applicability of deep learning in the detection of diabetes mellitus. From the study, the author documents the importance of various deep learning methods in diabetes detection like CNN, RNN and HAN. They highlighted the possibility of meeting the challenges of early detection of diabetes with the advancement in the of deep learning field and the availability of a huge volume of data from the medical industry. Tymchenko (2019) carries out a research study and proposes an automatic method in order to detect diabetic retinopathy based on deep learning by a single image of the human fundus. Tymchenko proposes the multistage concept for transfer of learning that makes the application of similar datasets with different labelling. They used an ensemble of three CNN architectures (EfficientNet-B4, EfficientNet-B5, SE- ResNeXt50). The method performs better in measures like accuracy, specificity and sensitivity when compared with other deep learning models and also enhances generalization with reduced variance. For future work the author proposes calculation of SHAP for the whole sample, instead of for only a particular network. Grzybowski et al. (2019) in a research study describe the modern artificial intelligence techniques suitable for diagnosis of diabetes retinopathy that have already been described in the literature. They compare the performance of various artificial intelligence algorithms and emphasize on future research in the field to address multiple challenges like medico legal implications, clinical deployment models and ethics. Kopitar et al. (2020) in a research study study and compare multiple machine learning models based on the prediction concept with widely used regression models used for prediction of Type 2 diabetes mellitus. The authors implemented multiple bootstrap iterations in various subgroups of data simulating new data and measure the performance in predicting the fasting plasma glucose level. The results show negligible improvement with more refined prediction models. In a research study Naito et al. (2021) developed a deep human leukocyte antigen deep learning model for ascribing HLA genotypes, and the validation was done using Japanese and European HLA reference panels. The developed model achieves better performance in terms of accuracy with substantial performance for low-frequency and infrequent alleles. They worked on the model that is less dependent on distance dependent linkage disequilibrium decay of the target alleles. They applied the model for the type 1 diabetes GWAS data collected by Biobank Japan and Biobank UK. Islam Ayon and Milon Islam (2019) in a research study propose a mechanism for designing a model using deep neural network for early diagnosis of diabetes by training the attributes of the PIMA dataset using fivefold and tenfold cross validation. They achieved the F1 score of 98, accuracy of 98.3% and MCC of 97 for fivefold cross validation. With the application the tenfold validation the accuracy was 97.11% with 96.25% sensitivity and specificity of 98.08%. Bora et al. (2021) carried out research and created two versions of deep learning systems for the prediction of diabetic retinopathy using a set of threefold or one-field color fundus images. The dataset of 57,5431 images was used, out of which 28,899 had already known results and 546,532 were used for the training process. The validation was performed on one eye per patient from two datasets.

The developed system was found independent and more informative than available risk factors and will thus help in optimization of screening intervals and ultimately in reduction of costs and improvement in vision-related outcomes. De Bois et al. (2021) in a research study propose a multi-source adversarial transfer learning framework for the learning of feature representations across the source that are similar. The authors applied the proposed model to glucose forecasting using a convolutional neural network. The results obtained show that the accuracy of the model can be improved by using adversarial training methodology that overperforms the existing deep learning technologies. The proposed model performs well when used with multiple datasets. Ashiquzzaman et al. (2017) propose a diabetes prediction system using the dropout method and also address the problem of overfitting. Using the deep learning neural network of fully connected layers followed by dropout layers, they devised a system that outperforms conventional neural network methods. As a future strategy, the author proposes the use of real-time data of wearable healthcare devices. Latchoumi et al. (2021) proposed a diabetes prediction solution called the Smart Diabetes Diagnosis System using machine learning, huge medical datasets and a large cloud of health intelligence to provide a more focused risk assessment and personal treatment schedule. The proposed system also provides patients with daily guidance for improvement of medication. The authors employed DT, SVM, ANN and Naïve Bayes machine learning algorithms. Ahmad et al. (2021) in a research study investigate the prediction of diabetes mellitus and study the role of HbA1c and FPG used as input features. The authors use five different machine learning classification algorithms and use the concept of feature elimination using hierarchical clustering and feature permutation. The authors evaluate the models and find good performance in terms of precision, recall, accuracy and f1-score on the dataset. The authors (Thaiyalnayaki 2021) carried out a research study to identify the problem of automatic prediction of diabetes and provide the solution using the deep learning algorithm and support vector machine for better therapeutic management. The authors achieved an accuracy of 65.1042%.

10.4 BIG DATA

The modern age is the age of data. Data is the main resource for success of organizations. A huge collection of data has a great statistical power. This term came into use in 1990, and the term was popularized by John Mashey. The term *big data* (Sidiq and Zaman 2020) refers to datasets with so large a number of records that they are beyond the capability of humans, and commonly software tools are used to collect, interpret, manage and process this data in a time-bound manner. Big data comprises data that is of unstructured, semi-structured and structured nature. The size of big data is continuously increasing ranging from few terabytes to multiple zettabytes. This enormous collection of data (Sidiq et al. 2017) contains much useful information that may prove fruitful for the organizations. However, for processing such big datasets there is a need for a set of techniques and technologies with the latest forms of integration so that the useful patterns can be revealed from the dataset. Different organizations describe big data differently. Some organizations use variety, veracity, velocity and value to describe the characteristics of big data. Variability is also an additional quality of big data.

Neha (2021) claimed that "big data is a dataset that requires parallel computing power to handle it." As per this definition, there is a clear change in computer science with parallel programming theories and loss of capabilities and rules made by Codd regarding the relational model. As stated earlier (Sidiq et al. 2019), the data with multiple fields has greater statistical power and may prove useful for the success of organizations; however, data with higher complexity may result in false discovery rate also. So it is a great challenge to handle big data properly. The main challenges include data collection, storage, analysis, searching, sharing, transfer, visualization, querying, updating, privacy security and data integrity (Neha et al. 2020). Traditionally big data was associated with volume, velocity and variety. The main challenge in big data processing was sampling. Big data was considered any data that exceed the capability of traditional database management systems and tools to process such data in a time limit.

In the present scenario big data means (Ashraf et al. 2017) use of predictive analytics, user behavior analytics and some more sophisticated data analytics methods that enable us to extract usable values from big data with least importance given to size of the dataset. Although the data available now are of large size, the size is no longer considered an important quality of a dataset. Big data has found applications in business, disease control in healthcare, crime control, weather forecasting and much more. With the use of smart phones and other mobile devices like smart watches, smart vehicles and other IoT devices, the size and number of available datasets has grown enormously. With the application of such devices the data generation approximately doubles every three years. By the year 2025, IDC predicts approximately 163 zettabytes of data to be available. In order to process and analyze such a huge volume of data, thousands of servers running massively parallel software will be required. Which data qualifies to be called big data depends on those analyzing it and their tools.

10.5   DEEP LEARNING

Deep learning is a subset of AI that imitates the function of the human brain for data processing, object detection, speech recognition, language translation and other decision making. With the help of deep learning, the computers can learn without the intervention of humans. With the advancement in the digital era, there is an explosion of data in various forms and from various regions. This huge data most commonly known as big data originates from multiple sources like social networks, search engines, online entertaining agencies, ecommerce etc. Big data is easily available and sharable through fintech applications like cloud computing. The unstructured data is so huge in volume that it could take hundreds of years for humans to understand the patterns in it and to extract useful information so that it can be used for decision making. Researchers realize the potential of artificial intelligence systems, machine learning and deep learning for processing such a huge volume of data and to extract relevant information for making smart decisions. Deep learning has the capability to learn without being supervised by humans. The learning capability is drawn from the data, which can be structured or unstructured. Deep learning has proved useful in detection of fraud and money laundering among many other applications.

10.5.1 CONVOLUTIONAL NEURAL NETWORK (CNN)

CNN is one of the most widely used deep learning algorithms. It is mostly used for analysis of image data. The CNN uses a technique called *convolution* for processing data. A convolution is a mathematical operation of two functions that generates another function that expresses how the shape of one is modified by other.

CNN is a composition of artificial neurons arranged in multiple layers. The neurons are similar to biological neurons. These neurons are actually mathematical functions that compute the weighted sum of inputs and give activation value as its output, which is passed onto the next layer. The task of the first layer is to extract basic features from its input and pass the extracted features on to the next layer. The job of the next layer is to extract more complex features and pass them on to the next layer as output. The deeper we move into the network, the more complex features get extracted. Depending upon the activation map of the final layer in the network, the classification layer finally outputs a set of confidence scores that is a value between 0 and 1. This confidence score is a final classification that specifies the class of the input entity.

Another layer is a pooling layer whose objective is to reduce the spatial size of the convulsed features. The main purpose of the pooling layer is to reduce computational power for processing the data by reducing the dimensions of input data. There are two types of pooling that can be applied known as Max pooling and Min pooling.

The final layer is the fully connected layer that has weights associated with each neuron and biases, and it connects the neurons between multiple layers. The fully connected layer is mostly placed before the output layer. Figure 10.1 shows basic steps of CNN.

10.5.2 RECURRENT NEURAL NETWORK (RNN)

This is another class of artificial neural network in which the nodes are connected with each other in a form of a directed graph in temporal sequence. This sequence of connected nodes gives a network temporal dynamic behavior. The RNN is basically derived from a feedforward neural network and uses the internal state memory for the processing of a variable length sequences of inputs. With this property the RNN is suitable for applications like handwriting recognition, speech recognition etc. In general terms the recurrent neural network is used to refer to two broad classes of networks that have similar general structure. The first is finite impulse and another one is infinite impulse. The temporal dynamic behavior is found in both classes of networks. A finite impulse recurrent network is in a form of directed acyclic graph that can be

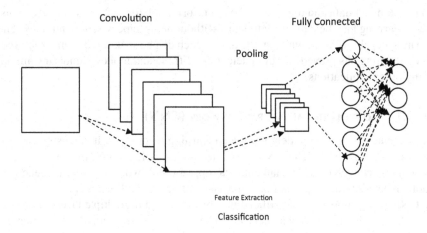

FIGURE 10.1   Convolutional Neural Network.

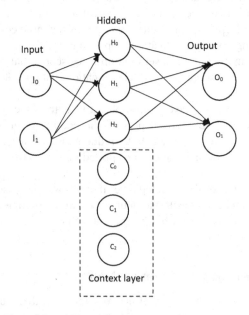

FIGURE 10.2   Recurrent Neural Network.

unrolled and replaced by a feed forward neural network. The infinite impulse recurrent neural network on the other hand is a directed cyclic graph that can't be unrolled. Both the neural networks may have additional stored states that are directly under the control of the neural network. This storage can further be replaced by another network or graph by incorporating time delays or feedback loops. The controlled states are also known as gated states or gated memory and form part of long short-term memory and gated recurrent units. Figure 10.2 depicts basic components of RNN.

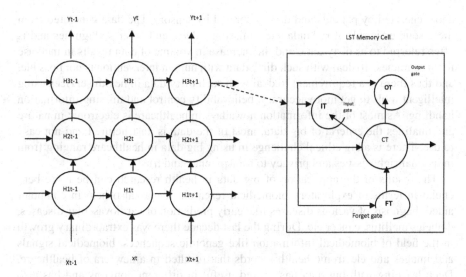

FIGURE 10.3 Long short-term memory network.

10.5.3 LSTM NETWORK

The LSTM known as long short-term memory network does not apply the concept of a typical neuron-based neural network, but it is based on the concept of a memory cell that has the ability to retain the value for a particular duration of time. The LSTM is mainly composed of three gates that control the flow of the information in and out of the cell. The gates used in the network are an input gate followed by a forget gate and finally the output gate. The first gate known as the input gate controls the flow of input/new information into the cell. This is followed by the second gate known as forget gate, which is concerned with controlling when to forget the existing information in the cell. Finally, the third gate controls the passing of information from the cell as output. The cell in LSTM is associated with weights that control each gate. The training algorithms are also used for the optimization of the weights that depends on the variation between the predicted and the expected value. Figure 10.3 shows the working of the LSTM network.

10.6 DEEP LEARNING WITH BIG DATA IN HEALTHCARE

Big data has vast scope in the field of healthcare as it may aid the medical staff by providing personalized medicine, accurate analytics, better medical risk intervention, correct predictions, waste and care variability reduction, automatic reporting of patient data and so on (Teli & Wani 2018). Some aspects of healthcare have already been implemented, but other areas of improvements are more aspirational. The data collected in healthcare so far is not small, but the use of mHealth, eHealth, and other IoT wearable devices may increase the volume of data even more. The data generated in healthcare includes electronic health records, data in the form of images, data generated by patients and data generated by sensors. The data generated from these sources is full of redundancies, missing values, and other ambiguities and is often referred to as dirty data, and the increase in volume of data results in increase in inaccuracies. To deal with such dirty data with human intervention is not possible, and thus there is a requirement to deal with such dirty data in health services using intelligent tools to get more accuracy, believability control and missing information handling. As most of the information nowadays in healthcare is electronic in nature and qualifies the criteria of big data, most of the data is unstructured and not easy to use. There is a big ethical challenge in using big data in healthcare ranging from individual rights issues and privacy to transparency and trust.

The results of the application of big data in health research can be more beneficial in terms of exploratory biomedical research. It may also help in computer aided diagnosis of various diseases like early prediction of cardiovascular diseases, diabetes mellitus, cancer etc. During the last decade there was extraordinary growth in the field of biomedical information like genomic sequences, biomedical signals and images and electronic health records that resulted in a new era of healthcare. Deep learning with big data has proved useful in different domains and has also opened up the latest prospects for handling a large quantity of biomedical data. The use of deep learning with big data is not an alternate for human doctors in the near future, as human physicians can never be replaced by any machine learning algorithm; however, the application of such technologies can certainly support them for better decision making in various fields of healthcare. Deep learning with big data is becoming popular in healthcare due to the ability to find the connections and relationships between risk factors and diseases. A huge collection of data is being generated worldwide every single day in the medical field; however, all the data cannot be processed due to some issues that include inconsistency, data quality, instability, multiple scales, legal issues and incompleteness. Besides, the processing of the data generated in the medical field requires expertise in multiple disciplines like physics, signal processing, mathematics, computer science, physiology, biology, instrumentation and medicine. With the application of deep learning with big data, such data can be managed in an efficient and cost-effective manner. Deep learning with big data finds applications in healthcare like personal disease diagnosis, disease prevention, disease prediction and disease prognoses as well as in suggesting better health treatments based on day-to-day habits of an individual. Google Flu Trends is an example of use of deep learning with big data in healthcare, as this enables us to predict influenza-like illness more than double the percentage of doctor visits using observation reports from multiple laboratories in the United States. Another example is Enlitic Company, which improves accuracy of disease diagnosis in less time and with reduced cost by application of deep learning with big data to analyze medical images like MRIs, CT scans and x-rays. One more contribution of deep learning with big data application in healthcare is the "Precision Medicine Initiative" project that was sponsored by the US Government in 2015 with the aim to plot the human genome of one million US citizens to find precise defects in the primary cause of disease for a group of people at a genetic level. This project assists in developing new generations of drugs and treats part of molecular complications common in patient groups having a given disease.

The success of the previously mentioned applications encourages more big research institution centers and fundraising agencies to make investments in this area. The deep learning with big data can be applied to hospital MIS to accomplish the objective of lower cost, fewer hospital stays, detection of insurance scams, change detection in disease patterns, better healthcare and efficient use of medical resources.

10.7 EXAMPLES OF DEEP LEARNING WITH BIG DATA IN HEALTHCARE

There are a number of applications of deep learning with big data in healthcare. Deep learning has achieved wonderful results in the prediction and diagnosis of various diseases like heart-related diseases, diabetes mellitus, gestational diabetes mellitus, kidney-related diseases and many more. Some of the examples where deep learning with big data has been used successfully are given in the next subsection.

10.7.1 BIOMEDICAL IMAGES

Biomedical images provide visual information about the body of humans with the aim of helping physicians in diagnosis and treatment of various diseases more efficiently. The most common use of deep learning that uses medical data includes medical image processing applications, and the success of CNN architecture in the field of image analysis proves very beneficial in this approach. Earlier the biomedical images were analyzed and studied by physicians using manual methods that suffer from the drawbacks of limited speed, fatigue, and limited experience. In addition to this, manual analysis may result in incorrect diagnosis that may result in harm to the patients. To avoid this there is a need for an accurate, efficient, and automated medical image analysis system to aid physicians in the process of diagnosis. Some of the milestones achieved in medical image processing applications are disease classifications, wounds or abnormalities and segmentation of regions like tissues and organs. Classification is the main issue in the field of medical image examination, and this can be done using computer-aided diagnosis using segmentation. Some other applications are localization, detection and registration. Localization works by defining a bounding box around a single entity in the image. Detection involves defining a bounding box around more than one entity from different classes in the image and classifying each entity. Finally, registration is the process of fitting one image onto another.

Magnetic resonance imaging (MRI), positron emission tomography (PET), computed tomography (CT), ultrasound (USG), x-ray, histology slides, retinal photography and dermoscopy images along with various small datasets and images that are multidimensional, multimodal, and multichannel makes medical imaging a field better suited to challenges. The deep learning solutions were used for medical image analysis in various topics related to classification and detection like identification of lung nodes into benign or malignant, multiple sclerosis detection, alcoholism identification using an MRI of the brain, categorization of chest x-rays into various diseases, differentiating patients with Alzheimer's disease from normal ones, detection of cancerous lung nodes and lung cancer stages on CT scans, diagnosis of diabetic retinopathy with the help of digital images of fundus of eye, discriminative kidney cancer histopathological pictures into tumor or non-tumor, malignant skin cells on dermatological images, mitotic images in breast histology pictures, and many more. In the field of image segmentation deep learning can help in various body parts like liver, prostate, knee cartilage, spine, etc.

10.7.2 ELECTRICAL BIOMEDICAL SIGNALS

Various biomedical equipment that is in use in various medical organizations produces electrical signals that are generated from sensors when positioned on the skin of patients. The characteristics of these signals are determined by the location of the sensors on the body. Such signals are very useful data sources for disease diagnosis and detection. With the application of deep learning it is possible to implement trustworthy mechanisms by applying physiological signals like electromyogram (EMG). RNN and CNN algorithms of deep learning are most widely used to process such physiological signals.

The electromyogram calculates the electrical actions of skeletal muscles like muscle activation, contraction, strength and state. The EMG signal contains all such activities in an overlapped manner, making it difficult to interpret manually for one particular problem. Deep learning can analyze such signals and help in interpretation of limb movement estimation, neuroprosthesis control, hand movement classification, gesture recognition, movement intention decoding and much more.

The EEG reads the electrical activity of brain and then evaluates it. The most common application of EEG is the brain computer interface that uses deep learning. The manual analysis of brain computer interface (BCI) signals is impossible on a real-time basis. Therefore the automated decision making is required and is performed by the application of deep learning algorithms. Deep learning also finds applications in identification of sleep state, decoding of excited movements, epileptogenicity localization and seizure detection.

The EOG counts the corneo-retinal potential in between the front and back of the human eye. Two electrodes are kept above and below each human eye, and a signal is generated that helps in ophthalmological diagnosis. The presence of noise and artifacts results in complications to interpret the signal. The application of deep learning algorithms can overcome these limitations to a large extent and can help in detection of driving fatigue, drowsiness and momentary mental workload.

10.7.3 OTHER NON-ELECTRICAL BIOMEDICAL DATA

A huge volume of data is generated from electrical biomedical signals; the same is the case with non-electrical sources as well. In the medical field various signals are generated that are of non-electrical nature like mechanical signals from mechanomyogram, chemical signals from oxygenation, acoustic signals from photoplethysmogram, and optical signals from photoplethysmogram. In addition to this, various biomedical information is obtained from other data sources like HER, medical imaging, sequence of genomes, and other results from the laboratory. This ever-growing medical data makes deep learning with big data a primary global actor in healthcare and medicine by helping in development of new applications that guide physicians in better and more accurate classification of diseases, prescribing better treatment to patients, and developing new and efficient therapies. Deep learning algorithms help in considering various features of patients by measuring big data sets from multiple data sources and ultimately helps in delivering better treatment to a specific patient at the right time.

The EHRs also produce a good volume of data and have gained importance in the last few years. The data includes a good amount of amorphous text like physical checkup, lab reports and operative summary notes. Deep learning finds applications to operate on such data as the natural language processing based on deep learning can help to extract useful patterns and information from the text data and helps in clinical decision making. The big data techniques have possible benefits not only in providing a large volume of data but also in developing deep learning algorithms that may help in solving multiple problems like dealing with heterogeneity, avoiding over-fitting of data, improving the generalization ability and handling the uncertainty due to missing data mostly found in HER data. Deep learning also has applications to deal with continuous time data like laboratory results to automate identification of particular phenotypes and to distinguish between uric acid signs of gout and acute leukemia by modelling longitudinal structures of serum uric acid measurements. Deep learning is also applied widely in genomics to improve detection of specific patterns from data. For example, a deep architecture centered on CNNs can help in prediction of RNA and DNA binding proteins and in classification of cancer cases using gene expression profiles.

10.7.4 Role of Deep Learning with Big Data in ECG Processing

The heart is the main organ that pumps blood to each and every organ within the body. A healthy heart means a healthy body. Some of the medical conditions like hypertension, diabetes and other associated factors like smoking, drinking etc. can disturb the health of the heart. Besides, unhealthy and modern lifestyles also contribute toward the risk of having heart problems. The most serious problem is ischemic heart disease in which one of more coronary arteries gets blocked, which results in nutritional loss of myocardial cells and may result in heart attack. The timely diagnosis and treatment of this disease results in better chances of existence and value of life. Another heart problem that causes alteration of the heartbeat variation and makes it either slower or faster is known as cardiac arrhythmia. There are many other problems associated with the health conditions of the heart like congenital heart disease, infections, valvular heat diseases, inflammations etc. The ECG is a physiological signal that is generated from the contraction and recovery of the heart and is very helpful in determining the clinical conditions of the heart. From the ECG signal required information may be extracted and then interpreted by physicians for better treatment. However, the human interpretation of these signals requires extensive training, enough time and lots of experience. Deep learning has a promising future in processing such signals and providing better interpretation. Computer-based interpretation of ECG is not a new concept and has been done for several decades. Deep learning has been proposed in order to achieve more effective examination of ECG signals, as it has proven effective in terms of tolerance to noise and unevenness in various pattern recognition applications and provides a classification of complex signals in real time.

10.8 EXPERIMENT

For the purpose of application of machine learning and deep learning in healthcare, various classifiers were implemented in python like Naïve Bayes, Random Forest, KNN, SVM, Decision Tree and ANN. The dataset that was used for the experiment was taken from National institute of Diabetes and Digestive and Kidney Diseases popularly known as PIMA dataset (www.kaggle.com/uciml/pima-indians-diabetes-database). First, the algorithms are run on a dataset without any preprocessing, and the mean, median and mode results are obtained from the experiment. Table 10.1 shows various measures of data. In the second experiment pre-processing techniques, namely Z-Score and Min-Max (normalization techniques), were applied on a dataset, and then the ANN algorithm was used to predict the early diagnosis of diabetes. The results obtained are shown in Table 10.2 with significant improvement in accuracy with the application of pre-processing techniques.

The dataset contains 768 records. Out of 768 records 70% of samples were used for training of model and 30% of samples were taken randomly for testing of the model.

Findings

TABLE 10.1
Accuracy of Various Machine Learning Algorithms for Prediction of Diabetes Mellitus

Model	Accuracy		
	Mean	Median	Most Frequent
Naïve Bayes	75.58	69.05	75.57
Random Forest	77.36	75.57	75.41
KNN	72.31	73.61	72.96
SVM	77.04	76.21	77.04
Decision Tree	70.36	67.43	75.57
ANN	71.66	58.50	62.89
CNN	93.6%	75.54%	82.04%

TABLE 10.2
Results Achieved from Artificial Neural Network after Application of Z-Score and Min-Max Scaler Pre-Processing Techniques

Missing Value Strategy	Z-score	MinMax Scaler
Mean	75.75%	84.77%
Median	60.89%	82.14%
Most Frequent	65.19%	82.79%

The artificial neural network used in the experiment a is fully connected and feed forward with three hidden layers implemented in Python using Keras after application of pre-processing techniques.

10.9 CONCLUSION AND FUTURE IN BIG DATA HEALTHCARE

Nowadays emerging data techniques are playing a primary role in the healthcare system. However, various issues are still open today, and it is a big challenge to machine learning technology to be implemented in our day to day life. One of the main challenges in the current healthcare system is the concept of interoperability, which means the ability of the healthcare mechanism of various entities, organizations, healthcare centers and regions to intercommunicate and work together. Multiple standards have been proposed for various aspects, and some standards have been accepted also, while many others have not. In the current scenario, the healthcare system around the globe is still fragmented, and presently no efforts are being made toward integration of the global healthcare system. The concept of interoperability based on clinical practice is most relevant in the near future as it may develop a consensus among main stakeholders of the healthcare system about the needs and availability of current technology in order to deliver to society an answer to those needs. On the other side the generation of a large volume of data is also a point of concern for personal data privacy, and steps have been taken in this regard to provide protection laws of personal data throughout the world.

With pre-processing techniques, the convolutional neural network can provide more accurate results when used for early prediction of various diseases. More efforts are needed like feature extraction, data preprocessing, and hybrid deep learning approach in order to make the prediction process more accurate.

REFERENCES

Ahmad, Hafiz Farooq et al. 2021. "Investigating Health-Related Features and Their Impact on the Prediction of Diabetes Using Machine Learning." *Applied Sciences* 11(3): 1173.

Ashiquzzaman, Akm et al. 2017. "Reduction of Overfitting in Diabetes Prediction Using Deep Learning Neural Network." *Lecture Notes in Electrical Engineering* 449: 35–43.

Ashraf, M., M. Zaman, M. Ahmed, and S.J. Sidiq. 2017. "Knowledge Discovery in Academia: A Survey on Related Literature." *International Journal of Advanced Research in Computer Science* 8(1).

Bora, Ashish et al. 2021. "Predicting the Risk of Developing Diabetic Retinopathy Using Deep Learning." *The Lancet Digital Health* 3(1): e10–e19.

Chaki, Jyotismita, S. Thillai Ganesh, S. K.Cidham, and S. Ananda Theertan. 2020. "Machine Learning and Artificial Intelligence Based Diabetes Mellitus Detection and Self-Management: A Systematic Review." *Journal of King Saud University—Computer and Information Sciences.* https://doi.org/10.1016/j.jksuci.2020.06.013.

De Bois, Maxime, Mounîm A. El Yacoubi, and Mehdi Ammi. 2021. "Adversarial Multi-Source Transfer Learning in Healthcare: Application to Glucose Prediction for Diabetic People." *Computer Methods and Programs in Biomedicine* 199: 105874.

Grzybowski, Andrzej et al. 2019. "Artificial Intelligence for Diabetic Retinopathy Screening: A Review." *Eye.* http://dx.doi.org/10.1038/s41433-019-0566-0.

Islam Ayon, Safial, and Md. Milon Islam. 2019. "Diabetes Prediction: A Deep Learning Approach." *International Journal of Information Engineering and Electronic Business* 11(2): 21–27.

Kopitar, Leon et al. 2020. "Early Detection of Type 2 Diabetes Mellitus Using Machine Learning-Based Prediction Models." *Scientific Reports* 10(1): 1–12. https://doi.org/10.1038/s41598-020-68771-z.

Latchoumi, T. P., J. Dayanika, and G. Archana. 2021. "A Comparative Study of Machine Learning Algorithms Using Quick-Witted Diabetic Prevention." *Annals of the Romanian Society for Cell Biology* 25(4): 4249–4259.

Naito, Tatsuhiko et al. 2021. "A Deep Learning Method for HLA Imputation and Trans-Ethnic MHC Fine-Mapping of Type 1 Diabetes." *Nature Communications* 12(1): 1–14.

Naz, Huma, and Sachin Ahuja. 2020. "Deep Learning Approach for Diabetes Prediction Using PIMA Indian Dataset Deep Learning Approach for Diabetes Prediction Using PIMA Indian Dataset." *Journal of Diabetes & Metabolic Disorders* 19(1): 391–403.

Neha, K. 2021. "A Study on Prediction of Student Academic Performance Based on Expert Systems." *Turkish Journal of Computer and Mathematics Education (TURCOMAT)* 12(7): 1483–1488.

Neha, K., and S. J. Sidiq. 2020. "Analysis of Student Academic Performance Through Expert Systems." *International Research Journal on Advanced Science Hub* 2: 48–54.

Parte, R. S. et al. 2020. "Non-Invasive Method for Diabetes Detection Using CNN and SVM Classifier." *International Journal of Scientific Research and Engineering Development* 3(3): 9–13.

Rahman, Motiur, Dilshad Islam, Rokeya Jahan Mukti, and Indrajit Saha. 2020. "A Deep Learning Approach Based on Convolutional LSTM for Detecting Diabetes." *Computational Biology and Chemistry* 88(December): 107329. https://doi.org/10.1016/j.compbiolchem.2020.107329.

Reddy, Thippa, Gadekallu Neelu, Khare Sweta, and Bhattacharya Saurabh. 2020. "Deep Neural Networks to Predict Diabetic Retinopathy." *Journal of Ambient Intelligence and Humanized Computing*. https://doi.org/10.1007/s12652-020-01963-7.

Sidiq, S. J., and M. Zaman. 2020. "A Binarization Approach for Predicting Box-Office Success." *Solid State Technology* 63(5): 8652–8660.

Sidiq, Zaman, M., and Ahmed, M. (2019). *How Machine Learning is Redefining Geographical Scien : A Review of Literature*. 6(1): 1731–1746, JETIR January 2019.

Sidiq, S. J., M. Zaman, M. Ashraf, and M. Ahmed. 2017. "An Empirical Comparison of Supervised Classifiers for Diabetic Diagnosis." *International Journal of Advanced Research in Computer Science* 8(1): 311–315.

Teli, T. A., and Arif Wani. 2018. "Fuzzy Logic and Medicine with Focus on Cardiovascular Disease Diagnosis." *5th IEEE International Conference on Computing for Sustainable Global Development*.

Thaiyalnayaki, K. 2021. "Classification of Diabetes Using Deep Learning and SVM Techniques." *International Journal of Current Research and Review* 13(1).

Tymchenko, Borys. 2019. "Deep Learning Approach to Diabetic Retinopathy Detection." *arXiv:2003.02261*.

Zhu, Taiyu, Kezhi Li, and Pantelis Georgiou. 2020. "Deep Learning for Diabetes: A Systematic Review." *IEEE Journal of Biomedical and Health Informatics* 25(7): 2744–2757.

11 Pattern Recognition in EMG Control of Bionic Hands

Munish Kumar and Dr. Pankaj Khatak

CONTENTS

11.1 INTRODUCTION

The human hand is a perfect example of a sophisticated machine in terms of functions and control. It makes an essential part of human body that enables us to perform a wide variety of daily tasks. It has given us immense capabilities to manipulate the world around us. The loss of a hand thus has a substantial impact on the life of an amputee, disconnecting him from the surroundings. According to a study, there are

DOI: 10.1201/9781003245469-11

about 23.6 million disabled persons in India; out of which nearly 20% are having disability in movement. This includes upper-limb amputees having unilateral/bilateral amputations due to accidental, pathological, or congenital causes. The population contributes to a significant part of society with the majority of people living in rural or under-developed parts of the country. This shows that the social and financial condition of the amputees is below satisfactory levels. A large part of the society is not able to perform tasks for daily living and/or livelihood, leading to decreased standard of life, confidence, and self-esteem with minimal social acceptance and loss of human resources. It necessitates the development of low-cost as well as highly functional devices to replace missing limbs (Cordella et al. 2016). This would enable the amputees to acquire adequate skills to perform daily living, working, and social activities so as to reduce dependency, enhance their quality of life through employment, and increase social acceptance. The amputees provided with affordable and multifunctional prosthesis would be able to address the societal challenges.

Functional replacement of a human hand is found in terms of bionic prosthetic devices that are a close replica of the hand. These bionic devices and their predecessors have a topic of research and development for several decades and even centuries. Although the developments in the past few decades have put the bionic devices in a direct comparison with the human hand, they are still very far from achieving the capabilities a natural hand offers (Clement 2011). Some examples of commercial bionic hands are: iLimb, Bebionic v2, Michaelengalo, Taska hand, etc. as shown in Figure 11.1.

Most of the bionic hands are controlled through electromyography signals captured from the muscles of an amputee. The signals are filtered, amplified, processed, and classified through machine learning and pattern recognition procedures. The user's intent is thus identified, and commands are directed to the terminal device, which performs movements similar to a human hand. The control scheme of bionic systems involves a signal acquisition unit, a processing unit, and an anthropomorphic terminal device.

11.2   ELECTROMYOGRAPHY (EMG)

The electromyography (EMG) signals acquired from remnant muscles of an amputee contain neural information regarding motor unit activation to perform a given task. The information is different from that collected from direct nerve recordings as an EMG recording involves combined action potential of motor units innervated by a nerve signal. As a result, the surface EMG signal is intrinsically linked to the neurological signal conveyed from the spinal cord to the muscles. It is acquired by connecting electrodes to the surface of a subject's hand. The new process of targeted muscle reinnervation emphasizes the concept of surface EMG recordings as sources of brain input (TMR). The nerves that reach the missing limb muscles are rerouted to innervate accessory muscles, from which the participants' motion intent may be determined.

The EMG interface categorizes voluntary-contraction-related muscle activity and connects it with the intended function of the device. The EMG control provides a simple and intuitive mode of communication with the bionic device, thus enabling

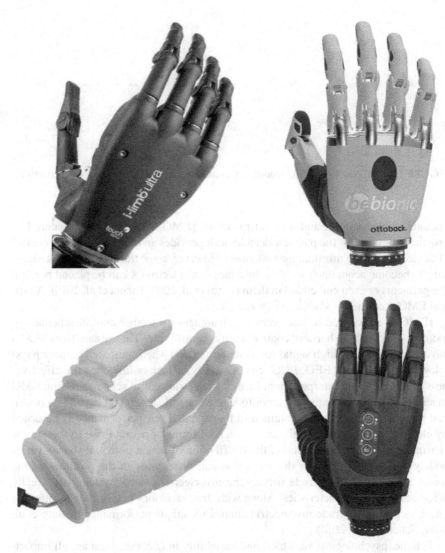

FIGURE 11.1 Commercial bionic hand.

the user to perform tasks without paying much attention. The only condition required from EMG control is a user's ability to flex or extend some skeletal muscles. EMG interfaces make use of this muscular information to control a bionic device achieving functional capabilities closer to a natural human hand.

EMG signals are collected using two methods, i.e., using surface electrodes or intramuscular electrodes. Surface electrodes capture the signal from the target muscle through skin contact whereas intramuscular electrodes penetrate through skin and capture signals through direct contact of the target muscle. The first method, also known as surface electromyography (sEMG), is simple and non-invasive. Intramuscular electromyography (iEMG) involves invasive procedures and is

FIGURE 11.2 Example of EMG recordings. Peaks are corresponding to muscle activation.

uncomfortable, painful, and difficult to set up. sEMG has been used in control of prosthetic devices for the past few decades as it provides an intuitive and easy control of devices requiring minimal user attention. However, some training is always necessary to become acquainted with the interface. The electrodes can be placed beneath the garments or even embedded in them (Finni et al. 2007; Farina et al. 2010). A typical EMG recording is shown in Figure 11.2.

sEMG is considered to have various advantages over other control schemes for bionic hands such as harness control, tactile control, etc. The signals from sEMG have a comparatively high signal-to-noise ratio (SNR) when compared to other physiological signals, e.g., EEG, ECG, etc. A variety of factors influence the effectiveness of sEMG signal interpretation. Most myoelectric control schemes rely on EMG macro-features (e.g., frequency, amplitude of electrical signal) that depend on neural and peripheral information contained in the signal. Their performance is governed by elements such as motor unit action potential forms or neural drive to the muscle (Farina et al. 2014). Furthermore, the distribution of action potential shapes varies widely between participants due to anatomical differences in the positions of the motor units inside the muscle tissue. The myoelectric control systems are greatly affected by these characteristics. Along with that, variations in electrode positioning and arm posture degrade myoelectric control technique performance (Young et al. 2011; Scheme et al. 2010).

Fatigue, psychological variables, and variability in task execution are all important factors influencing surface EMG and potentially causing alterations. Another crucial component affecting system performance is muscle crosstalk. It is defined as the signal captured from a target muscle that originated from other muscles nearby. In certain circumstances, the recorded EMG signal is a combination of the activities of many muscles that are physically and functionally adjacent, such as the forearm muscles, with at least as much crosstalk as the target muscle activity. Other variables such as muscle depth, innervation zones (IZ), quality of skin contact, skin impedance, and level of muscle contractions all contribute to the variability of sEMG recordings.

Despite the limitations of EMG data, these are the most widely used control methods for artificial hands. There are a number of different control schemes implemented for EMG control as discussed in further sections.

11.3   EMG CONTROL SCHEMES

To translate the information in the EMG, a broad range of control methods have been devised, which are commonly classed as amplitude-based control and feature-based control. In general, the following techniques are used to translate EMG signals:

  i. Amplitude-based controls—on-off control, proportional control, direct control, finite state machine, etc.
 ii. Feature-based controls—pattern recognition control, regression control, etc.

The amplitude-based methods use signal amplitude during muscle contraction to calculate an appropriate activation threshold, whereas the feature-based techniques extract and analyze signal attributes based on time and frequency for movement recognition. These control techniques can be used independently or in conjunction with others. The majority of control systems rely on EMG inputs from one or two muscle groups to set a threshold value for controlling the myoelectric limb. In the case of a single muscle group, altering the strength of muscular contractions is utilized to find appropriate activation thresholds that would activate a certain action or "state" of the end-effector.

Amplitude-based approaches enable movements such as a modest contraction to close the hand, a powerful contraction to open, and no muscular activity to stop the device (Battye et al. 1955). To govern hand opening and closing, the same approach might be used with two opposing muscle groups, such as flexors and extensors (Popov 1965). Different thresholds can be employed to identify whether EMG activity is significant and to distinguish it from background noise. Despite being both rapid and suitable to real-time control, the number of motions created is limited, and the control is sequential, as opposed to the smooth movements of the human hand. However, there have been developments in the past two decades to overcome the drawbacks of these systems.

11.3.1   PATTERN RECOGNITION CONTROL

EMG pattern recognition (EMG-PR) methods are developed to overcome the limits of traditional proportional control and to improve dexterity, controllability, and functions of myoelectric hands. Instead of depending only on EMG amplitude, EMG-PR extracts several properties from EMG signals (Naik et al. 2016). A well-developed design of bionic hands includes limb trajectories and corresponding patterns of movement, which need characteristics such as kinematic models, motion (Bi et al. 2019), and activity range (Samuel et al. 2019), incorporated in the control system. Researchers are exploring the notion that EMG patterns include a lot of information about desired movements. The prosthesis controller will get the instruction to implement the movement once the EMG patterns for desired motions have been detected using pattern recognition. As a result, the EMG pattern recognition technique enables amputees to manage their bionic hands more easily and with increased controllability.

It is difficult to model bionic hands with dexterity and intricacy like a human hand (Yang et al. 2017). Pattern recognition (PR) techniques have come through a number of developments in the past two decades for use in myoelectric hands (Radmand et al. 2016; Hahne et al. 2014; Strait 2006). They offer an intuitive control, making it easy for both the human and the machine to learn. It also allows for the independent control of multiple degrees of freedom (DOF) via simultaneous control or using sequential control, thus increasing the functional abilities of a device (Cordella et al. 2016).

Because of the stochastic nature of surface EMG signals, they is more difficult to analyze than other well-known bioelectrical signals (Padmanabhan & Puthusserypady 2004). Many extraneous variables have been proven to considerably affect the characteristics of the EMG signal and hence the effectiveness of EMG pattern recognition systems. Some of the issues are—changes in the signal's properties over time, electrode site movement, muscle fatigue, inter-subject variability, variations in muscle contraction intensity, and changes in limb position and forearm orientation (Scheme & Englehart 2011, 2013; Khushaba et al. 2016).

The majority of EMG-based PR research studies used in bionic hands are discrete methods. These methods use discrete data segments representing a class for training of classifiers and deployment for real-time control. Some researchers use continuous EMG data to model muscular tension or extension, also known as regression models. In these models, efforts are diverted from the goal of categorizing data patterns; to estimation of forces, angles, or any other physical parameter, which are then directly sent as a control command to actuators.

Pattern recognition in machine learning is implemented in different parts such as—signal acquisition and data preparation, feature extraction, and classification (Vidovic et al. 2016), which will be discussed in the following sections.

11.3.2 Signal Acquisition and Preparation

EMG signals originate from the electric potentials travelling across muscle fibers when stimulated by neural signals, which is different from electric potential in the nerves. There is, however, a substantial relationship between these two potentials (Farina et al. 2014). The EMG potential value is also connected to the level of muscle activation and the force generated in a non-linear manner.

EMG electrodes capture electric potential generated by motor units of a muscle when excited by motor nerves. This muscle potential correlates with the potential generated in motor nerves (Farina et al. 2014). An EMG signal is a super-positioned action potential of a number of individual motor units (MUAPs) fired at the same time within the range of electrodes. The higher the number of motor units engaged; the higher the electric potential generated. The amplitude and frequency of EMG changes with the level of contraction (Simao et al. 2019), a feature exploited in the control of prosthetic devices. The relation of EMG signal reading and muscle contraction level shows non-linear behavior, making it difficult to model (Simao et al. 2019).

11.3.3 sEMG Electrodes

The electrode is typically made up of two poles that are aligned along the muscle fiber path. There are also monopole sensors that compare the potential in relation to a reference electrode. Because any two poles may be joined to produce a reading,

FIGURE 11.3 A typical surface EMG electrode.

monopoles offer the benefit of allowing for more flexible configurations. The distance between the electrode poles as well as the diameter has an impact on the EMG signal (Merletti & Farina 2016).

The signal picked up by surface electrodes shown in Figure 11.3 is affected by factors such as material, electrode dimension, shape, electrode configuration, and type of skin interface (Merletti et al. 2009). Surface electrodes are generally made of noble metals (e.g., gold, silver, platinum, etc.), carbon, sintered silver, or silver chloride. They make either dry or wet contact with the skin. In wet interfaces, a layer of conductive gel or sponge is applied to the skin before attaching electrodes. These electrodes are self-adhesive and provide better signal quality due to low skin-contact impedance; thus, external interferences have low impact (Simao et al. 2019). But they can cause discomfort to the patient and hence are not recommended for long-term use. This makes dry electrodes more practical for sEMG-based prosthesis. The electrodes are also classified as polarized and non-polarized based on their electro-chemical behavior. For a detailed comparison of electrodes, see Merletti et al. (2009). There have been a few studies like Young et al. (2011) considering the size of electrodes. It was observed that a larger electrode size is more resilient to electrode shifts. However, the effect of electrode size on classification accuracy was not considerable.

There are three configurations of surface electrodes used in sEMG—bipolar, monopolar, and Laplacian. In bipolar electrodes, the EMG signal is measures as the potential difference between two electrodes placed over skin aligned with length of muscle. In monopolar configuration, the EMG signal is measured as the potential difference of an electrode attached to a muscle and a reference electrode (attached to a non-active site). The Laplacian configuration uses a middle electrode surrounded by other electrodes. Marletti et al. (2009) suggested monopolar electrodes to be optimal due to the entirety of information contained in the signal. Contrarily, Hakonen et al. (2015) recommended bipolar configuration due to higher noise tolerance compared to monopolar.

Many studies have explored the ideal number of electrodes for sEMG interfaces over forearm muscles (Hargrove et al. 2007; Farrell & Weir 2008; Young et al. 2012; Andrews et al. 2009, 2008). The classification accuracy is observed to reach higher values with three to four bipolar channels under constrained lab environments (Hargrove et al. 2007; Farrell & Weir 2008). The suggested configuration is four to six bipolar channels arranged in longitudinal and transverse directions relative to the muscle fibers (Young et al. 2012). However, the ideal configuration may vary with anatomy of subjects. The disadvantage of usinga few electrodes is that a single electrode malfunction might cause considerable reduction in classification accuracy. High-definition electromyography (HD-sEMG) might alleviate this problem. It makes use of spatial information and is robust against any change in number of electrodes.

11.3.4 ELECTRODE PLACEMENT

Electrode placement is vital in determining the signal quality acquired from muscle fibers. A layer of subcutaneous tissue and the skin separates the electrodes from the muscle of interest, functioning as a low-pass filter, which smooths the recorded action potentials, therefore lowering signal content. The signal to noise ratio (SNR) of EMG recordings can be improved when electrodes are placed closer to the source of EMG signals. Merletti et al. (2009) described lack of spatial selectivity as a limitation of EMG acquisition methods arising due to indetermination of source of signal within a muscle.

The bipolar channels are recommended to be positioned on a propagating section, that is, between the innervation zone (IZ) and the tendon region. The IZ in fusiform muscles is a relatively small area through which MUAPs propagate bidirectionally toward the tendons (Saitou et al. 2000). When bipolar electrodes are positioned on opposing sides of innervation zone, there may be significant cancellation of the signal, resulting in a reduction in amplitude (Merletti et al. 2004).

Therefore, it is important to identify the innervation zones of different subjects to get accurate and reproducible data (Falla et al. 2002). Despite the fact that the position of these zones has been documented for various muscles (Saitou et al. 2000; Falla et al. 2002; Castroflflorio et al. 2005), broad distributions of IZs are not always accurate for all persons.

11.3.5 SIGNAL PRE-PROCESSING

A typical issue with systems relying on signal acquisition through electrodes, such as EMG, is determining the way to prevent or cancel extraneous noise collected by the sensors. The EMG signals are largely affected by external noise and motion artifacts that need to be reduced before any useful feature can be extracted from signals. A typical EMG signal lies in the range of +−1 to 10 mV, which is amplified around 1,000 times and then filtered. This amplification makes it sensitive to even little disturbances in the action potential. The main sources of noises in EMG are—muscle crosstalk, power-line interference, electrode displacement, electrode-skin impedance, electromagnetic interference, etc. EMG signals are often dominated by electrical disturbances because of their low amplitude. Amplification of the signal, in the range of 500–2,000 gain, is required to increase the processing quality. Generally, a differential amplifier is used for this purpose.

Some criteria should be observed while designing data collecting systems to limit their negative impacts on the output signal, namely:

i. Electrode cables should be arranged in such a way that inductive and capacitive coupling between separate channels is reduced, for example, by isolating wires connecting different electrodes (Merletti & Farina 2016).
ii. Differential amplifiers and reference electrode should be used to enable common-mode rejection (Riillo et al. 2014).
iii. Preparing the skin to reduce impedance differences between electrode poles (Piervirgili et al. 2014).

There are several techniques to pre-process the sEMG signal. Most systems utilize high pass filters (HPF) ranging between 10–50 Hz and low pass filters (LPF) in the 500 Hz range. A notch filter is frequently used to reduce power-line noise at a frequency of 50 or 60 Hz (e.g., see Figure 11.4). Other methods such as regression-subtraction (Mewett et al. 2001) and spectrum interpolation (Mewett et al. 2004) are also used for this purpose. When using a battery as a power source, the cables may contribute noise into the output signal due to electrode pole impedance imbalances. Because of disruptions in the electrode-skin contact, these impedances fluctuate over time in sEMG electrodes.

11.3.6 SAMPLING AND FILTERING

The typically low amplitude of sEMG signals, i.e., below 10 mV, makes it susceptible to external disturbances. As a result, it is often amplified and filtered before

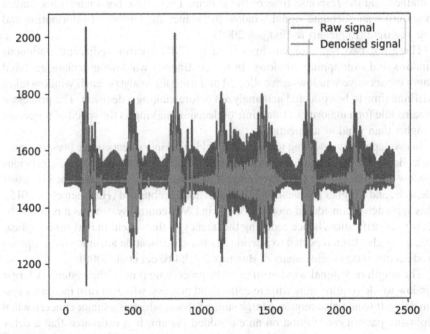

FIGURE 11.4 EMG signal filtering using wavelet denoiser.

performing digital conversion. In sEMG studies, a high pass cutoff range between 5–20 Hz is commonly utilized to reduce slow signal changes generated mostly by motion artifacts arising due to electrode movements and lying in the range of 0–20 Hz.

The major part ofenergy (about 95%) in myoelectric signal is constrained to a frequency range of 400–500 Hz, implying that components above 500 Hz are noise. Owing to the Nyquist theorem, sEMG recordings are sampled at 1,000 Hz, and a low pass filter is typically used at 400–450 Hz. Lower sampling rates are also used, as they may still retain significant information for pattern recognition.

One advantage of low sample rate is that it reduces computation time for the controller. It was evident from a study involving EMG data sampled at 500 Hz where it was found more optimal than the standard 1,000 Hz sampling rate. For sampling at 500 Hz, low pass filtration with less 250 Hz cutoff frequency is needed to avoid aliasing (Li et al. 2011).

In one study, a high pass cutoff at 60 Hz for sEMG interfaces was proposed by Li et al. (2011, 2010b), as the lower frequency components do not contribute much to movement classification due to the presence of information regarding firing rates of the active motor units (Stegeman et al. 2007).

11.3.7 SEGMENTATION

In bionic devices, signal processing and data analysis are performed on continuous EMG signals divided into segments called 'windows' or 'epochs'. A window is collected based on time (e.g., 100 ms) or based on number of samples (e.g., 200 samples). The size and selection of a window affects the degree of signal information contained and the response time of the system. There have been numerous studies on selecting an adequate signal window to balance the contained information and response time (Englehart & Hudgins 2003).

There are two types of windows used in EMG control—adjacent/continuous windows and overlapping windows. In the continuous windowing technique, fixed length consecutive windows are collected and used for analysis. Each window takes a definite time to be collected and analyzed before making a decision. The processor remains idle for a majority of total time of decision-making as the speed of processor is higher than window acquisition.

In overlapping windowing technique, fixed length increments to the present signal are collected and added to the present signal. Each window overlaps the previous window and slides along at relatively small increments. This decreases the processor idle time, and a higher frequency of class decisions is obtained (Hakonen et al. 2015). This approach is considered more favorable in EMG control systems as it reduces the delay in classification, hence reducing the latency of the system in real-time application. It has also been reported to provide a better classification accuracy as compared to adjacent windows (Englehart & Hudgins 2003; Oskoei et al. 2008).

The length of a signal window affects the processing time of the system. A larger window would require more time to collect and process, which in turn increases system delay. It would also require a large storage space, which is a major concern when deploying pattern recognition on an embedded system. It is estimated that a delay between user command and device actuation should be in the range of 50–400 ms.

Larger delays make a prosthetic device feel unresponsive and non-intuitive to the user. Generally, EMG signals are collected at a sampling frequency of 500 Hz or 1,000 Hz. At these frequencies, a signal window of 25–200 samples (for 500 Hz) and 50–400 samples (for 1,000 Hz) is considered valid (Farrell et al. 2011; Englehart & Hudgins 2003; Englehart et al. 2003). The length of a window also depends upon factors such as post-processing mechanism and type of feature vector (Oskoei et al. 2008).

11.4 FEATURE EXTRACTION

The pattern recognition system distinguishes the incoming EMG signal into various categories called 'classes'. Each class represent a different muscle movement based on the signal characteristics. In real-time application, such as a bionic prosthesis, continuous signal data is fed through a classifier performing pattern recognition. In the control methods, a set of features of signal, called a feature vector, is calculated from segments of signal (Hakonen et al. 2015). This feature vector is used for classifying the incoming signal into various classes.

In pattern recognition, it is essential that each pattern/class stays constant across studies and is capable of accurately differentiating each class. In order to identify the optimal set of all available features, one would have to examine every potential combination, which is impractical and, in the case of big datasets, impossible. Furthermore, the optimum combination for one application or environment is not always the greatest combination for another. In some cases, techniques for reducing the feature dimensions have been employed. Using these techniques such as principal component analysis (PCA), neighborhood component analysis (NCA) (Khan et al. 2021), or meta-heuristics like genetic algorithm (GA), particle swarm optimization (PSO), etc., redundant and low-performing features can be identified and removed (Karthick et al. 2018; Purushothaman & Vikas, 2018; Xi et al. 2018).

Feature selection has traditionally been the primary emphasis of EMG pattern recognition. On the other hand, feature learning, as illustrated by deep learning, is now enabling researchers to generate features that lead to higher classification accuracy. Deep learning techniques use large datasets of EMG signals to learn optimal features.

11.4.1 FEATURE SELECTION

The criteria for selection of proper features consist of properties like redundancy, class separability, robustness, and computational complexity (Hakonen et al. 2015). Class separability of a feature is its ability to separate each signal class with low overlap. Features with high class separability are more effective for use in classification. The features with low class separability are avoided as they can degrade the performance of PR systems, leading to lower classification accuracy.

Robustness is described as the capability of a feature set to preserve class separability in high noise conditions. Feature sets with high robustness remain less affected by noise, thus maintaining class separability. For real-time application of pattern recognition control, computational complexity of features should remain low. It includes the time required to calculate each feature and memory space required. Features that are computationally complex to calculate and store cannot be processed on embedded devices or microcontrollers.

Another important consideration in feature selection is the 'redundancy' of features. Redundancy is mostly observed in time-domain features where two or more features represent similar trends of class separability. Each of such features increases computational complexity without providing a distinctive criterion for signal classification. Redundant features should be avoided in the classification process.

11.4.2 Types of Features

The primary components in EMG pattern recognition are—data pre-processing, feature extraction, dimensionality reduction, and classification (Scheme & Englehart 2011; Phinyomark et al. 2011). Because of the EMG signal's stochastic and nonstationary nature, the instant value is inappropriate for typical machine learning techniques (Phinyomark et al. 2012a). Thus, before classifying the signals into classes, feature extraction is performed, which reduces the large signal data into short characteristics that provide similar information as the original signal. The reduced space is known as a feature vector, and it includes all the important information necessary for classifying different classes (Zardoshti-Kermani et al. 1995). For each signal window, features of the contained signal are calculated. The features used for pattern recognition are divided into different groups namely—i. time-domain (TD) features, ii. frequency-domain (FD) features, and iii. time-frequency domain (TFD) features. Some studies have also considered spatial domain (SD) features. In other words, EMG characteristics may be calculated using both linear and nonlinear analysis.

The first two feature groups are prominently used in EMG pattern recognition. TD features are the most widely used in real-time control of bionic prosthesis due to their low computational complexity and high performance in low-noise environment. However, these features are susceptible to noise and external interferences. On the other hand, FD features are considered less suitable for control applications and more for studying muscle fatigue and motor unit characteristics. It is reported that high dimensionality of TFD features has to be reduced for use in classification.

11.4.3 Time Domain (TD) Features

The time-domain features, as mentioned in Table 11.1, are often simple and easy to implement since they do not require any transformation (Hudgins et al. 1993; Tkach et al. 2010). These features are widely employed in various medical and technical studies. Features in this group consider EMG signals to be stationary. This is a significant disadvantage of features in this category as the EMG signals possess a non-stationary quality, i.e., they change in statistical properties with time. As a result, there are significant changes in the characteristics of features in this group due to dynamic recordings; the change of characteristics in this group is obtained to a significant extent. Furthermore, because their calculations are reliant on EMG signal amplitude values, excessive interference obtained while recording becomes a big disadvantage (Phinyomark et al. 2009). The ability of time domain features to provide

TABLE 11.1
Summary of Time Domain (TD) Features

Sr. No.	Feature Name	Abbr.	Sr. No.	Feature Name	Abbr.
1.	Integrated EMG	IEMG	12.	Zero Crossing	ZC
2.	Mean Absolute Value	MAV	13.	v-Order	v
3.	Modified Mean Absolute Value 1	MAV1	14.	Log Detector	LOG
4.	Modified Mean Absolute Value 2	MAV2	15.	Waveform Length	WL
5.	Simple Square Integral	SSI	16.	Average Amplitude Change	AAC
6.	Variance	VAR	17.	Difference Absolute Standard Deviation Value	DASDV
7.	Root Mean Square	RMS	18.	Myopulse Percentage Rate	MYOP
8.	Willison Amplitude	WAMP	19.	Multiple Trapezoidal Windows	MTW
9.	Mean Absolute Value Slope	MAVSLP	20.	Multiple Hamming Windows	MHW
10.	Slope Sign Change	SSC	21.	Auto-regressive Coefficients	AR
11.	Histogram	HIST	22.	Cepstral Coefficients	CC

high classification accuracy in low-noise environments and low computational complexity makes them ideal candidates for real-time implementation.

Phinyomark et al. (2012b) presented a comprehensive comparison of various features and reviewed their significance in classification accuracy. They divided the time-domain features in categories based on the information each feature contained, that is—energy, complexity, frequency, prediction model, and time dependence. For more details about the features, readers are referred to Phinyomark et al. (2012).

Several attempts have been made to find an optimal time domain feature vector. Prominently, the Hudgin's set consisting of mean absolute value (MAV), waveform length (WL), zero crossing (ZC), and signal slope changes (SSC) is commonly used with the LDA classifier (Li et al. 2010a; Oskoei et al. 2008; Englehart et al. 2003, 2001, 1999; Chan et al. 2000). The absolute value of the EMG signal amplitude is averaged to determine MAV, and the total length of the waveform across the time segment determines WL (Boostani & Moradi 2003; Phinyomark et al., 2012). It is a signal complexity metric that combines signal amplitude, frequency, and duration. The accuracy classification provided by Hudgin's set is significant as well as other advantages such as robustness for segment length and lower computational complexity (Oskoei et al. 2008). It has been observed that it is an appropriate time domain feature set for SVM, LDA, and MLP classifiers (Oskoei et al. 2008).

A feature commonly combined with Hudgin's set is autoregressive (AR) coefficients. This new set is known as TDAR feature set and provides high classification accuracy with LDA (Hargrove et al. 2007; Farrell & Weir 2008; Hargrove et al. 2008; Young et al. 2012; Al-Timemy et al. 2011; Yoshikawa et al. 2007), MLP, and Gaussian mixture models (GMM; Huang et al. 2005). Another feature based on the prediction model is Cepstral coefficients (CC). Both of these features, i.e., autoregressive and

Cepstral coefficients, are more difficult to calculate than other time domain features. As the order of a model is increased, time required to compute the coefficients also increases, necessitating use of simpler models. Some research has suggested that the third (Farrell & Weir 2005), fourth (Graupe & Cline 1975), and sixth (Liu et al. 2009) AR order models provide best results for classifying EMG data.

11.4.4 FREQUENCY DOMAIN (FD) FEATURES

The features in this group are generally used to analyze fatigue (Al-Mulla et al. 2012; Cifrek et al. 2009; Santos et al. 2008; Lindström et al. 1977), force generation (Bigland-Ritchie 1981), motor unit patterns, etc. These features are generally calculated using power spectral density (PSD). Mean frequency (MNF) and median frequency (MDF) are some of the PSD's most used variables (MNF and MDF), whereas mean power (MNP), peak frequency (PKF), and total power (TTP) are often used.

In a study, Phinyomark et al. (2012b) investigated characteristics of 37 time-domain and frequency-domain features. Time-domain features were found to outperform frequency-domain features as the latter are computationally more complicated than the former, in addition to having lower classification accuracy. On the other hand, combination of FD features with TD features may provide more robust and accurate classification than a feature vector consisting only of TD features. Mean frequency (MNF) was proposed by Phinyomark et al. as a viable option to combine with time-domain features since its discriminatory pattern differs from that of time-domain features (Phinyomark et al. 2012c).

11.4.5 TIME-FREQUENCY DOMAIN (TFD) FEATURES

The time-frequency domain (TFD) features generally used in EMG pattern recognition are shown in Table 11.2.

Out of these, DWT is one of the most frequent TFD features in sEMG interfaces because it is more computationally efficient than CWT. WPT has also received a lot of attention due to its capacity to offer frequency information in both the low- and high-frequency bands. TFD features are more difficult to calculate than TD features. However, provided suitable dimensional reduction and segmentation, they

TABLE 11.2

Summary of Time-Frequency Domain (TFD) Features

Sr. No.	Feature Name	Abbreviation	Reference
1.	Short Time Fourier Transform	STFT	Englehart et al. 2003, 2001
2.	Continuous Wavelet Transform	CWT	Englehart et al. 2003, 2001, 1999
3.	Discrete Wavelet Transform	DWT	Englehart et al. 2003; Zhao et al. 2006; Zhang et al. 2006; Kakoty & Hazarika 2011
4.	Wavelet Packet Transform	WPT	Englehart et al. 2003, 2001
5.	Stationary Wavelet Transform	SWT	Englehart et al. 2003
6.	Fractional Wavelet Transform	FrFT	Zahra et al. 2021

may be implemented for real-time sEMG classification (Englehart & Hudgins 2003; Englehart et al. 1999; Lucas et al. 2008). Wavelet transformations can improve the system's robustness as they can limit the analysis to the most important frequency bands by employing subsets of wavelet coefficients.

The time-frequency domain features produce a feature vector of high dimensions, which needs to be reduced in order to improve classification speed. Some of the reduction methods used are uncorrelated linear discriminant analysis (ULDA) and principal component analysis (PCA), neighborhood component analysis (NCA), etc.

11.5 CLASSIFICATION

After selecting an appropriate set of features, classification is used to separate various categories of the feature vectors. The pattern recognition-based technique assumes that the classifier can detect the input values and align them with a predefined set of classes. The classifier is fed a feature vector generated from the sEMG signals and classes corresponding to different control instructions sent to the bionic hand.

Several comparison studies, however, indicate that, using a sufficient number of channels and an adequate feature vector, most classifiers have comparable accuracy of EMG classification (Hargrove et al., 2007; Huang et al. 2005; Scheme et al. 2010, Scheme et al., 2011). Generally, the classifiers that are easy to implement, quick to train, and meet real-time constraints are employed. Some of the examples include linear discriminant analysis (LDA; Farrell & Weir, 2008; Boschmann et al. 2013; Farrell et al. 2011; Englehart & Hudgins 2003; Zhang et al. 2012; Chen et al. 2011; Scheme et al. 2010; Young et al. 2013), support vector machine (SVM; Yoshikawa et al. 2007), hidden Markov models (HMM; Lee et al. 2008; Chan et al. 2005; Scheme et al. 2011), artificial neural network (ANN), k-nearest neighbors (KNN), multi-layer perceptron (MLP), non-negative matrix factorization (NMF), Bayesian learning (Akira et al. 2021), etc. The most commonly used method in EMG pattern recognition is linear discriminant analysis (LDA), which has now become a general recommendation. Few researchers have examined the classifiers' capability to accurately distinguish EMG signals over time and in the presence of additive artifacts or noise. Because of superior capacity to generalize EMG signals, the linear classifiers such as LDA are expected to provide higher accuracy than nonlinear classifiers.

It is observed that it's difficult to address the issue of linear inseparability with an LDA classifier. The artificial neural network (ANN) is capable of describing nonlinear class borders across distinct categories. The two primary architectures of artificial neural networks are MLP and Cascade. The identification and categorization of distinct gait phases based on EMG data gathered is a standard linear inseparability problem. In a study that focused on this issue (Meng et al. 2010), three distinct MLPs were used to classify two major gait phases. Results showed that the three MLP models outperformed all popular models on average. LDA, on the other hand, can outperform ANN approaches when used in conjunction with methods for reducing dimensionality of feature space, such as PCA. Determining the appropriate size and structure of an ANN for a specific task may be difficult. Despite the advantage of obtaining latent features from raw EMG data, the extended training time is also an important factor to consider when employing the ANN.

The support vector machine (SVM) is considered to be among the most widely used algorithms in EMG pattern recognition. It is an efficient method where a low dimension dataset is projected by a kernel function to a high dimension feature space. During the training phase, SVM finds a hyperplane to identify various categories in high-dimensional space. It is thought that a good kernel function can aid in the reduction of indivisible linear data to linear separable sets in high-dimension space. As a result, it is clear that determining the best kernel function and its parameter values is the most important task while developing the SVM model.

Another approach utilized in the categorization of EMG signals is fuzzy logic (FL), which obtains definite results from imprecise data input in a straightforward manner. FL has been shown to have an advantage over other control approaches in EMG processing (Xie et al. 2014). Contrarily, it takes additional system memory and processing time due to the usage of fixed membership functions (Xie et al. 2014).

After classifying the incoming EMG signals into different classes, a control command is sent to the bionic hand or other terminal device, which then performs a particular movement. The movements are predefined as grip patterns or grasps similar to a natural human hand.

11.6   FUTURE DEVELOPMENTS

Current EMG PR development efforts are aimed toward enhancing classification reliability by predicting and correcting causes of variations due to muscle fatigue, fluctuating skin impedance, liftoff, electrode displacement, etc. The methods under consideration for future developments are based on the use of high-definition EMG electrode arrays to generate accurate mapping of muscle activity, signal conditioning, and preprocessing methods for reducing the effect of external disturbances and improving the robustness of classification models considering the dynamic nature of EMG signals. The feature-classifier combination continues to have a large impact on classification accuracy. Various time domain features are often used in EMG pattern recognition whereas novel time-frequency domain features are gaining more attention from researchers. LDA and its variants, SVM, and ANN in various versions are the most often used classifier models. All of these approaches have had some success with single muscle activation and may be employed in future human machine interfaces. The repeatability of the provided experiments is an essential aspect that is seldom discussed in most papers on this subject. Because datasets are rarely available, comparing different approaches is challenging, and it becomes difficult for researchers to replicate similar arrangements.

In recent years, studies have been carried out to determine alternatives to electromyography (EMG). Apart from acquiring electrical signals from muscles, other physical attributes are also being recorded, such as vibrations in mechanomyography (MMG), force in force myography (FMG), and sound in phonomyography (PMG). The new techniques are often used in combination with EMG leading to hybrid control schemes. These hybrid schemes have performed well by eliminating the drawbacks of conventional EMG systems and introducing new dimensions of neuromuscular information.

REFERENCES

Al-Mulla, M.R., F. Sepulveda, M. Colley, sEMG techniques to detect and predict localised muscle fatigue, in: M. Schwartz (Ed.), *EMG Methods for Evaluating Muscle and Nerve Function*, InTech, 2012, pp. 157–186, http://dx.doi.org/10.5772/25678.

Al-Timemy, A.H., G. Bugmann, N. Outram, J. Escudero, Reduction in classification errors for myoelectric control of hand movements with independent component analysis, in: *The 5th International Conference on Information Technology*, ICIT 2011, 2011.

Andrews, A.J. *Finger Movement Classification Using Forearm EMG Signals* (MS thesis), Queen's University, October 31, 2008.

Andrews, A.J., E. Morin, L. Mclean, Optimal electrode configurations for finger movement classification using EMG, in: *31st Annual International Conference of the IEEE EMBS*, September 2–6, 2009, pp. 2987–2990, http://dx.doi.org/10.1109/IEMBS.2009.5332520.

Battye, C. K., A. Nightingale, J. Whillis, The use of myo-electric currents in the operation of prostheses, *J. Bone Joint Surg. British Volume*, 1955, 37-B(3), 506–510.

Bi, L., A. Feleke, C. Guan, A review on EMG-based motor intention prediction of continuous human upper limb motion for human-robot collaboration, *Biomed. Signal Process. Control*, 2019, 51, 113–127.

Bigland-Ritchie, B. EMG/force relations and fatigue of human voluntary contractions, *Exerc. SportSci. Rev.*, 1981, 9, 75–117, http://dx.doi.org/10.1249/00003677-198101000-00002.

Boostani, R., M.H. Moradi, Evaluation of the forearm EMG signal features for the control of a prosthetic hand, *Physiol. Meas.*, 2003, 24(2), 309–319, http://dx.doi.org/10.1088/096 7-3334/24/2/307.

Boschmann, A., M. Platzner, Reducing the limb position effect in pattern recognition based myoelectric control using a high density electrodearray, in: *Biosignals and Biorobotics Conference*, BRC, February 18–20, 2013, pp. 1–5, http://dx.doi.org/10.1109/BRC.2013.6487548.

Castrofiflorio, T., D. Farina, A. Bottin, C. Debernardi, P. Bracco, R. Merletti, G. Anastasi, P. Bramanti, Non-invasive assessment of motor unit anatomy in jaw-elevator muscles, *J. Oral Rehabil.*, 2005, 32(10), 708–713, http://dx.doi.org/10.1111/j.1365-2842.2005. 01490.x.

Clement, R.G.E., K.E. Bugler, C.W. Oliver, Bionic prosthetic hands: A review of present technology and future aspirations, *Surgeon*, 2011, 9(6), 336–340.

Cordella, F., A.L. Ciancio, R. Sacchetti, A. Davalli, A.G. Cutti, E. Guglielmelli, L. Zollo, Literature review on needs of upper limb prosthesis users, *Front. Neurosci.*, May 2016, 10, 1–14.

Chan, A.D.C., K.B. Englehart, Continuous myoelectric control for powered prostheses using hidden Markov models, *IEEE Trans. Biomed. Eng.*, 2005, 52(1), 121–124, http://dx.doi. org/10.1109/TBME.2004.836492.

Chan, F.H.Y., Y. Yong-Sheng, F.K. Lam, Z. Yuan-Ting, P.A. Parker, Fuzzy EMG classification for prosthesis control, *IEEE Trans. Rehabil. Eng.*, 2000, 305–311, http://dx.doi. org/10.1109/86.867872.

Chen, L., Y. Geng, G. Li, Effect of upper-limb positions on motion pattern recognition using electromyography, in: *4th international congress on image and signal processing*, CISP, vol. 1, Shanghai, China, October 15–17, 2011, pp. 139–142, http://dx.doi.org/10.1109/CISP.2011.6100025.

Cifrek, M., V. Medved, S. Tonkovic, S. Ostojic, Surface EMG-based muscle fatigue evaluation in biomechanics, *Clin. Biomech.*, 2009, 24(4), 327–340, http://dx.doi.org/10.1016/j. clinbiomech.2009.01.010.

Farrell, T.R. Determining delay created by multifunctional prosthesis controllers, *J. Rehabil. Res. Dev.*, 2011, 48(6), xxi–xxxviii, http://dx.doi.org/10.1682/JRRD.2011.03.0055.

Farrell, T.R., R.F. Weir, Pilot comparison of surface vs. implanted EMG for multifunctional prosthesis control, in: *Proceedings of the 2005 IEEE 9th international conference on rehabilitation robotics*, ICORR, June 28–July 1, 2005, pp. 277–280, http://dx.doi.org/10.1109/ICORR.2005.1501101.

Farrell, T.R., R.F. Weir, A comparison of the effects of electrode implantation and targeting on pattern classification accuracy for prosthesis control, *IEEE Trans. Biomed. Eng.*, 2008, 55(9), 2198–2211, http://dx.doi.org/10.1109/TBME.2008.923917.

Finni, T., M. Hu, P. Kettunen, T. Vilavuo, S. Cheng, Measurement of EMG activity with textile electrodes embedded into clothing, *Physiol. Meas.*, 2011, 28(11), 1405–1419, http://dx.doi.org/10.1088/0967-3334/28/11/007.

Englehart, K., B. Hudgins, A robust, real-time control scheme for multifunction myoelectric control, *IEEE Trans. Biomed. Eng.*, 2003, 50(7), 848–854, http://dx.doi.org/10.1109/TBME.2003.813539.

Englehart, K., B. Hudgins, A. Chan, Continuous multifunction myoelectric control using pattern recognition, *Technol. Disabil.*, 2003, 95–103, ISSN:1055–4181.

Englehart, K., B. Hudgins, P.A. Parker, A wavelet-based continuous classification scheme for multifunction myoelectric control, *IEEE Trans. Biomed. Eng.*, 2001, 48(3), 302–310, http://dx.doi.org/10.1109/10.914793.

Englehart, K., B. Hudgins, P.A. Parker, M. Stevenson, Classification of the myoelectric signal using time-frequency based representations, *Med. Eng. Phys.*, 1999, 21(6–7), 431–438, ISSN:1350–4533.

Falla, D., P. Dall'alba, A. Rainoldi, R. Merletti, G. Jull, Location of innervation zones of sternocleidomastoid and scalene muscles—a basis for clinical and research electromyography applications, *Clin. Neurophysiol.*, 2002, 113(1), 57–63, http://dx.doi.org/10.1016/S1388-2457(01)00708-8.

Farina, D., T. Lorrain, F. Negro, N. Jiang, High-density EMG e-textile systems for the control of active prostheses, in: *32nd annual international conference of the IEEE Engineering in medicine and biology society (EMBC)*, EMBC, September 1–4, 2010, pp. 3591–3593.

Farina, D. et al., The extraction of neural information from the surface EMG for the control of upper-limb prostheses: Emerging avenues and challenges, *IEEE Trans. Neural Syst. Rehabil. Eng.*, July 2014, 22(4), 797–809.

Furui, A., Takuya Igaue, Toshio Tsuji, EMG pattern recognition via Bayesian inference with scale mixture-based stochastic generative models, *Expert Syst. Appl.*, 2021, 185, 115644, ISSN 0957–4174, https://doi.org/10.1016/j.eswa.2021.115644.

Graupe, D., W.K. Cline, Functional separation of EMG signals via ARMA identification methods for prosthesis control purposes, *IEEE Trans. Syst. Man Cybern.*, 1975, 5, 252–259.

Hahne, J.M., F. Biessmann, N. Jiang, H. Rehbaum, D. Farina, F.C. Meinecke, K.R. Muller, L.C. Parra, Linear and nonlinear regression techniques for simultaneous and proportional myoelectric control, *IEEE Trans. Neural Syst. Rehabil. Eng.*, 2014, 22, 269–279.

Hakonen, M., H. Piitulainen, A. Visala, Current state of digital signal processing in myoelectric interfaces and related applications, *Biomed. Signal Process. Control*, April 2015, 18, 334–359. https://doi.org/10.1016/j.bspc.2015.02.009.

Hargrove, L.J., K. Englehart, B. Hudgins, A comparison of surface and intra muscular myoelectric signal classification, *IEEE Trans. Biomed. Eng.*, 2007, 847–853, http://dx.doi.org/10.1109/TBME.2006.889192.

Hargrove, L., K. Englehart, B. Hudgins, A training strategy to reduce classification degradation due to electrode displacements in pattern recognition based myoelectric control, *Biomed. Signal Process. Control*, April 2, 2008, 3, 175–180, http://dx.doi.org/10.1016/j.bspc.2007.11.005.

Huang, Y., K.B. Englehart, B. Hudgins, A.D.C. Chan, A Gaussian mixture model-based classification scheme for myoelectric control of powered upper limb prostheses, *IEEE Trans. Biomed. Eng.*, 2005, 1801–1811, http://dx.doi.org/10.1109/TBME.2005.856295.

Hudgins, B., P. Parker, R. Scott, A new strategy for multifunction myoelectric control, *IEEE Trans. Biomed. Eng.*, 1993, 40(1), 82–94.

Kakoty, N.M., S.M. Hazarika, Recognition of grasp types through principal components of DWT-based EMG features, in: *IEEE International Conference on Rehabilitation Robotics*, ICORR, 2011, pp. 1–6, http://dx.doi.org/10.1109/ICORR.2011.5975398.

Karthick, P., D.M. Ghosh, S. Ramakrishnan, Surface electromyography based muscle fatigue detection using high-resolution time-frequency methods and machine learning algorithms, *Comput. Methods Prog. Biomed.*, 2018, 154, 45–56.

Kaufmann, P., K. Englehart, M. Platzner, Fluctuating EMG signals: Investigating long-term effects of pattern matching algorithms, in: *Proceedings of Annual International Conference of IEEE Engineering in Medicine and Biology Society*, 2010, 6357–6360.

Khan, S.M. et al., Pattern recognition of EMG signals for low level grip force classification, *Biomed. Phys. Eng. Express*, 2021, 7(6), 065012.

Khushaba, R.N., A. Al-Timemy, S. Kodagoda, K. Nazarpour, Combined influence of forearm orientation and muscular contraction on EMG pattern recognition, *Expert Syst. Appl.*, 2016, 61, 154–161.

Lee, K.-S. EMG-based speech recognition using hidden Markov models with global control variables (author abstract), *IEEE Trans. Biomed. Eng.*, 2008, 55(3), 930–940, http://dx.doi.org/10.1109/TBME.2008.915658.

Li, G., Y. Li, L. Yu, Y. Geng, Conditioning and sampling issues of EMG signals in motion recognition of multifunctional myoelectric prostheses, *Ann. Biomed. Eng.*, 2011, 39(6), 1779–1787, http://dx.doi.org/10.1007/s10439-011-0265-x.

Li, G., A.E. Schultz, T.A. Kuiken, Quantifying pattern recognition-based myoelectric control of multifunctional transradial prostheses, *IEEE Trans. Neural Syst. Rehabil. Eng.*, 2010a, 18(2), 185–192, http://dx.doi.org/10.1109/TNSRE.2009.2039619.

Li, G., Y. Li, Z. Zhang, Y. Geng, R. Zhou, Selection of sampling rate for EMG pattern recognition-based prosthesis control, in: *Annual international conference of the IEEE engineering in medicine and biology society (EMBC)*, EMBC, August 31–September 4, 2010b, pp. 5058–5061, http://dx.doi.org/10.1109/IEMBS.2010.5626224.

Lindström, L., R. Kadefors, I. Petersén, An electromyographic index for localized muscle fatigue, *J. Appl. Physiol.: Respir. Environ. Exerc. Physiol.*, 1977, 43(4), 750–754, ISSN: 01617567.

Liu, X., R. Zhou, L. Yang, G. Li, Performance of various EMG features in identifying arm movements for control of multifunctional prostheses, in: *IEEE Youth Conference on Information, Computing and Telecommunication, 2009*, YC-ICT'09, Beijing, China, September 20–21, 2009, pp. 287–290, http://dx.doi.org/10.1109/YCICT.2009.5382366.

Lucas, M.-F., A. Gaufriaua, S. Pascuala, C. Doncarlia, D. Farina, Multi-channel surface EMG classification using support vector machines and signal-based wavelet optimization, *Biomed. Signal Process. Control*, 2008, 3(2), 169–174, http://dx.doi.org/10.1016/j.bspc.2007.09.002.

Meng, M., Z. Luo, Q. She, Y. Ma, Automatic recognition of gait mode from EMG signals of lower limb, in: *Proceedings of the 2010 The 2nd International Conference on Industrial Mechatronics and Automation*, Hong Kong, China, 30 May 2010, pp. 282–285.

Merletti, R., A. Botter, A. Troiano, E. Merlo, M.A. Minetto, Technology and instrumentation for detection and conditioning of the surface electromyographic signal: State of the art, *Clin. Biomech.*, 2009, 24(2), 122–134. https://doi.org/10.1016/j.clinbiomech.2008.08.006

Merletti, R., D. Farina, *Surface Electromyography: Physiology, Engineering, and Applications*, Wiley, 2016.

Merletti, R., P. Parker, H.J. Hermens, Detection and conditioning of the surface EMG signal, *Electromyogram.: Physiol. Eng. Non-Invasive Appl.*, 2004, 107, http://dx.doi.org/10.1002/0471678384.ch5

Mewett, D.T., H. Nazeran, K.J. Reynolds, Removing power line noise from recorded EMG, in: *Proc. Conf. 23rd Annu. Int. Conf. IEEE Eng. Med. Biol. Soc.*, vol. 3, Oct. 2001, pp. 2190–2193.

Mewett, D.T., K.J. Reynolds, H. Nazeran, Reducing power line interference in digitised electromyogram recordings by spectrum interpolation, *Med. Biol. Eng. Comput.*, 2004, 42(4), 524–531.

Naik, G.R., A.H. Al-Timemy, H.T. Nguyen, Transradial amputee gesture classification using an optimal number of sEMG sensors: An approach using ICA clustering, *IEEE Trans. Neural Syst. Rehabil. Eng.*, 2016, 24, 837–846.

Oskoei, M.A., H. Hu, Support vector machine-based classification scheme for myoelectric control applied to upper limb, *IEEE Trans. Biomed. Eng.*, 2008, 55(8), 1956–1965.

Padmanabhan, P., S. Puthusserypady, Nonlinear analysis of EMG signals—a chaotic approach, in: *Proceedings of the 26th Annual International Conference of the IEEE Engineering in Medicine and Biology Society*, San Francisco, CA, USA, 1–5 September 2004, Volume 1, pp. 608–611.

Phinyomark, A., C. Limsakul, P. Phukpattaranont, A comparative study of wavelet denoising for multifunction myoelectric control, in: *International Conference on Computer and Automation Engineering*, ICCAE'09, Bangkok, Thailand, March 8–10, 2009, pp. 21–25, http://dx.doi.org/10.1109/ICCAE.2009.57.

Phinyomark, A., C. Limsakul, P. Phukpattaranont, Application of wavelet analysis in EMG feature extraction for pattern classification, *Meas. Sci. Rev.*, 2011, 11(2), 45–52, http://dx.doi.org/10.2478/v10048-011-0009-y.

Phinyomark, A., A. Nuidod, P. Phukpattaranont, C. Limsakul, Feature extraction and reduction of wavelet transform coefficients for EMG pattern classification, *Electr. Electr. Eng.* 2012a, 122(6), 27–32, http://dx.doi.org/10.5755/j01.eee.122.6.1816.

Phinyomark, A., P. Phukpattaranont, C. Limsakul, Feature reduction and selection for EMG signal classification, *Expert Syst. Appl.* 2012b, 39(8), 7420–7431, http://dx.doi.org/10.1016/j.eswa.2012.01.102.

Phinyomark, P., S. Thongpanja, H. Hu, P. Phukpattaranont, C. Limsakul, The usefulness of mean and median frequencies in electromyography analysis, in: R.G. Naik (Ed.), *Computational Intelligence in Electromyography Analysis—A Perspective on Current Applications and Future Challenges*, InTech, 2012c, ISBN 978-953-51-0805-4, pp. 220–295, http://dx.doi.org/10.5772/50639.

Piervirgili, G., F. Petracca, R. Merletti, A new method to assess skin treatments for lowering the impedance and noise of individual gelled Ag—AgCl electrodes, *Physiol. Meas.*, 2014, 35(10), 2101.

Popov, B. The bio-electrically controlled prosthesis, *J. Bone Joint Surg.* 1965, 47(3), 421–424. https://doi.org/10.1016/0022-3468(66)90118-7

Purushothaman, G., R. Vikas, Identification of a feature selection based pattern recognition scheme for finger movement recognition from multichannel EMG signals, *Aust. Phys. Eng. Sci. Med.* 2018, 41, 549–559.

Radmand, A., E. Scheme, K. Englehart, High-density force myography: A possible alternative for upper-limb prosthetic control, *J. Rehabil. Res. Dev.* 2016, 53, 443–456.

Riillo, F. et al., Optimization of EMG-based hand gesture recognition: Supervised vs. unsupervised data preprocessing on healthy subjects and transradial amputees, *Biomed. Signal Process. Control*, 2014, 14, 117–125.

Saitou, K., T. Masuda, D. Michikami, R. Kojima, M. Okada, Innervation zones of the upper and lower limb muscles estimated by using multichannel surface EMG, *J. Hum. Ergol. (Tokyo)*, 2000, 29(1–2), 35–52, ISSN:0300–8134.

Santos, M.C.A., T.A. Semeghuini, F.M. De Azevedo, D.B. Colugnati, R.D. Negrao, N. Alves, R.M. Arida, Analysis of localized muscular fatigue in athletes and sedentary subjects through frequency parameters of electromyographic signal, *Rev. Bras. Med. Esp.*, 2008, 14(6), 509–512, ISSN: 1517–8692.

Samuel, O.W., M.G. Asogbon, Y. Geng, A.H. Al-Timemy, S. Pirbhulal, N. Ji, S. Chen, P. Fang, G. Li, Intelligent EMG pattern recognition control method for upper-limb multifunctional prostheses: Advances, current challenges, and future prospects, *IEEE Access*, 2019, 7, 10150–10165.

Scheme, E., K.B. Englehart, Electromyogram pattern recognition for control of powered upper-limb prostheses: state of the art and challenges for clinical use, *J. Rehabil. Res. Dev.*, 2011, 48(6), 643–660.

Scheme, E., K. Englehart, B. Hudgins, A one-versus-one classifier for improved robustness of myoelectric control, in: *The 18th Congress of the International Society of Electrophysiology and Kinesiology*, Aalborg, Denmark, 2010.

Scheme, E., K.B. Englehart, B.S. Hudgins, Selective classification for improved robustness of myoelectric control under non-ideal conditions, *IEEE Trans. Biomed. Eng.*, 2011, 58(6), 1698–1705.

Scheme, E., K. Englehart, Training strategies for mitigating the effect of proportional control on classification in pattern recognition—based myoelectric control, *JPO J. Prosthet. Orthot.*, 2013, 25, 76–83.

Simão, M., N. Mendes, O. Gibaru, P. Neto, A review on electromyography decoding and pattern recognition for human-machine interaction, *IEEE Access*, 2019, 7, 39564–39582, doi: 10.1109/ACCESS.2019.2906584.

Stegeman, D.F., H.J. Hermens, Standards for surface electromyography, in: *The European Project Surface EMG for Non-Invasive Assessment of Muscles (SENIAM)*, 2007.

Strait, E. *Prosthetics in Developing Countries*, American Academy of Orthotists & Prosthetists, 2006.

Taghizadeh, Zahra, Saeid Rashidi, Ahmad Shalbaf, Finger movements classification based on fractional Fourier transform coefficients extracted from surface EMG signals, *Biomed. Signal Process. Control*, 2021, 68, 102573, ISSN 1746–8094, https://doi.org/10.1016/j.bspc.2021.102573.

Tkach, D., Huang, H., Kuiken, T.A. (2010). Study of stability of time-domain features for electromyographic pattern recognition, *J. Neuroeng. Rehabil.*, 7(21). doi:10.1186/1743-0003-7-21.

Vidovic, M.M.C., H.J. Hwang, S. Amsuss, J.M. Hahne, D. Farina, K.R. Muller, Improving the robustness of myoelectric pattern recognition for upper limb prostheses by covariate shift adaptation, *IEEE Trans. Neural Syst. Rehabil. Eng.*, 2016, 24, 961–970.

Xi, X., M. Tang, Z. Luo, Feature-level fusion of surface electromyography for activity monitoring, *Sensors*, 2018, 18, 614.

Xie, H.-B., T. Guo, S. Bai, S. Dokos, Hybrid soft computing systems for electromyographic signals analysis: A review, *Biomed. Eng. Online*, 2014, 13, 8.

Yang, D., W. Yang, Q. Huang, H. Liu, Classification of multiple finger motions during dynamic upper limb movements. *IEEE J. Biomed. Health Inform.*, 2017, 21, 134–141.

Yoshikawa, M., M. Mikawa, K. Tanaka, A myoelectric interface forrobotic hand control using support vector machine, in: *IEEE/RSJ International Conference on Intelligent Robots and Systems*, IROS 2007, San Diego, CA, USA, October 29–November 2, 2007, pp. 2723–2728, http://dx.doi.org/10.1109/IROS.2007.4399301

Young, J. Aaron, L.J. Hargrove, T.A. Kuiken, Improving myoelectric pattern recognition robustness to electrode shift by changing interelectrode distance and electrode configuration, *IEEE Trans. Biomed. Eng.*, 2012, 59(3), 645–652, http://dx.doi.org/10.1109/TBME.2011.2177662.

Young, A.J., L.J. Hargrove, T.A. Kuiken, The effects of electrode size and orientation on the sensitivity of myoelectric pattern recognition systems to electrode shift, *IEEE Trans. Biomed. Eng.*, 2011, 2537–2544, http://dx.doi.org/10.1109/TBME.2011.2159216.

Young, A., L. Smith, E. Rouse, L. Hargrove, Classification of simultaneous movements using surface EMG pattern recognition, *IEEE Trans. Biomed. Eng.*, 2013, 60(5), 1250–1258.

Zardoshti-Kermani, M., B.C. Wheeler, K. Badie, R.M. Hashemi, EMG feature evaluation for movement control of upper extremity prostheses. *IEEE Trans. Rehabilit. Eng.*, 1995, 3, 324–333.

Zhang, D., A. Xiong, X. Zhao, J. Han, PCA and LDA for EMG-based control of bionic mechanical hand, in: *International Conference on Information and Automation*, ICIA, Shenyang, China, June 6–8, 2012, pp. 960–965, http://dx.doi.org/10.1109/ICInfA.2012.6246955.

Zhang, Q., Z. Luo, Wavelet de-noising of electromyography, in: *International Conference on Mechatronics and Automation*, Luoyang, Henan, China, June 25–28, 2006, pp. 1553–1558, http://dx.doi.org/10.1109/ICMA.2006.257406.

Zhao, J., Z. Xie, L. Jiang, H. Cai, H. Liu, G. Hirzinger, A five-fingered underactuated prosthetic hand control scheme, in: *The first IEEE/RAS-EMBS international conference on biomedical robotics and biomechatronics*, BioRob, 2006, February 20–22, 2006, pp. 995–1000, http://dx.doi.org/10.1109/BIOROB.2006.1639221.

11.7   ABOUT AUTHORS

Munish Kumar

He is a research scholar in Mechanical Engineering at GJUST, Hisar (India). He is working on development of bionic hand for upper limb amputees at the bionics lab at the institute. He holds an M.Tech degree in mechanical engineering. His research interests include robotics, automation, electromyography, and bionic devices.

Email: munish.nc@gmail.com

Dr. Pankaj Khatak

He holds a doctorate degree in mechanical engineering from GJUST, Hisar (India) and an M.Tech degree from IIT, Delhi (India). He is currently a professor at the Department of Mechanical Engineering at GJUST, Hisar (India) and in charge of CAD/CNC and the bionics lab at the institute. His research interests include computer numerical control, robotics and bionic systems, and rehabilitation technologies.

Email: pankajkhatak@gmail.com

12 Application of IoT in the Food Processing Industry

Himanshi Garg, Vasudha Sharma and Soumya Ranjan Purohit

CONTENTS

12.1   INTRODUCTION

Most people enjoy chocolates, but as consumer awareness regarding palm oil and the complexities its use and production offer to the environment grows, the demand for food traceability increases. Even though sustainability has been a global approach since 2015, the mismatch in demand and supply has generated heaps of waste. Moreover, consumers' relationship with food is changing—they want to know what they are

DOI: 10.1201/9781003245469-12

buying, how it has been prepared, and from where it originated—for good reasons. According to the World Health Organization, contaminated food sickens one out of every ten people annually, resulting in 42k deaths. These sobering statistics put pressure on manufacturers to provide full transparency of food to their consumers from farm to plate, which undoubtedly necessitates the use of leading technologies. Furthermore, according to the *Harvard Business Review*, the global population has quadrupled since 1915, estimated to be around 1.8 billion. However, the United Nations predicts it will reach 9.7 billion by 2050, upsurging the food demand from 58% to 98% (Elferneek & Schierhorn, 2016), putting pressure on the agriculture industry to boost crop production and yield with limited land–water supply in a sustainable manner.

Food recalls are detrimental to consumer health and harm producers' economies. In 2017, the USDA issued 131 recalls resulting in the waste of over 20 million pounds of food. One of the prominent areas that requires improvement is the cold chain and temperature monitoring to maintain food hygiene and a lower decomposition rate. Moreover, manufacturers, distributors, and restaurant operators must transition from paper-based to digital records.

Another primary concern in the food industry is food fraud. "Food fraud is characterized as the purposeful alteration, deception, mislabelling, swapping, or tampering with any food product at any point along with the supply-chain." Fraud can occur in ingredients used, the final product, and the product's packaging. Annually food fraud incidents are estimated to cost between 10 billion and 15 billion dollars (Johnson, 2014). The rising cargo theft and improper supply chain are the reasons for food fraud. Moreover, the anonymity of how food is produced and where it came from has eased suppliers' food fraud. Hence, there is a need for a digital, real-time monitoring and tracking system from farm to store to prevent such fraud, as it will create a digital footprint that can help trace the fraudster.

However, to increase economic benefits while maintaining transparency, the food industry needs to be efficient. In this regard, the food industry must adopt environmentally friendly, sustainable or long term, high-tech production methods. In addition, there are growing consumer demands for healthy, customized, and fresh food. Hence, production techniques and trends must evolve. So, processing industries are rapidly transitioning from mechanized (Industry 1.0) to digital (Industry 3.0) and now to the Internet of Things (Industry 4.0).

12.1.1   WHAT IS INDUSTRY 4.0?

The internet is now a ubiquitous part of the world, with a significant impact on people's lives. In addition, IoT and other advanced technologies are helping food industries from farm to plate. But before proceeding with IoT and its application, let's explain the term "Industry 4.0." As Zhou and his colleagues explain, "the industry 4.0 concept is the integration of information and communications technologies with industrial technology" (Ben-Daya et al., 2019; Zhou et al., 2016). It is also called "the digitalization of manufacturing, produce, process, factory, and product." More specifically, it signifies the rise of smart factories. New manufacturing technologies empower machines and machine-level information handling and processing, enabling manufacturers to adopt new production prerequisites instantly.

Additionally, linking data frameworks and sharing throughout the supply chain expands productivity. Moreover, it prospects efficient energy consumption with real-time data and process optimization, which can be mined from the heaps of information generated. It is an outlook of lean and green simultaneously.

12.1.2 WHAT IS IoT?

Kevin Ashton coined the term "Internet of Things" (IoT) to depict an innovation worldwide that envisions a tremendous link of devices and machines (Misra et al., 2020). IoT is a network of interconnected devices, appliances, or people that can communicate and exchange data via the internet between devices without the need for humans (Lokers et al., 2016). Standardized data classification, communication protocols, and storage capacity are required for efficient deployment of IoT devices. However, IoT is also mentioned as the "Internet of Everything or the Industrial Internet." Therefore, IoT utilization in the industry is often stated as "Industrial Internet of Things (IIoT)." A supply chain manager defines IoT as "a network of physical objects that are digitally connected to sense, monitor, and interact within a company and between the company and its supply chain enabling agility, visibility, tracking and information sharing to facilitate timely planning, control and coordination of the supply chain processes" (Ben-Daya et al., 2019).

Slowly but steadily the food processing industry is becoming acquainted with this. If statistical data is taken into account, IoT usage is expected to increase from around a billion machine-to-machine connections in 2017 to approximately 4 billion connections by 2022 (S.J., 2019). As a result, IoT's market scenario in the food and beverage industry is expected to grow from 4.08 billion USD in 2017 to 8.43 billion USD in 2025 with a compound annual growth rate (CAGR) of 9.5 percent (Kayalvizhi & Sivagnanaprabhu, 2020).

IoT has been expected to connect processes, data, things, and people to provide more value and relevance. As a result, countless applications, food providers, processors, and retailers see promising functional and financial augmentation opportunities. IoT network has reduced waste, risk, and production costs in the food supply chain. Likewise, it advances traceability and transparency throughout the supply chain, ensuring better food handling. IoT is credited with enabling Industry 4.0, which resulted in the fourth industrial revolution. Intelligent products, services, and machines such as quality-controlled coordination and support could all be considered "things" in Industry 4.0 (Ben-Daya et al., 2019). As a result, this chapter will outline the application of IoT-based technologies and other emerging trends that have recently yielded promising outcomes in reinforcing food processing and safety frameworks or that can answer numerous forthcoming food-related concerns in the future. Finally, the chapter will discuss the opportunities and challenges associated with full-fledged IoT implementation and industry 4.0 acceptance in the food processing sector.

12.2 ARCHITECTURE OF IOT

In simple terms Internet of things (IoT) is defined as the network where technology communicates with the technology via the internet and without human interventions. As previously stated, IoT is also called IIoT. Thus, a typical framework or architecture as described by a research group (Bouzembrak et al., 2019) is depicted in Figure 12.1, which includes several layers:

a) *The device layer* encompasses all devices and communication channels implemented in the environment, such as temperature-, location-, light-, movement-, etc. based sensors that exchange information directly or through gateways like receptors or transmitters over the communication network. It may also involve energy supply devices (batteries, solar panels) and devices that manage functionalities and gateways. This layer also includes all relevant wired and wireless communication technologies such as CAN bus, Wi-Fi, Bluetooth, Zigbee, and so on (Bouzembrak et al., 2019). It boosts efficiency, encourages better decision-making, and generates competitive advantages regardless of industry size.

b) *The network layer* includes the features required for device data and related protocol conversion to network-layer protocols. In addition, it provides network functionalities like authentication, connectivity, authorization, mobility, and accounting, and it is referred to as the transport layer in the open systems interconnection (OSI) protocol (Bouzembrak et al., 2019).

c) The *service support layer* presents the features that enable IoT applications and services to function (Bouzembrak et al., 2019). It includes features such as information handling and processing using machine learning and algorithm techniques to produce noteworthy perceptions and particular functionalities depending on the application because the requirement of emerging services varies.

d) IoT applications and services are included in the *application layer*.

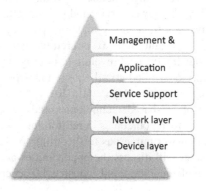

FIGURE 12.1 Architecture of IoT.

e) *Management and security* refer to the characteristic arrangement, geography, asset performance, accountability, and record management and security.

As a result, it was noted that IoT is a method of producing and sending a lot of data that contains valuable information. In the next section, key IoT technologies will be discussed, which will enable a better understanding.

12.3 IOT TECHNOLOGIES

According to various research groups, numerous IoT technologies will benefit the food processing industry. This section will discuss the critical IoT techniques that are helping the industry or are under research.

12.3.1 RADIO FREQUENCY IDENTIFICATION (RFID)

It enables the identification, chasing, and transmission of data throughout the supply chain. Figure 12.2 shows inventory management using RFID. It has evolved, but the basic categories are as follows:

a) Read/write passive memory tags.
b) Passive tags with enhanced security features.
c) Battery and sensor-powered semi-passive tags.
d) Battery-powered active tags that canlink with other labels of the same type.
e) Labels that can activate other tags and connect to logistics directly (Ben-Daya et al., 2019; López et al., 2011).

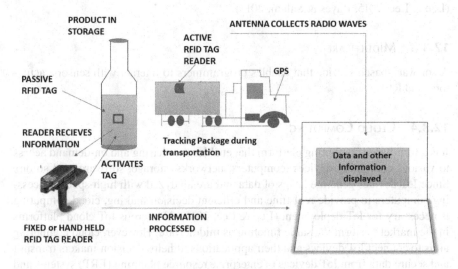

FIGURE 12.2 Inventory and logistics management using RFID tags.

Today, products can be linked to RFID tags combined with IoT-based sensors and a corresponding machine learning model to increase the monitoring efficiency of food safety systems. A group of researchers from Korea, Indonesia, and Pakistan proposed such a model in 2020. A machine learning model was integrated with RFID gates through which tagged perishable products are moved in and out (Alfian et al., 2020). This proposed system has proven to help gather and manage real-time product information and record absolute temperature and humidity histories. Fixed RFID readers installed at the manufacturer's and supplier's gates in the cold storages will function by automatically identifying products that move in and out of the facility. Transporters can use handheld RFID readers. IoT-based sensors in this area and at the end of suppliers and transporters can collect critical environmental conditions (humidity, temperature, etc.). The on-site host computer's integrated tag direction module will make it easier to keep track of the direction in which the product moves. This module is built on machine learning concepts and will distinguish between products moving in and out of the facility. It is simple to distinguish between received and shipped products in a system like this. A server database can be updated with information collected by RFID tags and IoT sensors. This setup has been extensively tested for the kimchi supply chain to determine if it can carefully and appreciably monitor the product's freshness, quality, and acceptability. The system yielded promising results throughout the supply chain. In COVID times, such a setup can increase consumer acceptance and assurance greatly.

12.3.2 WIRELESS SENSOR NETWORKS (WSN)

These are the sensors-based networks that screen and track the status of different devices like location, displacement, temperature, etc. Aside from that, it can collaborate and communicate with RFID tags to perform various other functions like pressure-flow levels, pollution levels, imaging, propinquity, moisture, and swiftness (Lee & Lee, 2015; Rayes & Salam, 2017).

12.3.3 MIDDLEWARE

A software-based service that enables programmers to interact with sensors, actuators, and RFID tags.

12.3.4 CLOUD COMPUTING

It is a web-based computing platform that enables the sharing and on-demand access to various computing devices (computers, networks, storage, software, and so on). Since IoT devices generate heaps of data and are analyzed with high-speed processing computers to provide real-time and efficient decision-making, cloud computing is necessary for IoT deployment (Lee & Lee, 2015). Numerous IoT cloud platforms in the market perform the same function as middleware. However, cloud computing aims to connect IoT devices and their applications. It helps decision-makers transmit and secure data from IoT devices to enterprise resource planning (ERP) systems and business intelligence by providing real-time information.

It is a cost-effective option for owning and managing data centers (Bonomi et al., 2014). However, businesses may require local, on-premise storage, computing, and communication capabilities due to latency-sensitive applications. (Bonomi et al. 2014). It is a cost-effective option for owning and managing data centers (Bonomi et al., 2014).

Fog computing can be considered a great answer to all the cloud computing problems and versatile supports required by IoT arrangements (Ben-Daya et al., 2019). Therefore, fog computing can be defined as a hybrid of local and cloud-computing services that comprises "a highly virtualized platform that provides compute, storage, and networking services between end devices and traditional cloud computing data centers" (Bonomi et al., 2012).

12.3.5 BLOCKCHAIN

One of the best ways to provide product visibility throughout the supply chain is to use Blockchain. Figure 12.3 shows tracing of food with help of block chain technology. Blockchain-based ledgers can be used to keep a digital record of the history of transactions in a food production network. Such a digital record cannot be layered and contributes significantly to food transparency. Furthermore, this efficient record keeping is one of the most promising ways to identify and inform actions that can be taken. There are many unique features in the functioning of blockchain networks that make this technology stand out in terms of promoting greater trust and transparency in food safety. According to Frank Yiannas, these distinguishing elements of Blockchain technology include:

a) Decentralized (multiple locations for same data combats the risk of single-point network failure)
b) Immutable (due to encryption data that cannot be altered without detection)
c) Consensus (data requires consensus from each element of the supply chain; hence one part cannot be dominant over others)
d) Democratic (a diverse group of stakeholders can have an equal say in data ownership, sharing, and protection) (Yiannas, 2018).

Food supply chains can be studied more efficiently in the event of an outbreak of a problem due to these specific outfitting features, and information on future prevention efforts can be obtained.

Beyond monitoring supply chains and preserving transparency, it has a wide range of applications like preventing food frauds and counterfeit. Globally there has been a significant gap between demand and production of food items due to COVID-19. The international scenario, which has swiftly shifted the focus from choice to availability, has consequently disrupted production and transportation. Under such circumstances, committing food fraud is easier than ever. However, to check on such fraudulent activities, digital footprints can be of great help. The real-time ability to trace and track these footprints from farm to store can help monitor food suppliers, which otherwise go unnoticed due to anonymity.

Similarly, producing safer food is critical because frequent product recalls and batch discards are costly for manufacturers with limited production. Hence, reliable and

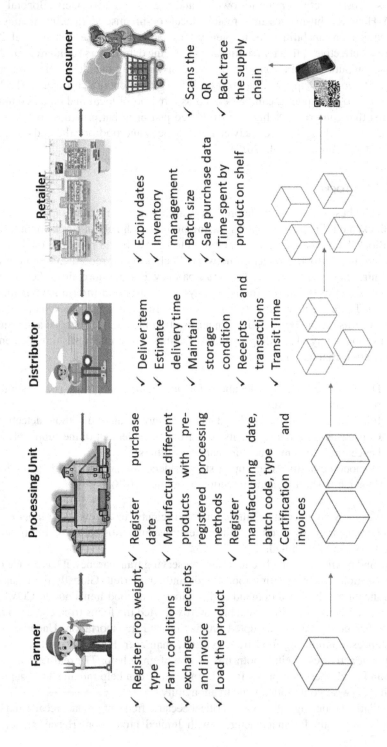

FIGURE 12.3 Food traceability using blockchain technology.

sufficiently analyzed data is required to streamline product recalls and avoid wastage of safe and unaffected products. In addition to this, post-purchase consumer wastages are on the rise as people have begun doubting the safety of everything around them, which for obvious reasons also includes food items. Post-purchase discernment could be curbed by sending out more explicit and more assuring messages about a product's safety. Blockchain technology can boost food transparency and can help to improve accountability and security while also lowering the risk of food fraud and wasteful product recalls.

12.4 APPLICATIONS OF IOT

After understanding the Internet of Things and its basic architecture, the benefits of using it in the food industry could probably be imagined. No doubt, the introduction of new techniques is not easily adopted. However, IoT is quickly becoming a necessity. It can find unimaginable applications in the food and beverage industry. According to the European Commission Information Society (Bouzembrak et al., 2019), IoT integration in the food processing industry is an upcoming revolution. IoT will bring precision farming, storage, production, preparation, distribution, detectability, consumption, permeability, and controllability challenges. Soon, the new technologies based on IoT, on the other hand, are expected to bring more safe and effective, sustainable, and affordable food to plates. The use of information technologies simplifies the process of monitoring and real-time visualization of how food products are handled and managed. It also makes the procedure more transparent and the stakeholders better prepared to deal with food emergencies and crises. As a result, relying on intelligent systems, either by using IoT technologies to fabricate food supply chain tracking and tracing or by using IoT to monitor existing ones better, is the modern answer to keeping food safe.

Furthermore, today's customers care about food characteristics, origin, freshness, microbiological, and chemical elements of product safety. It's challenging to establish a quality standard for all of these requirements. In this section, various applications of IoT in the food processing industry will be mentioned.

12.4.1 Inventory Management, Sustainability, and Wastage Reduction

One of the critical purposes of the food and beverage sector is to provide excellent and safe food to consumers, which is subject to how a food product is preserved before arriving at the buyer's plate. One of the most critical considerations in keeping perishable foods like meat or dairy is maintaining proper temperature and humidity levels. IoT can be employed for continuous monitoring and functioning of equipment. Further, food quality is hampered if food storage conditions are compromised, resulting in significant losses owing to spoiled or expired inventory. It can also aid in demand tracking and ordering raw materials as needed, resulting in better production efficiency.

12.4.2 Food Safety

Food safety is becoming more critical in the food and beverage (F&B) industry. In this context, IoT-empowered solutions help F&B businesses be flexible while delivering the best products and adhering to food safety management laws.

Temperature sensors and humidity sensors, for example, are utilized to screen the vital production state, freight time, and container temperature to ensure that the proper environment is provided to food products. Moreover, to ensure active cold chain management, food manufacturers can intently monitor food safety data points using real-time temperature tracking sensors. When data from sensors indicate potential concerns or violations, this prompts employees to conduct and verify mandated food safety inspections.

Additionally, in food packaging right now, tagged sensors and linking of supply-chain aids investigators in the following and tracing process, saving time and exertion.

Besides, Automated Hazard Analysis and Critical Control Points (HACCP) specifications are used in the food processing industry, enabling production or quality managers to access reliable information that helps manufacturing units implement food safety arrangements.

12.4.3 LOGISTICS

IoT technology can leverage the F&B industry in its operations much like any other logistics department. These technologies can assist F&B teams in following stocks in realtime, in addition automating shipments depending on solicitations or forecasts for a refill. It would be better to understand "Kagome," a tomato processing industry, as an example. Kagome achieved previously unthinkable levels of transparency across the whole processing cycle by using RFID tags and GPS technology, among other technologies, to collect data. This also allowed Kagome to mitigate threats to the tomato supply. Businesses can benefit from such a strategy in various ways, including immediately identifying the source of contamination in the supply chain, determining who is responsible, and preventing repeat outbreaks.

12.4.4 TRANSPARENT SUPPLY CHAINS

Traceability and transparency have been implemented in the global supply chain, which helps to increase customer loyalty and develop confidence between businesses and customers. Both customers and companies can track products as a result of this. The Internet of Things (IoT) can make this process more advantageous and productive.

Consequently, the safety for food (S4F) initiative has been adopted by many firms. Let's take an example for better understanding: A customer goes to a store and chooses a chips packet; he may scan the QR code on the package to learn how the chips were manufactured with manufacturing date, the ingredients used, and how it came to store racks. All this will assist customers in making better-educated purchasing decisions and establishing confidence among purchasers and manufacturers.

12.4.5 SMART OPERATIONS

Manufacturing's buzzword is operational efficiency. Every food processing industry aspires to attain it by coordinating workflows and diminishing manual methodology, linking corporate frameworks, optimizing and adopting advanced processes for better decision making, growth, and profits. Let's understand this with examples of sensors notifying engineers about the wear and tear happening inside the machine. Also, the predictive software could alert the operation handler about the maintenance of the machinery before it fails. This will further eliminate the frequent inspection requirements.

Similarly, innovative vending machines have been introduced, which allow these machines to accept digital payments. The utilization of IoT in such systems has helped boost automatic vending and diminish functional expenses. These systems could recognize and recommend any issues raised in food or beverage, predict user behavior, send reminders for removing obsolete food, and send cautions to refill machines quickly and efficiently.

Sensors can significantly improve quality control, product tracking, workers' activities, and production by real-time analysis. For example, installed sensors persistently observe the color and specks in the flour production unit with quick action against any inaccuracies. Additionally, it helps in measuring moisture and protein content in the case of flour production for real-time process optimization.

12.4.6 ENERGY SAVINGS

Restaurant appliances, especially outdated ones, account for the significantly higher electricity bill, making it challenging for restaurant proprietors to monitor or reduce related expenditures. Additionally, it jumbled up the maintenance schedule because of untimely breakdowns.

In this case, energy utilization can be perceived and normalized for a piece of equipment by following energy use at the production level; this permeability can limit extensive energy utilization by fixing equipment with higher power utilization. However, now, Energy Management Solution (EMS) has started forming a part of QSRs that reliably and intelligently collect, analyze, and actuate the restaurant's energy consumption. In fact, according to American West Restaurant Group, "Pizza hut expects an 18% reduction in average monthly energy usage and save $2million in energy costs between 2018 to 2022, all because of IoT" (Insights, 2018).

Moreover, restaurants can rely on intelligent appliances to regulate energy consumption. For example, intelligent fryers can define the oil temperature and alert the cook when the meal is done correctly. Also, they can notify managers to change or filter the oil, ensuring better food hygiene.

Additionally, integrating smart fridges can collect data for on-demand sale and analysis, which could help kitchen managers prepare a checklist of what to buy in what quantity, therefore reducing food waste and minimizing energy loss from the thermal transfer. The restaurant staff may analyze the cause-and-effect scenario by obtaining real-time information from various departments and areas, which will help gain prominence in energy consumption and cost analysis.

The previously mentioned are some of the application areas of IoT In the food processing industry. Moreover, we are aware of E-Nose application in the food control system, which mimics the human nose and gives precise sensory results quickly and cost-effectively. However, recent approaches architectures for IoT in the food processing industry have been developed or are being studied for enhanced performance. Table 12.1 shows IOT-based technologies in the food industry, and Table 12.2

TABLE 12.1

IoT Based Technologies Paving the Way in the Food Industry

Technology Developed	Material and Software Used	Specialties	Reference
Sensor-based smart food trays	Stainless steel Aluminum Arduino software	• Detects open and closed tray lid action • Ensures food hygiene • Prevents contamination • Reduces environmental exposure and human contact	(Rohani et al., 2020)
Sensor-based digital cold storage	DHT-11 sensor and ESP-8266 Node MCU module	• Relatively more robust than an Arduino board in terms of humidity and temperature monitoring, with an automated switch to turn on/off in the event of an emergency • It can function well on mobile phones or via PCs	(Mallik et al., 2018)
Smart meat product tracking	AI, IoT, and Blockchain	• Optimized supply chain provenance system • On the table, supply chain and origin details could be obtained	(Khan et al., 2020)
Pocket-sized immunosensor system	Wireless network + lens-free CMOS image sensors	• Prevents contamination • Observed on fish for *V. parahaemolyticus*	(Seo et al., 2016)
Real-time food traceability and transportation	Raspberry Pi (RPi) Based on the power supply	• Transparency in supply chain • Low cost and small size • Real-time access and automatic cargo identification	(Vujovic et al., 2015)
Food adulteration monitoring system	RPi with sensors and Zigbee	• Detect the presence of adulterants in the food product	(Gupta & Rakesh, 2018)
Traceable agriculture product (TAP) mechanism model	RFID	• Increased consumer trust owing to improved traceability and transparency • Improved fresh produce management • Reduced inventory loss	(Chen et al., 2021)
Automated fruit quality checking	Arduino based Microchip + Probe	• Economical and efficient apple ripening index was checked	(Ray et al., 2016)
Smart log	Deep learning	• Automatically monitors the nutrition with 98.6% accuracy.	(Sundaravadivel et al., 2018)
Time-temperature indicator	Glucose-based biosensor	• Intelligent food packaging system with integrated TTIs, smart radio frequency identification (smart-RFID), and electrical signal.	(MijanurRahman et al., 2018)

TABLE 12.2

IoT-Based Various Food Traceability Models

Technology Name	Technology Involved	Applications	References
Tilapia Products Traceability System	RFID +ONS (Object Name Service) and EPCIS (EPC Information Service) server	Aqua Supply chain tracking, traceability, and recall.	(Yan et al., 2012)
Anti-counterfeiting Model	RFID and NFC (near field communication) with Zigbee	Improves bi-directional pork traceability.	(Zou, 2016)
Milk Fraud Prevention	RFID based gas, temperature, viscosity, and salinity sensors	Monitor milk spoilage.	(Rajakumar et al., 2018)
Smart Wine	Cloud-based system	Monitors sustainable wine production.	(Smiljkovikj & Gavrilovska, 2014)
Prepackaged Food Supply chain traceability	Integrated RFID and QRcode with extensible markup language (XML)	Economical and effective tracking and tracing of food. Efficient data sharing throughout the stakeholders.	(Li et al., 2017)
FiSpace	Cloud based	Production, shipment, logistics, cargo search, and market activities are all virtualized and optimized in realtime.	(Verdouw et al., 2016)
Kanban System	-	Reverse logistics Waste Management by signaling time and quantity	(Thürer et al., 2016)
Blackberry Supply chain traceability	Fog computing	Real-time quality monitoring of blackberries using email and SMS services.	(Musa & Vidyasankar, 2017)
Gastrograph AI by Analytical Flavour Systems	Genetic algorithm using AI	Predicts consumer taste and flavor profile. It saves time and cost.	(Shah et al., 2017)
Drones	A.I. + Sensors	Monitor land, soil-crop performance, and weather conditions. Help in improvising management strategy at a farm.	(Farooq et al., 2019; Navarro et al., 2020)
Bacterial Spoilage indicator	Machine learning +AI	Post-pasteurization milk safety from contamination.	(Murphy et al., 2021)

shows food traceability models. Furthermore, since IoT has been a term in use for the last few years, there are some market players. Following the tables, the text discusses the market players and recent trends in IoT application in the food processing industry.

However, machine learning correlation models were trained with six months of historical data on 29 different processing variables, including the amount of starter culture, mixing times, and raw milk composition, in another case study from a US-based cheese manufacturer to classify impacts on the final moisture content. The manufacturer was able to boost average moisture content by up to 0.6 percent within regulatory compliance limitations as a consequence of the optimization research, saving more than $1 million in one year.

The aforementioned are the various researches conducted on IoT techniques with their application. But, before mentioning the market players, let's introduce to some of the latest projects related to food safety management:

a) **Musetech** project is a multi-sensor device where photoacoustic spectroscopy, quasi-imaging spectrometry, and temperature sensors are responsible for real-time monitoring of food quality and safety from chemicals.
b) **IQ-FRESHLABEL** is another project intended to create a temperature and oxygen-sensitive smart label for frozen and packaged food.
c) The **rabbits** aim at integrating IoT with enterprises system by developing such technologies and processes. This will further help in interoperable businesses' accountability and smoothness.
d) Human eyes could not achieve the same accuracy, reliability, consistency, and quantitative data as **image processing techniques** when it comes to grading the fruit size. Hence, **the histogram of oriented gradients (HOG)** is intended to evaluate fruit quality in image processing on the basis of looks and shape, which is an intensity distribution mechanism connected with gradients.

Henceforth the companies that play a crucial role in connecting the supply chain effectively are discussed:

a) **Elitech** has introduced the RCW-360 series, which combines IoT technology. "RCW-360 is a 2G/4G/WIFI-based IoT temperature and humidity data trimmer. It is a perfect choice for a huge number of monitoring applications." "Its function is to alert about any change in temperature and humidity, provide real-time solutions and data preservations." It consumes very low power and also has a built-in rechargeable battery, which makes it work for a longer duration even after being turned off (Kayalvizhi & Sivagnanaprabhu, 2020).
b) **STELLAPPS:** IoT-based dairy firm; SMARTMOO IoT router controller gathers data from sensors built into the milking equipment, animal wearables, and milk cooling accessories, all of which are linked to automated milking controls. This technique's ultimate goal is to create Smart Farms in order to accomplish expected cattle value increases, effective productivity

management tactics, and fundamental healthcare explanations for improved performance levels.

c) **TAG BOX:** It is a solution for the cold chain supply business. The main idea behind it is to reduce cold storage wastage quantitatively while maintaining the quality of the higher cold chain, end-to-end traceability of the stock retaining units, and eliminating existing general limitations such as delays and breakdowns.

d) **Wal-Mart** is a multinational operator of hypermarkets that prefers warehouses for product storage. Various other major sellers and distributors, like Wal-Mart, prefer warehouses for product storage. Any increase in the demand or supply of the food products would compel these major distributors to stock their warehouses with those specified food products in order to meet the demand. As mentioned earlier, monitoring the movement of each and every product appears to be difficult in realtime, which clearly proves to be a significant challenge. Moreover, the massive sizes of the warehouses make it difficult to keep track of an inventory. Companies today rely on the benefits of pressure-sensitive sensors for efficient stock monitoring in order to improve inventory management. The sensor's function is to alert the inventory manager when stock runs low. Furthermore, in 2020, Walmart introduced robots in approximately 1,000 stores that checked inventory and price accuracy. Companies can dwell on integrating artificial intelligence and IoT to understand consumers in better ways in terms of purchasing style.

e) **Siemens** uses a variety of IoT components and digitalization tools for various processes, including brewing, sugar production, dairy processing, baking, and agricultural production. Another example is a multi-sensor system that assesses fouling in chocolate spread using ultraviolet fluorescence imaging and acoustic sensors as part of the decision-making process for clean-in-place (CIP) systems.

f) **Tetra Pak** uses a variety of sensor networks and data analytics tools to improve business productivity, manufacturing sustainability, and the human centricity of digital transformation in its own 53 production factories, as well as on the processing and filling equipment of its customers around the world. Tetra Pak machines are outfitted with sensors, which are then linked to the Microsoft Azure IoT Hub and displayed in the Power B.I. Data visualization platform to improve performance and drive continuous improvement activities within the total predictive maintenance framework. As part of the predictive maintenance services, routine action alerts based on real-time data and models obtained from manufacturing assets are monitored by a global quality and performance management center and automatically sent to field-based staff and consumer maintenance personnel.

g) Cleaning costs, water, and downtime were reduced by using real-time ultrasonic sensors and imaging data analyzed by AI algorithms to monitor residue levels and optimize cleaning schedules. A team from **Martec of Whitwell**, in collaboration with universities working on sustainability and recycling, created the aforementioned self-optimizing CIP (Simeone et al., 2016).

h) Food safety, as an important component of food processing, will also be a major focus of AI-based models. **Kan Kan by Remark Holdings** created an AI solution equipped with face and object recognition technology to track operator movement and food safety compliance practices.

i) Besides companies, startups have also started contributing to this area. One such example is the company **Rachio**, which has added intelligence to water usage. This is a mobile app-based system. The company's product provides solutions like calibrating slope, soil quality and sunlight exposure, vegetation, and types of nozzles to customers, which enables customers' visibility and good control over cultivational land.

j) **Lumens** is another sensor-based platform for chemical and biological monitoring that aims at providing safe and better food, air, and water. This system monitors and automatically maintains liquid and gas requirements in realtime.

k) Grab and go is a convenience offered by fast-food chains, and the COVID-19 pandemic has changed the consumer approach to quick serving restaurants (QSR). An AI is currently taking drive-thru orders in a Chicago suburb. The idea is that AI will handle orders, freeing up human employees to focus on accuracy and food quality.

Finally, if applications of IoT are tried, then IoT integration in the industry with other emerging techniques will help the industry in bringing transparency throughout the supply chain with less burden on workers, more consumer trust, less paperwork, better traceability, and improved confidence or assurance among manufacturers and customers.

12.5 CHALLENGES AND OPPORTUNITIES IN IOT IMPLEMENTATION

Despite the success stories and the opportunities, the food industry's digitalization has been slow. According to a 2017 survey, food suppliers from different countries reported vulnerability against potential attacks in IoT infrastructure (21 percent), inability to analyze generated data (16 percent), data privacy concerns (18 percent), lack of standards for IoT arrangements (15 percent), and regulatory reporting changes driven by the technology (13 percent) as major reasons for not adopting new technologies and processes (Boz, 2021; Rentokil, 2017).

Furthermore, the right IoT system must take into account factors such as operating costs, power, internet connectivity, data analysis, and a count of connected devices. Consequently, incompatibility of equipment from different vendors, digitalization costs, old equipment that cannot be manually retrofitted, and a lack of expertise could all be some other challenges.

Additionally, data-driven models require a vast amount of data training and validation. The data requirements for machine learning and AI models involve multiscale, multimodal, openness, and standardization considerations, which will require trained professionals who understand food industry dynamics as well as technological advancements. Moreover, the exclusion of the human factor in digitalization efforts can have negative consequences for employees, businesses, and society (Neumanna et al., 2021). For example, the jams or food pastes manufacturing industry, which involves boilers, evaporators, etc. to achieve a targeted paste consistency to ensure the product quality while processing (Zhang et al., 2014) are detrimental to key performance indicators (KPIs) at the plant and product requirements. IoT aims at real-time data collection from sensors like refractometers (in case of jam and paste forming industry and other sensors) and analyzing it to provide actuation to the machinery (like evaporator or boilers in paste processing). Likewise, it can help in achieving targeted paste consistency. However, the quality of experience, in this case, could be if the target set is achieved or not. Hence, quality of experience (QoE) is a necessity here for system designers and developers since it gives them an approach to evaluate the results provided by the application they are creating (Fizza, Banerjee, Mitra, Prakash Jayaraman, et al., 2021). Typically, it is a subjective analysis involving human feedback. But it could be integrated with machines in order to achieve the target set by KPIs (Fizza, Banerjee, Mitra, Jayaraman, et al., 2021). However, this presents a significant challenge in evaluating how individual components impacted the final quality of the product demanding an objective approach in the future.

Considering opportunities, human-centric and AI-enabled tools in Industry 4.0 can create a more balanced structure of human-machine interfaces because this vast store of experience captured in datasets will increase in power as more data is accumulated and made available and as machine learning and artificial intelligence mine and pinpoint relevant information and solutions.

Additionally, collaborative teams of humans and smart robots will reduce mental and physical stress on workers, lower manufacturing costs, increase product quality, and allow manufacturers to quickly respond to changing customer demands. By eliminating the need for several special-purpose tools, advanced robotic systems that are flexible and perform multiple tasks will reduce capital investment and increase manufacturing agility. Robot-based production systems can also enable mass customization through efficient batch-of-one production.

Through object recognition and computer vision, artificial intelligence (AI) can enable further optimization, automation, and rapid decision-making without necessarily shedding light on the underlying complex physical mechanisms. Although the jury is still out on the benefits and drawbacks of black-box models versus mechanistic or combination approaches, AI can enable rapid and real-time decision-making by replacing mundane or difficult tasks in processing operations.

12.6 CONCLUSION AND FUTURE REMARKS

IoT is rapidly replacing the mechanized world. However, the food industry is growing in this field at a slow pace. But with the changing consumer taste, the industry needs to advance. Hence, IoT found a privileged application in the food processing industry where machine talks to machine without human intervention. Moreover, IoT is a solution to many other manufacturer problems like logistics, food wastage, inventory problems, traceability, food recalls, and tracking. Most importantly, transparency throughout the chain could be easily achieved using IoT integration in the food industry. It could be predicted from recent applications and research in this field that sensor-based technology is advancing to AI and robotics. The best example of it has been presented by Walmart, the biggest supply chain in the USA. Also, a Chicago-based fast-food chain involved A.I. to take drive-through orders.

This technique came with associated challenges along with a complete transformation of industry like adding cost, requiring objective analysis for quality of experience studies, and data loss, but it is the best solution that provides data actuation with most prominent results ensuring manufacturer and consumer confidence. Even the COVID-19 pandemic has enhanced the value of this.

The chapter concludes that, in the future, machine learning, fuzzy cognitive studies, artificial intelligence, and robotics integration and collaboration of human expertise will bring a paradigm shift into society that could be called "industry 5.0."

REFERENCES

Alfian, G., Syafrudin, M., Farooq, U., Ma'arif, M. R., Syaekhoni, M. A., Fitriyani, N. L., Lee, J., & Rhee, J. (2020). Improving the efficiency of RFID-based traceability system for perishable food by utilizing IoT sensors and machine learning model. *Food Control, 110*, 107016. https://doi.org/10.1016/J.FOODCONT.2019.107016

Ben-Daya, M., Hassini, E., & Bahroun, Z. (2019). Internet of things and supply chain management: A literature review. *International Journal of Production Research, 57*(15–16), 4719–4742. https://doi.org/10.1080/00207543.2017.1402140

Bonomi, F., Milito, R., Natarajan, P., & Zhu, J. (2014). Fog computing: A platform for the internet of things and analytics. *Studies in Computational Intelligence, 546*, 169–186. https://doi.org/10.1007/978-3-319-05029-4_7

Bonomi, F., Milito, R., Zhu, J., & Addepalli, S. (2012). Fog computing and its role in the internet of things. *Proceedings of the First Edition of the MCC Workshop on Mobile Cloud Computing—MCC'12*. https://doi.org/10.1145/2342509

Bouzembrak, Y., Klüche, M., Gavai, A., & Marvin, H. J. P. (2019). Internet of things in food safety: Literature review and bibliometric analysis. *Trends in Food Science & Technology, 94*, 54–64. https://doi.org/10.1016/J.TIFS.2019.11.002

Boz, Z. (2021, July). Moving food processing to industry 4.0 and beyond. *Food Technology Magazine*. www.ift.org/news-and-publications/food-technology-magazine/issues/2021/july/columns/processing-food-processing-industry

Chen, X., Chen, R., & Yang, C. (2021). Research and design of fresh agricultural product distribution service model and framework using IoT technology. *Journal of Ambient Intelligence and Humanized Computing 2021*, 1–17. https://doi.org/10.1007/S12652-021-03447-8

Elferneek, M., & Schierhorn, F. (2016). Global demand for food is rising: Can we meet it? *Harvard Business*.

Farooq, M. S., Riaz, S., Abid, A., Abid, K., & Naeem, M. A. (2019). A survey on the role of IoT in agriculture for the implementation of smart farming. *IEEE Access, 7*, 156237–156271. https://doi.org/10.1109/ACCESS.2019.2949703

Fizza, K., Banerjee, A., Mitra, K., Jayaraman, P. P., Ranjan, R., Patel, P., & Georgakopoulos, D. (2021). QoE in IoT: A vision, survey, and future directions. *Discover Internet of Things, 1*(1), 1–14. https://doi.org/10.1007/S43926-021-00006-7

Fizza, K., Banerjee, A., Mitra, K., Prakash Jayaraman, P., Ranjan, R., Patel, P., & Georgakopoulos, D. (2021). Discover internet of things QoE in IoT: Avision, survey and future directions. *Discover Internet of Things, 1*, 4. https://doi.org/10.1007/s43926-021-00006-7

Gupta, K., & Rakesh, N. (2018). IoT-based solution for food adulteration. *Smart Innovation, Systems, and Technologies*, *79*, 9–18. https://doi.org/10.1007/978-981-10-5828-8_2

Insights, E. E. (2018). Advanced IoT solutions expected to serve up millions in energy savings for pizza hut franchisee. *Eco-Energy Insights*. www.prnewswire.com/news-releases/advanced-iot-solutions-expected-to-serve-up-millions-in-energy-savings-for-pizza-hut-franchisee-300698566.html

Johnson, R. (2014). *Food fraud and economically motivated adulteration of food and food ingredients*. www.fredsakademiet.dk/ORDBOG/lord/food_fraud.pdf

Kayalvizhi, M., & Sivagnanaprabhu, K. K. (2020). *A Survey on Applications of IoT in Food Processing Industry*, International Journal of Advanced Science and Technology, *29*(6), 4177–4195.

Khan, P. W., Byun, Y.-C., & Park, N. (2020). IoT-blockchain enabled optimized provenance system for food industry 4.0 using advanced deep learning. *Sensor*. https://doi.org/10.3390/s20102990

Lee, I., & Lee, K. (2015). The internet of things (IoT): Applications, investments, and challenges for enterprises. *Business Horizons*, *58*(4), 431–440. https://doi.org/10.1016/J.BUSHOR.2015.03.008

Li, Z., Liu, G., Liu, L., Lai, X., & Xu, G. (2017). IoT-based tracking and tracing platform for the prepackaged food supply chain. *Industrial Management & Data Systems*, *117*(9), 1906–1916. https://doi.org/10.1108/IMDS-11-2016-0489

Lokers, R., Knapen, R., Janssen, S., van Randen, Y., & Jansen, J. (2016). Analysis of big data technologies for use in agro-environmental science. *Environmental Modelling and Software*, *84*, 494–504. https://doi.org/10.1016/j.envsoft.2016.07.017

López, T. S., Ranasinghe, D. C., Patkai, B., & McFarlane, D. (2011). Taxonomy, technology, and applications of smart objects. *Information Systems Frontiers*, *13*(2), 281–300. https://doi.org/10.1007/S10796-009-9218-4

Mallik, A., Karim, A. Bin, Md, Z. H., & Md, M. A. (2018). Monitoring food storage humidity and temperature data using IoT. *MOJ Food Processing & Technology*, *6*(4). https://doi.org/10.15406/mojfpt.2018.06.00194

MijanurRahman, A. T. M., Kim, D. H., Jang, H. D., Yang, J. H., & Seung, J. L. (2018). Preliminary study on biosensor-type time-temperature integrator for intelligent food packaging. *Sensors*, *18*(6), 1949. https://doi.org/https://doi.org/10.3390/s18061949

Misra, N. N., Dixit, Y., Al-Mallahi, A., Bhullar, M. S., Upadhyay, R., & Martynenko, A. (2020). IoT, big data, and artificial intelligence in agriculture and the food industry. *IEEE Internet of Things Journal*, *4662*(c), 1–1. https://doi.org/10.1109/jiot.2020.2998584

Murphy, S. I., Reichler, S. J., Martin, N. H., Boor, K. J., & Wiedmann, M. (2021). Machine learning and advanced statistical modeling can identify key quality management practices that affect post-pasteurization contamination of fluid mill factors affecting post-pasteurization contamination. *Journal of Food Protection*, *84*(9), 1496–1511. https://doi.org/https://doi.org/10.4315/JFP-20-431

Musa, Z., & Vidyasankar, K. (2017). A fog computing framework for blackberry supply chain management. *Procedia Computer Science*, *113*, 178–185. https://doi.org/10.1016/J.PROCS.2017.08.338

Navarro, E., Costa, N., & Pereira, A. (2020). A systematic review of IoT solutions for smart farming. *Sensors*, *20*(15), 4231. https://doi.org/10.3390/S20154231

Neumanna, W. P., Winkelhaus, S., Grosse, E. H., & Glock, C. H. (2021). Industry 4.0 and the human factor—a systems framework and analysis methodology for successful development. *International Journal of Production Economics*, *233*, 107992. https://doi.org/10.1016/j.ijpe.2020.107992

Rajakumar, G., Ananth Kumar, T., Samuel, T. S. A., & Kumaran, E. M. (2018). IoT based milk monitoring system for detection of milk adulteration. *International Journal of Pure and Applied Mathematics*, *118*(9), 21–32. www.ijpam.eu

Ray, P. P., Pradhan, S., Sharma, R. K., Rasaily, A., Swaraj, A., & Pradhan, A. (2016). IoT-based fruit quality measurement system. *Online International Conference on Green Engineering and Technologies (IC-GET)*, 1–5. https://doi.org/10.1109/GET.2016.7916620

Rayes, A., & Salam, S. (2017). The things in IoT: Sensors and actuators. *Internet of Things-From Hype to Reality*, 57–77. https://doi.org/10.1007/978-3-319-44860-2_3

Rentokil. (2017). *The impact of the internet of things from farm to fork*. www.ift.org/news-and-publications/food-technology-magazine/issues/2021/july/columns/processing-food-processing-industry

Rohani, M. Z., Nor, M., & Ab, H. (2020). Integration of automation technology in food industry: A case study. *January*, 471–475.

S.J.C. (2019). *Cisco visual networking index: Global mobile data traffic forecast update, 2017–2022*. SJC.

Seo, S. M., Kim, S. W., Jeon, J. W., Kim, J. H., Kim, H. S., Cho, J. H., Lee, W. H., & Paek, S. H. (2016). Food contamination monitoring via the internet of things is exemplified by using a pocket-sized immunosensor as a terminal unit. *Sensors and Actuators B: Chemical*, *233*, 148–156. https://doi.org/10.1016/J.SNB.2016.04.061

Shah, D., Ahn, R., & Cohen, J. (2017). *Predicting consumer preference from reviews of professional tasting panels on the gastrograph sensory system*. De Gruyter.

Simeone, A., Watson, N., Sterritt, I., & Woolley, E. (2016). A multi-sensor approach for fouling level assessment in clean-in-place processes. *Procedia CIRP*, *55*, 134–139. https://doi.org/10.1016/j.procir.2016.07.023

Smiljkovikj, K., & Gavrilovska, L. (2014). SmartWine: Intelligent end-to-end cloud-based monitoring system. *Wireless Personal Communications*, *78*(3), 1777–1788. https://doi.org/10.1007/S11277-014-1905-X

Sundaravadivel, P., Kesavan, K., Kesavan, L., Mohanty, S. P., & Kougianos, E. (2018). Smart-log: A deep-learning based automated nutrition monitoring system in the IoT. *EEE Transactions on Consumer Electronics*, *64*(3), 390–398. https://doi.org/10.1109/TCE.2018.2867802

Thürer, M., Pan, Y. H., Qu, T., Luo, H., Li, C. D., & Huang, G. Q. (2016). Internet of Things (IoT) driven kanban system for reverse logistics: Solid waste collection. *Journal of Intelligent Manufacturing*, *30*(7), 2621–2630. https://doi.org/10.1007/S10845-016-1278-Y

Verdouw, C. N., Wolfert, J., Beulens, A. J. M., & Rialland, A. (2016). Virtualization of food supply chains with the internet of things. *Journal of Food Engineering*, *176*, 128–136. https://doi.org/10.1016/J.JFOODENG.2015.11.009

Vujovic, V., Omanovic-Miklicanin, E., & Maksimović, M. (2015). *A low-cost internet of things solution for traceability and monitoring food safety during transportation competence-based software engineering curriculum development view project system of systems-the collaborative framework development view project a low*. www.researchgate.net/publication/285055479

Yan, B., Hu, D., & Shi, P. (2012). A traceable platform of aquatic foods supply chain based on RFID and EPC internet of things. *International Journal of R.F. Technologies*, *4*(1), 55–70. https://doi.org/10.3233/RFT-2012-0035

Yiannas, F. (2018). A new era of food transparency powered by blockchain. *Innovations: Technology, Governance, Globalization*, *12*(1–2), 46–56. https://doi.org/10.1162/INOV_A_00266

Zhang, L., Schultz, M. A., Cash, R., Barrett, D. M., & McCarthy, M. J. (2014). Determination of quality parameters of tomato paste using guided microwave spectroscopy. *Food Control*, *40*(1), 214–223. https://doi.org/10.1016/J.FOODCONT.2013.12.008

Zhou, K., Liu, T., & Zhou, L. (2016). *Industry 4.0: Towards future industrial opportunities and challenges*. 2015 12th International Conference on Fuzzy Systems and Knowledge Discovery, FSKD 2015, 2147–2152. https://doi.org/10.1109/FSKD.2015.7382284

Zou, X. (2016). Design and realization of pork anti-counterfeiting and traceability IoT system. *Acta Technica*, *61*(4B), 281–290. http://journal.it.cas.cz

13 Emerging Technologies for Healthcare during the COVID-19 Pandemic

Imran Aslan and Sam Goundar

CONTENTS

13.1 INTRODUCTION

The first cases of COVID-19 were reported at the end of 2019 in Wuhan/China, and the World Health Organization (WHO) announced COVID-19 caused by SARS-CoV-2 (Severe Acute Respiratory Syndrome Coronavirus 2) as a global pandemic in March 2020 (Smith et al., 2020). The COVID-19 virus is highly contagious and can spread by coughing or sneezing. Coronavirus is a respiratory virus showing fever, cough, and myalgia, diarrhea, and dyspnea. Respiratory distress, sepsis, and septic shock have been seen in critically ill patients requiring intubation and intensive care treatments. Other symptoms are hemoptysis, anorexia, weakness, fatigue, sputum removal, headache, vomiting, anosmia, normal or decreased leukocyte count, lymphopenia, abdominal pain, and ground-glass opacities (Kutlu, 2020; Smith et al., 2020; Sheikhzadeh et al., 2020). People are categorized into four groups during the COVID-19 pandemic: healthy without infection, sick from COVID-19, healthy after recovery or after (approximately) symptom-free infection, and dead. It is assumed that COVID-19 infections lead to symptoms in only about 20%, therefore only about half of the sick people are reported. The other half is categorized as a normal cold, or people don't go to a doctor for some reason, or no COVID-19 tests are carried out, so just 6% of them have been reported as being sick (Donsimoni et al., 2020). The time between the exposure and the onset of symptoms is estimated to be 5 days on average or can be 2–14 days. For potential exposures, 14 days' quarantine is suggested (Mofti & Vasudevan, 2020).

Increasing human migration, extreme weather events, food shortage, unemployment, etc. have affected the whole world and have impacts on human well-being as seen during that pandemic (Berchin & Guerra, 2019). Hence, low- and middle-income countries, weaker health systems, limited resources, and lower socioeconomic status can make the life of people challenging (Vieira et al., 2020). Elderly persons in nursing homes, indigenous populations, homeless people, populations of lower socioeconomic status, migrants, and people in prison are risk groups (Vieira et al., 2020). Food and nutritional deficiency were the main problems among the vulnerable poorest people due to loss of livelihood (Shammi et al., 2020). Furthermore, shifts in food habits have decreased people's immunity and also increased sources of infection (Razvi & Prasad, 2020). Increased psychological morbidities were seen in workers at the frontlines of the outbreak due to significant threat to life and extreme vulnerability and uncertainty. They have a higher rate of anxiety, depression, suicide, and burnout in normal life and are nervous due to uncooperative patients resisting adherence to safety instructions and fear of contracting the virus themselves. Personal protective equipment (PPE) difficulties such as not breathing well, limited accessibility to water and toilets, and PPE shortages are other reasons for the additional physical and mental fatigue (Rooney and & McNicholas, 2020).

Diagnosis, treatment, and prevention of diseases, injuries, and other physical and mental disorders are provided in healthcare (Pashazadeh & Navimipour, 2018). Governance, smart technology usage, privacy concerns, lockdown, activism, information sharing, and infodemic are the main constructs during the COVID-19 pandemic to support decision making and to take strategies against the pandemic. Identifying, isolating, and quarantining to reduce virus transmission are methods applied by all countries according to World Health Organization criteria (Kummitha, 2020). Limiting travel, lockdowns, closing education facilities, emergency investments in health care, increasing testing, and using data and technology to trace contacts are precautions taken against the COVID-19 pandemic around the world.

Cloud computing, big data, the Internet of Things (IoT), mobile internet, artificial intelligence (AI), blockchain, 5G, telehealth, sensors, robots, and other digital technologies have been used to develop monitoring and to control the system during the pandemic. Epidemic situation dynamics and prevention knowledge can be accessed by mobile phones. Material allocation and monitoring of personnel movement can be done with big data technologies. Intelligent diagnosis of medical imaging and temperature measurement technology based on computer vision and infrared technology can be supported with AI (Ye et al., 2020). Nanomaterial-based coatings for antibacterial applications, manufacturing of antiviral coated masks, and fumigation chambers that last five seconds allowing a quick disinfection of a person while visiting hospitals, universities or airports are some new improvements aimed at decreasing the spread of COVID-19 (Goel et al., 2020). Decreased waiting times; tracking patients, staff, and inventory; preventing drug allergies; and identifying and authenticating possible harmful mistakes to patients are the main advantages of new smart technologies (Nasrullah, 2020). Many in-person didactics and examinations have been cancelled or rescheduled due to COVID-19. However, some of them have been handled with telehealth (Shebrain et al., 2020).

Healthcare medical data and other real-world data categorization can be done for efficient sharing. To fight COVID-19, large-scale public data is generated and collaborated by many countries to make common decisions and strategies. A large number of COVID-19 patients' data can be used to make therapeutics, for drug repurposing, and for clinical studies so that epidemiological characteristics, host susceptibility and host immune responses, risk factors for severe disease, and routes of transmission can be determined. Open data (clinical and epidemiological data) can help many strategymakers to make decisions or make models. More learning can be done with the help of on-time and reliable data. During the 2016 Zika and Ebola outbreaks, open data coming from leading national agencies and science organizations has helped decision-makers to carry out plans and studies. Generating data platforms to be reused in a long time can make countries more aware for the next pandemics. Furthermore, open source tools, workflows, and computational resources can help in handling COVID-19 data. Thus, the severity of disease can be determined from these models, and COVID-19 research projects can be carried out. The origin of pandemics and preventing the risk in the future can be done with "one health approaches."

Preserving trust and respecting the privacy of patients and participants are important considerations in data management (Blomberg & Lauer, 2020).

This chapter aims to explain the usage of informatics and technologies through focusing on the COVID-19 pandemic. The most widely used technologies during the pandemic and their prominent applications are explained. Later, some successful smart technology countries are elucidated. Finally, the most important challenges of using smart technologies and some solutions are mentioned.

13.2   INFORMATICS AND TECHNOLOGIES IN HEALTHCARE MANAGEMENT

Innovative health technology for early detection and patient-centered treatment focus on medical equipment such as production of ventilators for intensive care units and protective innovations such as vaccines on health system preparedness for COVID-19 handling due to rising COVID-19 infections (Aminullah & Erman, 2021). Also, technologies during the pandemic have been used for prediction of epidemic trends, tracking of close contacts, and remote diagnosis (Ye et al., 2020). Categories for digital health interventions are emergency planning and preparedness; public health response as surveillance; risk assessment and response; risk communication; and non-pharmaceutical interventions such as protective measures, travel-related measures, and clinical services such as prevention through healthcare (Chen et al., 2021). Quality of care, consent and autonomy, access to care and technology, laws and regulations, clinician responsibilities, patient responsibilities, changed relationships, commercialization, policy, information needs, and evaluation of ethical, legal, and social issues have been changed with these new emerging technologies that are to be adapted and re-evaluated (Kaplan, 2020).

Artificial intelligence, 5th generation (5G) telecommunication networks and the Internet of Things (IoT) have created new opportunities in handling the COVID-19 pandemic. Cloud-based ubiquitous computing, big data analytics, machine learning, autonomous systems, smart robots, 3D printing, and virtual and augmented reality (VR and AR) are other advanced technologies (Lee & Trimi, 2021). Digital technologies usage as shown in Figure 13.1 can be permanently normalized with the help of the pandemic (Doyle & Conboy, 2020). Ophthalmology based on image-based investigations is another successful major application used during the pandemic as an example of new normal healthcare management.

Clinical informatics, public health informatics, consumer health informatics, and clinical research informatics are types of health information technology. Clinical experts in China and around the country have provided healthcare service through mobile internet and 5G technologies. In addition, online mental health services have been given through internet hospitals due to increased physiological problems among medical staff and the public. Also, internet hospitals have helped patients with chronic illness (Ye et al., 2020).

Figure 13.1 shows integrated usage information technologies in handling COVID-19.

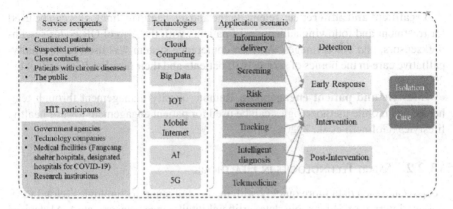

FIGURE 13.1 Health information technology framework for the COVID-19 pandemic (Ye et al., 2020).

13.2.1 Usage Purposes of Technology in the Healthcare Management During the COVID-19 Pandemic

Diagnosis, prevention, treatment, adherence, lifestyle, and patient engagement have changed from the traditional way with smart technologies during the COVID-19 pandemic (Golinelli et al., 2020). Advanced technologies are necessary for testing, contact tracking, and treating people during the pandemic. Restructuring supply chains, remote working, and remote education are other noteworthy topics related to the pandemic in healthcare. Due to limited resources for testing and treatment, innovative solutions are required to prevent shortages in the supply chain of healthcare (Lee & Trimi, 2021).

Diagnosis: COVID-19 can be diagnosed with artificial intelligence methods based on the use of computed tomography (CT) data. An integrated deep-learning framework on chest CT images for auto-detection of novel coronavirus pneumonia is an example of that technology. Risk of radiation exposure and operator or machine-type dependence are drawbacks while nucleic acid test kit shortages can be solved with that technology. A low cost blockchain and AI-coupled self-testing and tracking systems for COVID-19 requesting a user's personal identifier before opening pre-testing instructions can be applied in low-income countries. Moreover, chatbots can be used for patient triage, clinical decision support for providers, directing patients and staff to appropriate resources, and mental health applications (Golinelli et al., 2020).

Prevention and surveillance: Coronavirus symptom checkers can be used for self-triage, self-scheduling, and avoidance of unnecessary in-person care, including personalized recommendations. A contact-tracing app such as a quick response (QR) code-screening to follow people's movement creates a temporary record of proximity events. Also, a contact-tracing app gives recent close contacts of diagnosed cases for self-isolation. For instance, WeChat, a Chinese social media, was used to follow daily cases. Electronic personal protective equipment (ePPE) can be used for preventing healthcare workers' infections (Golinelli et al., 2020).

Treatment and adherence: Telemedicine and telehealth technologies can be used for treatment and following adherence of patients with the help of QR codes screening, sensors, and video images. A video conference system was used for specialty-palliative care in the homes of seriously ill patients and their families (Golinelli et al., 2020).

Lifestyle and patient engagement: Remote obesity management through telehealth and mobile methodologies can be used for patient engagement toward healthy lifestyles (Golinelli et al., 2020).

13.2.2 SMART TECHNOLOGIES IN HEALTHCARE

People can reach data more quickly with the help of smart technologies (cloud technology, Internet of Things, big data, artificial intelligence, sensors, etc.). AI devices include intelligent machines working autonomously without human interference that can self-learn and improve their performance from current data. IoT devices, sensors, processors, wearables, electronics, software, actuators, vehicles, cell phones, and computers can help in collecting data, and this data can be analyzed by AI in seconds without the intervention of humans. Different government departments can be connected with IoT devices, and algorithms are used to analyze and report data patterns (Kummitha, 2020). Furthermore, the body area network can get data from the human body with the help of sensors (Pashazadeh & Navimipour, 2018).

13.2.2.1 Cloud Technology

Different devices are integrated without needing to be near each other, and data is stored in an internet server with cloud computing (CC). Optimal resources allocation can be done by means of data analytics with minimal management efforts and more efficiently. Smart hospitals or healthcare concepts are developed based on new sophisticated technologies. Service (IaaS), software (SaaS), and platform (PaaS) are categories of CC infrastructure. Private cloud, public cloud, community cloud, and hybrid cloud are types of CC and the platform, hardware, software, and infrastructure are parts of the CC model. CC applications have increased the volume of data in the form of structured, semi-structured, or unstructured data, stored in different formats that cannot be easily integrated at high speeds. Different formats of data cannot be easily integrated, and CC applications can help in overcoming this problem. QR codes have been used in China to track millions of people and have increased internet use and data volume. Virtual work, virtual meetings, and distance education are some applications of the pandemic, creating more data (Alashhab et al., 2020).

Lowering of costs without purchasing the hardware and servers, ease of interoperability, gaining easy access to the patient data collected from numerous sources, giving out timely prescriptions and treatment protocols, access to high powered analytics, extracting relevant patient information, patient's ownership of data, allowing patient participation in decisions, and telemedicine capabilities are ways CC is affecting the health industry through having the patient's data in the cloud as a way of seamless transfer between the different stakeholders (Kamani, 2019).

13.2.2.2 The Internet of Things (IoT)

A large number of interconnected devices are engaged together in Internet of medical things (IoMT) to make a wide healthcare management system, helping healthcare providers with appropriate decision-making. A disease can be tracked and alerts for the safety of the patient can be given by IoMT (Intawong, Olson and Chariyalertsak et al., 2021). Interconnection between devices and machines maintaining connections without deliberate human intervention is a new concept developed in coordination with IoT. Embedded sensors can send information about machines and people in an automated manner. Smart watches, living in smart homes with connected fridges and heating systems, wireless metering, mobile payments, and commuting in smart cities in driverless cars are some types of these technologies.

Increasing the treatment efficiency and decreasing healthcare costs can be realized through intelligent and improved prediction capabilities of IoT-based healthcare systems bridging between sensor infrastructure network and internet as applied around the world in many countries shown in Figure 13.2 (Rahmani et al., 2018; Aslan, 2021). 24/7 expert support, smartphone-integrated devices, and medical history are properties of that system in healthcare (Kim & Kim, 2018). Unnecessary visits, utilizing better-quality resources, and improving allocation and planning are the main advantages of IoT in healthcare. Moreover, governments and private institutions have become a part of this new state-of-the-art development for decreasing costs and getting more benefits over the management of services. Appropriate medical record-keeping, device integration, and determination of cause of sickness can be carried out with help of IoT based on sensors, machine learning, real-time testing, and embedded systems through the internet. Intelligent medical equipment linked through a smartphone can communicate health data to the required place. Erroneous and premature health information have caused the deaths of many patients, and IoTs can prevent that by means of instantly detecting health problems with the use of sensors so that many patients with diabetes, heart failure, asthma, and abnormal blood pressure can be saved. Furthermore, identifying changes in COVID-19 by IoT in vital parameters can help physicians and facilitate the work of the surgeon through reducing risks and enhancing overall performance to improve hospital treatment systems such as tracking patients' calorific intake and therapy with COVID 19 asthma, diabetes, and arthritis (Mukati et al., 2021; Aslan, 2021).

Figure 13.2 shows how information technologies worldwide performing COVID-19 data collection with phones, robots, and health monitors. Data transfer, data analytics, and data storage in a central cloud are the stages of IoT (Al-Ogaili et al., 2021). A Bluetooth-enabled weight scale and blood pressure cuff, together with a symptom-tracking application for cancer treatment, smart continuous glucose monitoring, connected inhalers helping people with asthma and chronic obstructive pulmonary disease, ingestible sensors monitoring adherence to medicine taking, connected contact lenses measuring tear glucose and providing an early warning system for diabetics, the Apple watch app monitoring depression, coagulation testing allowing patients to check how quickly their blood clots, Apple's Research Kit, and Parkinson's disease monitoring symptoms are some examples of the IoT in healthcare (Econsultancy, 2019). Adverse drug reaction detection; medication management;

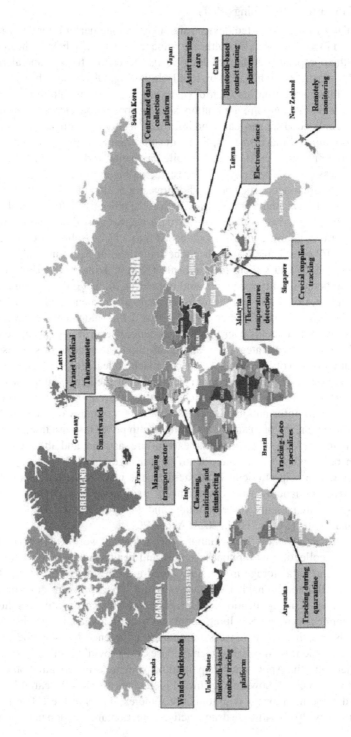

FIGURE 13.2 A worldwide map of IoT technologies during the COVID-19 pandemic (Al-Ogaili et al., 2021).

community healthcare monitoring a municipal hospital, a residential area, or a rural community; children's health information; glucose level sensing; electrocardiogram monitoring; blood pressure monitoring; body temperature monitoring; oxygen saturation monitoring with a wearable pulse oximetry method based on a Bluetooth system; rehabilitation system as an effective remote consultation; wheelchair management with different sensors; imminent healthcare solutions for skin infection; peak expiratory flow; abnormal cellular growth; hemoglobin detection; cancer treatment; eye disorder; remote surgery; and IoT healthcare by smart phones for managing cancer, diabetes, and mental health; and fitness controlling for diet, nutrition, weight loss, etc. are other IoT applications based on m-IoT healthcare services (Dhanvijay & Patil, 2019). Detecting allegations of fraud for a subscription for insurance firms; tracking medical equipment such as wheelchairs, defibrillators, nebulizers, oxygen pumps, etc. against theft; IoT-enabled hygiene monitoring, control of pharmaceutical stock inventories, and environmental monitoring such as refrigerator temperature in the supply chain of the vaccination are other important usages of this novel technology (Mukati et al., 2021).

13.2.2.3 Sensors

Microphones, cameras, and other sensors are used to collect data (Matheny et al., 2019). Accelerometer, gyroscope, heart rate and heart rate variability sensors, galvanic skin response sensors, skin temperature sensors, location sensors, and orientation sensors are kinds of wearable sensors (Bhatia & Sood, 2017). Environment sensors (measuring light conditions, temperature and humidity, noise, and air quality), objects sensors with RFID antenna as used for collision detection or fall protection, biometric sensors based on a smart wristband-biometric information, RFID, and GPS are types of sensors used widely. Skin conductance can detect sodium or potassium concentration in sweets (Qi et al., 2017b). Light and noise levels, temperature and humidity rates, and air quality can be measured by sensors with help of RFID and GPS combination (Aslan, 2019; Qi et al., 2017b). Lying, standing, and sitting static posture positions can be determined by inertial sensors, detecting falling of a patient. Barometric pressure sensors can detect stairs and falling. Magnetic field sensors can detect the location and recognize watching TV. On-object sensors with RFID can detect watching television, preparing a meal, or washing clothes. Wristband devices record step counts, distance, and calories burnt (Bhatia & Sood, 2017).

Hearables for people suffering hearing loss interacting with the world, ingestible sensors for monitoring the medication in the body and warning in case of any irregularities as for a diabetic patient, moodables used for improving mood, computer vision technology with artificial intelligence for visually impaired people, and healthcare charting capturing the patient's data used for reducing doctors' manual work are major applications of sensors and IoT in healthcare (Nasrullah, 2020). Sensor information can be collected by means of smartphones, and the collected data can be used for community healthcare and national communication as shown in Figure 13.3. Coronavirus-identifying sensors embedded in face masks and ultraviolet (UV) light being able to kill coronavirus are two new innovative ways to fight COVID-19 (Goel et al., 2020; Nasrullah, 2020). Bio-signals (e.g., ECG, EEG, and EMG) collected from users' bodies can be filtered (Rahmani et al., 2018).

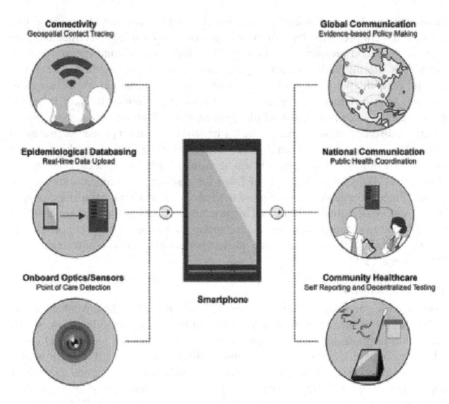

FIGURE 13.3 Collection of data with sensors (Goel et al., 2020).

Figure 13.3 highlights components of the COVID data collection strategy using sensors.

Novel developed technologies for SARS-CoV-2 detection-biosensors are used as a fast and reliable alternative for clinical diagnosis, real-time detection, and routine measurements. Dual-functional plasmatic biosensors applied for the detection of various viral sequences including RdRp-COVID, ORF1ab COVID, and E genes from SARS-CoV-2 and discriminating between SARS-CoV and SARS-CoV-2 viruses; cell-based potentiometric biosensor detecting the SARS-CoV-2 S1 anti-genand field effect transistor detecting SARS-CoV-2 spike protein are applied for diagnosis of infectious diseases (Sheikhzadeh et al., 2020). Furthermore, nano-biosensor-based technology using RNA or DNA sequences, gel electrophoresis, and an enzyme-linked immunosorbent assay can be used for diagnostic purposes within a few seconds. The high-resolution CT was found to be the best alternative at an early stage for detection of SARS-CoV-2 (Kumar et al., 2020a).

13.2.2.4 Telehealth

People under self-isolation unable to attend clinic visits with diabetic and other chronic conditions can be monitored by telemedicine (Intawong et al., 2021).

Communications, education for users, and training for staff and feedback mechanisms are to be established in that system. Telehealth appointments can be managed by physicians, nurse practitioners, and psychologists (Farrugia & Plutowski, 2020). In-person and virtual encounters can be developed in telehealth (Rodin, Lovas and Berlin et al., 2020). Existing infrastructure (platform, scheduling, and electronic medical record software) can be integrated into telehealth systems. Different stakeholders and actors can be integrated in this system to prevent bureaucratic hurdles and improve processes. Telehealth regulations can enable virtual care with fast-tracking application processes for remote patients instead of face-to-face visits. Telemedicine can help doctors to check their patients remotely in distant areas where there are shortages of doctors; it is a beneficial and cost effective way of reducing travel and the associated carbon footprints. Waiting times can be decreased, and access to care can be improved with this method. Smart phones with slit-lamps enabling ocular biomicroscopic videography can show the patient's examination features to doctors, and also image sharing of data can be done easily by just sharing the screen (Farrugia & Plutowski, 2020). Telepsychiatry, a subset of telemedicine, has psychiatric services for patients remotely using information and communication technologies (Khan et al., 2021).

Real-time telemedicine can be an efficient way to care for patients during the COVID-19 period as a kind of a new standard of care to prevent unnecessary visits and decrease the exposure risks. Moreover, video consultations with innovative service can improve the quality of treatment. Deploying a wearable device to transmit a patient's respiratory rate, heart rate, and pulse oximetry for "high risk" patients with COVID-19 at discharge controlled by the virtual health center is an application that technology (Lin et al., 2020). Screening for symptoms can be used to identify symptomatic patients earlier. Virtual home visits for high-risk individuals and remote intensive care unit monitoring methods are applied to patients with suspected or confirmed cases of COVID-19 in order to prevent spreading. In addition, remote triaging is applied in telemedicine to decrease hospital visits by video counseling the patients, virtual patient monitoring, and mobile applications for symptom observation. Decreasing the rate of discharge without complete treatment and proper follow-up and reducing the number of in-person visits and subsequent risk of transmission of infection are advantages of the telemedicine triage (Reeves et al., 2020). Unbillable phone calls can be beneficial from the users' side to make free calls (Lin et al., 2020).

Health literacy is defined as "the ability to access, understand, evaluate, and communicate information as a way to promote, maintain and improve health in a variety of settings across the life-course," which is associated with older age, lower levels of education, and minority status and is to be developed due to fast-moving situations through telemedicine-targeted education besides virtual care and remote patient monitoring (Brahmbhatt et al., 2021). Clinicians posting new ideas and experiences, nurses using an instant messaging system within the epic secure chat to communicate with bedside nurses, physical therapists, and social workers involved in the care of hospitalized patients with COVID-19 are methods of communication in that system (Lin et al., 2020). Asynchronous messaging, nurse-led virtual clinics, and virtual peer-navigation in post-COVID-19 cancer care can be used in the future as a part of standard-of-care (Rodin et al., 2020).

13.2.2.5 Artificial Intelligence (AI) and Machine & Deep Learning

13.2.2.5.1 Artificial Intelligence

Health assessment, health monitoring, disease prevention and management, medication management, and rehabilitation are areas where AI is used in healthcare (Matheny et al., 2019; Nuffield Council on Bioethics, 2018). With the artificial neural network model, overall health severity can be determined in order to prevent serious health loss (Bhatia & Sood, 2017). Medical imaging detecting conditions such as pneumonia, breast and skin cancers, and eye diseases; echocardiography detecting patterns of heartbeats and diagnosing coronary heart disease; screening for neurological conditions and monitoring symptoms of neurological conditions; and surgery knots to close wounds are applications of clinical AI.

Diagnosing illnesses; remote surgery; precision medicine; patient safety; identification of risk of individuals for suicide by using social media data; genomics (analysis of tumor); population health by controlling air pollution, water, and epidemiologic data; fraud detection in billing; cybersecurity; classification of diseases; and drug discovery are major applications of AI in health. Diagnosing retinopathy, identifying cartilage lesions, detecting lesion-specific ischemia, predicting node status after positive biopsy for breast cancer, image recognition techniques, guiding clinicians in radiotherapy and surgery planning, and microorganism detection can be done with AI by helping radiologists, dermatologists, pathologists, and cardiologists (Matheny et al., 2019).

Wireless and satellite communication channels, machine learning, imaging technology, artificial intelligence, and tele-surgical robotics can be used to carry out remote surgery in different regions by supervising the surgical robots (Zeadally & Bello, 2021). Remote-controlled robotic surgery can decrease the risks and increase precision so that consequences of surgical decisions can be assessed with patient risk factors (Matheny et al., 2019). Heart failure before symptoms even arise can be found by machine learning and neural networks (Farrugia & Plutowski, 2020). Personalizing chemotherapy dosing and mapping patient response, determining the best surveillance intervals for colonoscopy exams, and best treatment options for complex diseases are types of personalized management and treatment; a patient-centered medical homes model (Matheny et al., 2019).

Performing mundane, repetitive tasks in a more efficient, accurate, and unbiased fashion can decrease administrative work. The best possible action in a situation can be taken to maximize the performance with help of AI (Vetrò et al., 2019; Matheny et al., 2019). Tasks can be performed by AI without receiving instructions from people. Intelligent algorithms and decision-making models are used to reduce human involvement in processing (Zeadally & Bello, 2021). Decreasing failure of care delivery, care coordination of overtreatment or undertreatment, pricing failure, fraud and abuse, and administrative difficulties and automating highly repetitive tasks and workflows are the main advantages of AI in healthcare (Matheny et al., 2019).

Building early warning systems for adverse drug events, fall detection, and air pollution systems can be developed. Microscopes to detect bacterial contamination in treatment plants can decrease costs. AI-enabled robots for handling hazardous materials, handling heavy loads, working at elevation or in hard-to-reach places, and completing tasks that require difficult work postures, risking injury, can decrease work accidents. Hand gestures and facial expressions of patients can be communicated to the system by Apple's Siri, Amazon's Alexa, or Microsoft's Cortana tools. Depression, smoking cessation, asthma, and diabetes can be addressed by conversational agents from health monitoring. Combating loneliness and isolation can be done through social support improving treatment outcomes with a conversational agent: conversational AI instead of professional or personal emotional domain (Matheny et al., 2019).

COVID-19 and other pneumonias can be differentiated by AI, and new treatment pathways can be identified by that technology. Furthermore, the risk profiles of non-hospitalized patients can be calculated. Credit card information can also be used to find the activities of citizens for tracking (Farrugia & Plutowski, 2020). Also, tracking disease outbreaks using search engine query data can be used for public health. Digital epidemiological surveillance can be developed from news and online media, sensors, digital traces, mobile devices, social media, microbiological labs, and clinical reporting relying on AI's capacities in spatial and spatiotemporal profiling, environmental monitoring, and signal detection (Matheny et al., 2019).

13.2.2.5.2 Deep and Machine Learning

Deep learning (DL) uses neural networks, similar to the human brain, to make predictions about the health status of patients. Image recognition and speech recognition are some fields of DL. Detecting pulmonary tuberculosis from chest radiographs and malignant melanoma from digital skin photographs are applications of DL. DL algorithms are prepared to analyze optical coherence tomography images to detect the early stage of disorders such as diabetic macular edema. Deep learning can detect heart failures better than doctors, decreasing the need for doctors (Tuli et al., 2020).

Machine learning is constructed with algorithms in order to make predictions. Abnormal behaviors can be predicted with the help of collected data (Saheb & Izadi, 2019). Activity recognition, behavioral pattern discovery, anomaly detection, and decision support are developed with ambient intelligence algorithms in healthcare (Qi et al., 2017a). Machine-learning algorithms can be trained to categorize patterns from the raw data as indicators of an individual's behavior and health status. Learning systems can personalize treatment for users over time. Devices can communicate with one another such as a glucometer receiving feedback from refrigerators about frequency and types of food consumed. Giving information, disease natural history, patient values and cost through a DL model, helping the surgical team in the operating room, reducing risk, and making surgery safer are advantages of machine learning (Matheny et al., 2019).

Novel drugs can be discovered with help of artificial intelligence and deep learning (Ye et al., 2020). Machine learning techniques with mature brain-image analysis tools can give better results to extract chemical information from large compound databases and to design drugs with important biological properties and comprehensive assessment of cellular systems so that potential drug effects can be determined by AI. Predicting synthesis, drug selectivity, pharmacokinetic and toxicological profiles, modeling polypharmacy side effects, designing de novo molecular structures, and structure-activity models are advantages of AI in drug discovery. Biobanks, radiology images, and notes are sources of data (Matheny et al., 2019).

Virtual reality devices such as IrisVision™ with a headset hold a smartphone displaying images in the peripheral vision magnifying the image, and NuEyes™ used to magnify images are used for eye treatment. The Oculenz™ AR headset can enable one to view the image normally by using image remapping strategy. It can help alignment of the eye gaze and the projected image with patented eye-tracking technology. Augmented reality headsets can improve visualization in the operating room. Alcon (Ngenuity™) and Zeiss (Artevo™) can help surgeons with ophthalmic surgery by more comfortably operating by wearing 3D glasses to view surgery on a large monitor positioned beside the patient's surgical bed.

13.2.2.6   Big Data Management

Vehicles, mobile phones, and camera data are used to give information, and warning can be given through analyzing this data. Patients' backgrounds and underlying health conditions can be analyzed to give treatment methods and warning. Weather changes can be estimated based on big data. Facial recognition technology can be used on the streets to penalize people going out, and alerts can be given to authorities with the help of connected systems categorized as continuous monitoring, allowing one to make quick decisions (Kummitha, 2020). Automatically registering the arrival of a patient from their personal devices to minimize wait times and streamlining their journey by organizing their visits to department and giving priority to departments with fewer queues, uploading images automatically to the patient's electronic health records (EHR), integration with automation to prepare alerts for diagnosing, cataloguing patients' drug histories to prevent drug interactions, dispensing new prescriptions, integrating healthcare records from different providers, integrating lifestyle trackers into patients' records, retinopathy screening, and reducing errors by minimizing human intervention are current topics benefiting patients. EHR-integrated secure messaging, tracking of persons, tracking COVID-19-related infection in EHR, and utilization of EHR for training can improve the efficiency of healthcare (Reeves et al., 2020).

13.2.2.7   5ᵗʰ Generation (5G) Telecommunications

6G is the next stage of technology enabling extended reality (XR) and high-fidelity mobile holograms with higher security and privacy. Complex network connection problems can be solved with a higher speed of data transmission by 5G wireless communications. Less latency, better download speeds, and reduced energy consumption are major advantages of that new technology. Ultra-high-definition (UHD) multimedia streaming is another useful property of the technology with user experience. Better quality and reliable video consultations can create an improved physician-patient relationship, and the eye can be examined remotely. A 5G telemedicine network was implemented in Sichuan province/China, and a single hospital selected as the center resulted in lower case fatality ratio.

13.2.2.8 Health QR Codes

Health QR codes and big data technology have been used whether a person has been in contact with a COVID-19 patient or not. An infected person can be located and traced by governments to prevent spreading (Ye et al., 2020). A QR code on phones can be scanned when necessary at work and public places in different colors (green code—unrestricted, yellow code—quarantine for seven days, and red—a two-week quarantine), making control easier in that red or yellow codes are to be reported quickly. At the point QR code is scanned, the location address is sent to the system's servers so that the authorities can track movements of people over time. According to NY (2020), 900 million users are registered in Alipay, and WeChat working with authorities has more than a billion monthly users to build their own health code system across China (NY, 2020).

13.2.2.9 Technologies Limiting Travelling: Home Monitoring Devices, Augmented and Virtual Reality

The remote patient monitoring devices market has expanded during the COVID-19 pandemic because patients try to avoid hospitals due to the high risk of being infected. Chronic conditions are attempted to be managed remotely by means of understanding aspects of their health (Medical Technology, 2020). Remote patient monitoring is the new concept developed in healthcare. Up-to-the-minute information of patients; patient-reported information (symptoms, activities, self-quarantine etc.); physical measurements (temperature, pulse, etc.); and ongoing metrics can be given by digital technologies from homes (Mayo, 2020). Physiological parameters can be decided through wearable patient monitoring devices for reduction of healthcare providers during the pandemic (FDA, 2020).

Virtual reality (VR) and augmented reality (AR) can be used for educational purposes in healthcare. Creating virtual reality simulations can be more efficient for students, and more students can reach classes in this way, making education more economical and automatable. Virtual tours can be prepared by a VR/AR developer during the period of lockdown and isolation (Pillai et al., 2022). More personalized care can be realized by limiting the physical contact. VR and AR technologies are used in operating rooms and classrooms for helping surgeons (Marbury, 2020).

13.2.2.10 Blockchain Technology

Untrusted different users' data is dispersed in a secure way for data reliability and record keeping safeness by a blockchain so that copies of information are included in all hubs in a decentralized system to protect data against changes (Kumar et al., 2020a). All members of the network can see the data on the blockchain, making it more transparent, and in the case of any change, all members will be alerted with authenticating participants. Reliable and trustworthy information from governments can be provided through a blockchain-based tracking system. Impenetrable information infrastructure, transparency, and cryptographic encryption tools are used in blockchain technology. Cryptography technology can track data from its origin and prevent changes in data blocks with a unique cryptographic signature with help of Hash label. All members share equal rights to overcome security problems (Marbouh et al., 2020). Blockchain technology or distributed ledger technology (DLT) can enable patient-centered interoperability, putting patients at the center of the eco-system. Increased patient trust, accessing trusted data, developed transparency, decreased operational costs, monitoring who shares data and with whom in privacy, automated payments with smart contracts, bringing all transactions onto a platform, private messaging, tracking medicines, preventing thefts, and enabling proof of docu-ment existence are advantages of that technology. MEDREC, BURSTIQ, FACTOM, MEDICALCHAIN, GUARDTIME, ROBOMED, and BLOCKPHARMA are some types of blockchains in different fields of healthcare (Attaran, 2020). Radio-frequency identification (RFID) and blockchain technology prioritizing service, security, and timeliness can be used in the multi-sport event, especially during and post pandemic for implementing health protocols issued by the World Health Organization (WHO) for safety to break the chain of COVID-19 infection in order to optimize the event through tagging with a unique unchangeable numbering system and UHF RFID gate reader (Nugraha et al., 2021).

The blockchain records anonymously in clinical trials including participants or patients related to COVID-19 to protect the identity of patients. People can be identified if they leave their home in that system. Blockchain applications such as Ethereum smart contracts and oracles can be used in combating the COVID-19 pan-demic for collected data from different external sources. Fraudulent mobile apps such as fake coronavirus tracker apps can be used by attackers to get names, pass-words, and credit card numbers. Data available online is susceptible to data manip-ulation. Imperfect and incomplete data is a major problem for decision makers. Data verification and validity are required in pandemic management for recommendations to the public from reported data statistics. Determining and predicting the disease spread is a crucial dynamic to measure the performance of the countermeasures. Misinformation and spam of fake news can be prevented by blockchain technology. The aim is to get trusted and reliable data about new cases, deaths, and recovered cases of COVID-19. Algorithms are developed between stakeholders in the network for secure data transfer. World Health Organization (WHO) has stated that detecting the newly infected cases correctly is a problem for predicting coronavirus infection risks (Marbouh et al., 2020). Decentralization, transparency, and immutability of blockchain technology can be effective to control this pandemic by early detection of outbreaks. Public hospitals and clinical laboratories can give data not collected according to the settled guidelines. Enormous energy consumption is a challenge of that technology due to needing powerful hardware resources for each transaction. Validation requirements for some time, the complexity of blockchain, and privacy preservation are other challenges of that technology (Sharma et al., 2020).

13.2.2.11   Social Distancing Technologies

Maximizing physical distance can reduce the chance of getting coronavirus. Wireless, Wi-Fi, cellular, networking, and artificial intelligence technologies can help to create a safe area for social distancing practices. Measuring the distances among people and alerting them in case they are in less than 1.5–2 meters' distance of one another can be done by artificial intelligence. Global navigation satellite systems (GNSS) technology can locate in outdoor environments for social distance. Most smartphones having GPS can be used to track locations of mobile users. People suspected of having COVID-19 and returning from an infected area need to be isolated. GPS-based positioning devices are effective at monitoring people who are not to leave their residences during the quarantine. GNSS technology can follow people everywhere and anytime compared to Wi-Fi or infrared-based solutions. On Violation of social distancing norms, mobile application through server, can send warnings to concerned persons automatically. Also, shopping online and receiving items or medicines can be done with drone delivery services to decrease contact as used already by Amazon and DHL, called GNSS-based autonomous services (Nguyen et al., 2020).

The technologies used for tracking are shown in Figure 13.4. Cellular technology or SIM card data for tracking foreigners is widely used for location tracking with 5G cellular networks for real-time monitoring and infected movement data scenarios by sharing cellphone locations so that quarantined cases can stay in their homes. For instance, Taiwan has deployed an "electronic fence" with the cellular-based network for ensuring that the quarantined cases stay in their homes. If the phones are turned off, they are visited, and they need to carry their phones with them. Russia has used SIM card data for tracking foreigners that can be used during COVID-19 for social distancing. Zigbee is a technology used for social distancing based on a standard wireless communication technology and low-power wireless networks such as wireless sensor networks (Nguyen et al., 2020).

A Zigbee-enabled device such as ID card or member card can be equipped with a person. Zigbee central hub can determine the people in the place they enter. A network coordinator and Zigbee-enabled devices are used for this purpose so that the location of the user can be determined by that technology and the central hub can detect crowds and notify the local manager about that. RFID technology can also be used for tracking, monitoring crowds from the RFID tag, and the RFID reader to determine the location of the user named as an RFID-based localization system for indoor environments. The number of people inside the place can be determined by an RFID reader, used in supermarkets or workplaces where people's member/staff ID cards are equipped with RFID tags. Ultra-wideband (UWB) technology for precise indoor positioning systems (IPSs) is another solution for social distancing for detection (e.g., tracking users' location), public place monitoring, and access scheduling at the centimeter level. Dimension4 sensor can be used for tracking with a built-in GPS module. UWB-supported phones or personal belongings equipped with tags can be used in that technology supported with AI/machine learning algorithms. Smartphones or smartwatches can be used to measure distance between persons. Bluetooth-enabled devices and Bluetooth technology detecting and connecting automatically to other nearby devices can be used for the same purposes as these technologies used by Apple and Google detecting other smartphones nearby using Bluetooth technology, and a notification can be sent to a person in close contact with the positive case (Nguyen et al., 2020). Figure 13.4 shows how social distancing can be managed with the help of different IT devices.

Satellite signals are important to make tracking effective. Interference signals such as being in a building or in crowded areas can prevent the location accuracy

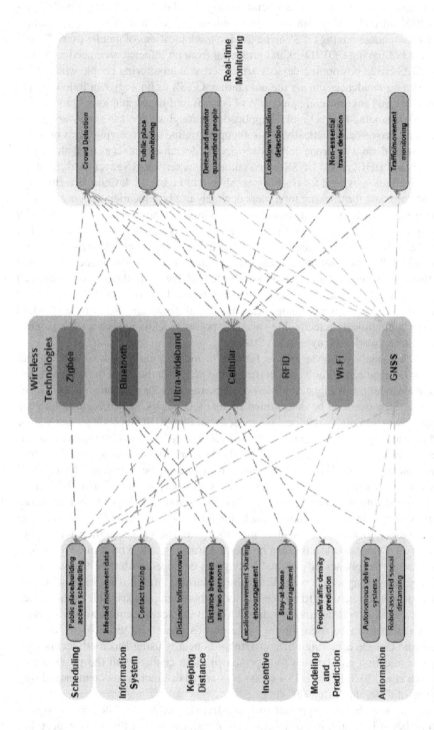

FIGURE 13.4 Application of technologies to different social distancing scenarios (Nguyen et al., 2020).

determination. Ground-based transceivers in a place where satellite signals are weak can be used to overcome these problems to transmit GNSS signals enhancing the positioning accuracy of the GPS. High price and strict time synchronization requirements are further challenges of these pseudolites. RFID tag and the RFID reader problems and connections are also problems in transmitting data. The accuracy of localization techniques and turning on Bluetooth mode are other problems of social distancing technologies (Nguyen et al., 2020).

13.2.2.12 Other Technologies

- Geographical information systems can be used for spatial mapping COVID-19 hotspots in the country to control lockdowns, intercity travels, and distributions of masks and sanitizers. Rapid visualization of epidemic information, spatial tracking of confirmed and suspected cases, using contact-tracing applications and tracking movements of patients, and mapping immigration mobility are uses of GIS.
- Autonomous robots and drones are used for collecting samples of throat swabs, controlling social distancing, disinfecting and sterilizing coronavirus, monitoring social distancing, delivering health equipment, and jamming.
- Social media platforms can be used for creating awareness, reporting shortage and distribution of COVID-19 and personal protective equipment, reporting COVID-19 suspected cases and contact-persons, tracking people's mobility patterns, providing real-time COVID-19 updates, and clarification of uncertainties. Facebook, TikTok, and web applications such as Line (https://lin.ee/dAEig3e) and Twitter (https://twitter.com/thaimoph) have been used for informing people about the outbreak and giving correct information such as the number of COVID-19 cases, areas of high risk, and reports in Thailand (Intawong et al., 2021).
- Noncontact 3D scanning helps the thoracic chest scanning for COVID-19 so that 3D scanning can be used to detect and quantify the COVID-19 pandemic, and 3D printing can be used for mask production and production of personal protective equipment such as additive manufacturing.
- Wireless thermometers within some distance to measure temperature without touching, using feet to open doors, using wireless devices or cards to use elevators, digital and contactless payments, and online entertainment are other innovative technologies.
- Medications and other essential health supplies can be supplied with IoT-enabled drones as done in Rwandan and Ghana rural health institutions during the epidemic. Disinfection or detection of symptoms related to COVID can be also done by drones.
- Nano biotechnology applications: self-disinfecting surfaces, nanomaterial-based vaccine development, nanomedicine, enzyme-linked immunosorbent assay (ELISA) and chemiluminescence immunoassay (CLIA), implantable medical devices, nano robots, and the use of biological molecules in nano devices are other future novel ways of combating COVID-19 (Kamaraj, 2020).

13.3   APPLICATIONS OF DIGITAL TECHNOLOGIES

13.3.1   TELEMEDICAL INTERVENTIONAL MONITORING IN HEART FAILURE

Telemedical interventional monitoring in heart failure (TIM-HF) collects daily weights and symptoms and provides daily coaching. A cloth-based nanosensor array incorporated into an undergarment is used to monitor cardiac output and stroke volume. Smartwatches with photo plethysmography (PPG) technology coupled with algorithms can detect irregular rhythms for stroke prevention in the screening of atrial fibrillation. Smartphone video-capture to estimate blood pressure with the use of transdermal optical imaging is an emerging technology still being developed. The Eko (Oakland, CA, USA) digital stethoscope developed with artificial intelligence and machine learning can detect echocardiographic abnormalities and predict heart murmurs with high accuracy. Continuous blood glucose monitoring devices are developed to measure capillary blood glucose levels for better diet and self-care. Low adherence to that technology, poor data accuracy, and either delays or lack of actionability on data received were found from a study done in Canada. Not knowing when to seek assistance as their condition is deteriorating was another problem of that technology (Brahmbhatt et al., 2021).

13.3.2   DIABETIC RETINOPATHY (DR)

Patients with diabetes can have sight-threatening consequences. Fundoscopy for patients with diabetes mellitus is aimed to prevent sight loss. An AI-based DR screening algorithms can be used for diagnosing. The "Prevention first" principle is applied through screening processes in China. Hence deep learning in telescreening for DR can help non-eye professionals to perform DR screening and make recommendations. Smartphone-based imaging devices such as Remidio fundus on the phone camera are used in teleophthalmology screening.

Glaucoma explained with structural changes in the optic nerve head, intraocular pressure, retinal ganglion cell death, and loss of visual field are the main causes of irreversible blindness. It is diagnosed according to intraocular pressure measurements, fundus photographs, and changes in nerve. Experienced ophthalmologists are required to detect this disease early. A test by the Humphrey Visual Field Analyzer and Oculus Field Analyzer is used to detect optic nerve damage, and telemedicine can be used to screen glaucoma with combined technologies.

Age-related macular degeneration is a major cause of visual loss and blindness of elderly persons. Telemedicine with other findings can help to detect that illness in face-to-face examination for screening purposes. Detecting myopia, anterior segment diseases, cataract screening, infectious keratitis with blue light-emitting diode (LED) and smartphone light sources, corneal abrasions, ulcers, scars, and dry eye disease by smartphone app DryEyeRhythm are other applications of smart technologies for detection and timely management to decrease costs and ensure quality of care is measured. Safer and more secure applications can be used in place of telephones, messaging, and video calling. Losing some of the non-verbal communication and impeding empathy are to be investigated so that acceptability and accessibility for all potential users can be improved for digital literacy.

13.3.3 CANCER TREATMENT

Patients in rural and remote areas have the problem of getting cancer treatment in good hospitals. It is suggested that cancer follow-up can be done effectively with smart technological solutions by considering patient safety and satisfaction. Outpatient volumes can be decreased with well-established telehealth programs (Rodin et al., 2020). The risk of developing cancer, risk of disease recurrence or risk of treatment complications, predicting prognosis, radiation treatment planning, etc. are common applications of machine learning in healthcare. AI was used for predicting the risk of reoccurrence of breast cancer surgical resection (Khan et al., 2017).

13.3.4 PSYCHOLOGY AND REHABILITATION OF PATIENTS

A tablet with a remote translator service was placed in every non-English-speaking patient's room. An in-room tablet allowed a single member of a team to perform a physical exam while the remainder of the team interacted remotely from the hallway. Psychology and rehabilitation patients use tablets and video visits to resume group therapy sessions (Lin et al., 2020). Employees' physiological conditions can be improved with wearable devices. Psychosocial risks related to chronic diseases for elderly people can be detected to improve their life conditions (Aslan & Aslan, 2019).

Facial gesture recognition software and voice analytics can be used to determine the mood of patients with mental health disorders. As a patients' emotional state is determined, intervention may be required. Smart watches and glasses are types of wearable devices that can use sound recognition (Zeadally & Bello, 2021). Personal digital assistants (PDA) and smartphones can be used for pain assessment as a kind of self-management where facial expressions can be used as a kind of pain indicator (Prada, 2020).

13.3.5 INTELLIGENT DIAGNOSIS FOR CHEST COMPUTED TOMOGRAPHY IMAGES

Computer-assisted diagnostic products for COVID-19 have been developed by scientists. Some results of real-time reverse transcriptase-polymerase chain reaction (RT-PCR) test results can be wrong, and chest computed tomography (CT) features with RT-PCR can give more reliable results. CT abnormalities were detected in 96% of COVID-19 patients. CT using x-ray probes (providing > 90% sensitivity) has been found better than the reverse-transcription polymerase-chain-reaction (RTPCR) test (Goel et al., 2020; Ye et al., 2020). The procedure of RT-PCR can take up two hours, but lesions of COVID-19 can be detected with high accuracy by an AI-based diagnostic system in two seconds so that the risk of cross-infection in healthcare facilities can be decreased. Clinical informatics have been developed for information exchange and rapid response of electronic health records for emergencies (Ye et al., 2020).

13.3.6 M-HEALTH

Displaced populations who live in informal settlements or in crowded camps such as tents and abandoned buildings are often without access to essential resources like food, soap, and water, making them susceptible to COVID-19. A novel mobile health (m-health) platform named HERA App, an open-source mobile application, was adapted for health access for vulnerable displaced populations to improve their health and to provide preventive maternal and child health services for Syrian refugees in Turkey. It was used for performing symptomatic assessment, disseminating health education, and bolstering national prevention efforts as an innovative, cost-effective, and user-friendly approach. It is beneficial for receiving healthcare appointment reminders such as for antenatal visits and childhood immunizations; accessing health-related communication, storing medical records, contacting emergency services, and navigating the Turkish healthcare system (Narla et al., 2020).

Clinical decision support provides screening criteria, information on specimen acquisition, requirements for personal protective equipment, and expectations on a test. Generating real-time alerts during clinic visits to aid in case identification was developed in Taiwan within a national health insurance database to quarantine high-risk individuals at home and for remotely monitoring for the development of symptoms with mobile device applications (Reeves et al., 2020).

13.3.7 MONITORING AND SUPPORT FOR ELDERLY PEOPLE

Mobile networks (3G/4G and 5G), Bluetooth devices, wireless networks, and sensors can communicate with each other. Long-Range (LoRa) technology wireless communication networks (radio frequency) and connection to the things network with various sensors are used to monitor elderly people, known as remote monitoring. LoRa wide area network (LoRaWAN) gateway collecting data from sensors sends data to the internet, and then it is sent to the IoT. Data servers control all data, and results of IoT analytics and artificial intelligence can transfer data in a shorter time with machine learning algorithms (Lousado & Antunes, 2020).

13.4 HEALTH INFORMATICS AROUND THE WORLD

IOT; AI; augmented reality and virtual reality; following travel history and cell phone areas in Taiwan, in China, and in South Korea; a portable application *"Trace Together"* in Singapore that works by trading IDs through Bluetooth; cell phone information of COVID-19 patients in Israel; Aarogya Setu in India for giving advice about the hazards of COVID-19; robots; and drones are the main applications and technologies most widely used around the world during the pandemic (Kumar et al., 2020a).

13.4.1 EUROPEAN UNION (EU)

Mobile applications have been developed by many countries for proximity and contact tracing. News, general alerts, general instructions to avoid infections, and maps to avoid hotspots are information given by these apps. Medical support for self-diagnosis, reporting, and information access are also provided. Maps of hotspots and alerts to populations are sent to people in some countries through cell phone messages or mail. Websites with health questionnaires; use of bracelets measuring skin temperature, pulse, respiration and blood flow; use of smart cameras allowing for facial recognition, thermal scans, and innovative solutions to fight COVID-19; remote control by drones and robots; and mandatory virus testing are digital solutions and tools applied in different EU countries (Council of Europe, 2020). Drones were used in some European countries to implement full lockdowns (Kummitha, 2020). Identification and movement data have been carried out by telecommunication providers by sending SMS warnings to end users (Council of Europe, 2020).

To bring experts of data management to single platform, ELIXIR Project has been launched by EU countries. ELIXIR not only builds a strong networks experts but also develops toolkits for data management. COVID-19 workflow and fast-tracked instances were developed in some countries in the European Union. Bringing all bioinformaticians together on COVID-19 and defining guidelines for data sharing efficiently are the main achievements of this project (elixir-europe.org/services/covid-19-resources). Collecting GPS location for 28 days can be done by Safe Paths and GeoHealth. People who tested positive for the virus can share their data with these application to health officials. The Corona Data Donation smart watch App getting anonymous data from volunteers was developed in Germany to track down the infections (Kummitha, 2020).

Travelling from one city to another requires coordination among cities. Different uncoordinated regional and national healthcare institutions can be a problem for sharing data on time by not allowing them to use big data effectively. Decentralized administrative regimes in the EU have the problem of collecting data together, preventing creation of effective strategies against COVID-19 (Kummitha, 2020).

13.4.2 CHINA

Healthcare informatics have been used in China for epidemic monitoring, detection, early warning, prevention, and control. Internet hospitals and WeChat, big data analyses (including digital contact tracing through QR codes or epidemic prediction), cloud computing, Internet of things, artificial intelligence (including the use of drones, robots, and intelligent diagnoses), 5G telemedicine, and clinical information systems were used in China (Ye et al., 2020). Sophisticated technologies such as the techno-driven approach have helped Chinese governments to get the COVID-19 pandemic under control in the first three months. AI technology is used to diagnose people infected with COVID-19 in 10 seconds instead of CT scan results in 15 minutes. Drones equipped with cameras are used by operators to detect people not wearing masks. Instructions and warnings to obey the rules are sent to people not wearing masks. Not allowing people to leave quarantined houses can be controlled by cameras as there is one camera for every six residents in Chongqing in China (Kummitha, 2020).

AI body temperature screening system: an AI-based system screening 15 patients every second from a maximum distance of 3m is used to measure body temperature to detect people having COVID-19, as the basic symptoms are identified as contactless temperature detection. These systems were put in metro stations in China. Baidu, another AI firm, have developed a system to scan 200 people per minute. Infrared systems were established in famous public areas, and as the people pass these systems, their body temperatures are shown on machines. Smart helmets are used by police to detect people having high temperatures within a 5-m radius. Thermal scanners were installed in all major train stations in China. When an infected person is detected, he/she is put in a local isolation room and then his/her travel history is determined to find potential infected people. Body and face identification is combined with AI to help officers to identify patients quickly in China. Drones are also used to control the full lockdown (Kummitha, 2020).

A QR code is given to people to track their movements. If they use buses or other transportation systems or are going to markets, they need to scan their QR code. If an infected person is found, a quarantine procedure starts for him and people in contact. A red color code for high risk, a yellow code for self-isolate for two weeks, and green code meaning access to the city are given. IoT devices can be used to determine the travel history of citizens. If someone has traveled to red zones or places with COVID-19 infections, AI applications can make reports. Alipay and WeChat apps are used track down people infected with COVID-19. All data is collected in a central system to follow virus testing potential and plan resources in Wuhan, China. Graphics processing unit-0 accelerated AI is used in hospitals to diagnose COVID-19 patients. The app Close Contact Detector is used to identify whether a person has a close contact with a COVID-19 patient. The travel history of positive patients is sent by social media platforms to find citizens for the self-isolation process. A matured smart city-based ecosystem combines governments and firms with mobile applications and helps citizens whether they are in contact with infected persons or not. Citizens register their health conditions by using mobile applications such as DingTalk and Alipay Apps for not visiting hospitals to prevent the spread of the virus (Kummitha, 2020).

13.4.3 SOUTH KOREA

Minimizing mortality, flattening the epidemic curve, and limiting the socio-economic burden are the success of South Korea against the COVID-19 pandemic by using the latest digital technologies. The cellular broadcasting service has been used to send emergency alert text messages as short emergency alerts and guidelines about the COVID-19 outbreak in the coordination of mobile telecom base stations. Provincial and municipal governments can also send individual or public messages in that system. An AI-based chatbot in municipal authorities is used for patients under active monitoring to check for fever or any respiratory symptom by sending check-up results automatically via email including AI speakers in a patient's home. The number of the confirmed cases and deaths, prevention guidelines, and information are sent automatically in that system. A smart working system was applied by providing the private companies' software solutions of remote working with free solutions and related infrastructural and technical supports through governments. Remote education has been started by supporting digital access in the vulnerable class and lending computers and smart tablets. Smartphone data and subscription charges are excluded while accessing the Korea Educational Broadcasting System website. Online simulation software and platforms were created for university students (Heo et al., 2020).

A self-diagnosis mobile application is developed for monitoring any possible COVID-19 symptoms of inbound passengers to report their health once a day during

14 days of mandatory self-quarantine. A self-quarantine safety mobile application was developed for monitoring people under self-quarantine. A GPS-based location tracking is used to prevent violations of self-quarantine orders. A smart quarantine information system (SQIS) for identifying inbound passengers from high-risk regions and monitoring them during the latent period of the infection and an epidemiological investigation support system (EISS) for identifying the transmission routes and places are other digital applications of Korean governments. Telemedicine service, information sharing with open public data, government's fast approval and immediate funding, AI installation for chest radiography detection, and innovative walk-through testing centers (K-Walk-Thru) are smart applications of Korean governments (Heo et al., 2020).

13.4.4 JAPAN

Digitalization and the adoption of AI have been accelerated by Japan's government during the COVID-19 period known as the "new normal" age after the pandemic. Electronic contract services (digital documents, records of transmissions and receipts, and digital signatures) instead of traditional written contracts decreasing the amount of paperwork have been developed to decrease cost and contacts mainly in administrative and judicial sectors during the pandemic. Introduction of digital currencies is a new project of Japan's government (Lida, 2020). Restricting contact with hospitalized patients is aimed to decrease the spread of infections within hospitals. Digital therapeutics with noninvasive personal sensors is a new category of digital medicine (Uschamber, 2020). Online fitness, research and development support for AI drug discovery, and online practice are three major areas of digital health investments in healthcare in Japan. Telemedicine integrated to HER for medical examinations, diagnosis, and prescriptions is part of digital care available to all patients in Japan. AI-based medical devices for detecting lesions in the colon, discerning pulmonary nodules, lung cancer diagnoses, and COVID-19-related lesions have been put into use as AI hospital concepts (Cosmo, 2020).

High-sensitivity cameras and laser beam guidance experiments have been carried out at Toho University in Japan to show saliva spray during a sneeze containing thousands of viruses such that droplets fall off due to their heavier weight while aerosols could remain for up to three hours in the air due to relatively small size depending on air flow rate, humidity, and dryness. Other body fluids of infected people such as feces, blood, oral fluids, anal secretions, tears, and urine can contain SARS CoV-2, attacking mainly the respiratory system (Goel et al., 2020).

13.4.5 TURKEY

The hospital information management system project integrating all systems and all medical and non-medical information and communication systems have started to digitalize the healthcare services in Turkey. The central physician appointment system, pharmaceutical tracking system, decision support system, and e-Pulse application facilitating communication between the patient and the physician and enabling access health data over the internet are some developments in digital transformation in Turkey (Sahiner & Özer, 2020). "Life fits inside the house" (HES) app is a pandemic isolation tracking project giving real-time location data in Turkey. It follows the movement of people for self-isolation. A warning via SMS is sent to people if they leave their homes to return to isolation. A HES code is used mandatory for people wanting to travel by train and plane (Council of Europe, 2020).

13.4.6 MALAYSIA

The Internet of Things was widely used to combat COVID-19 in intelligent transport systems for e-parking, taxi reservations, bus transportation information, travel and train information tips, smart cities and energy sector, etc. Mapping IoT mobile data can be used for mapping purposes from infected patients with a geographic informa-tion system. Medical IoT (MIoT) has been initiated in Malaysia for monitoring the patients and improving their wellness. Smart thermal detection relying on thermal cameras within the infrared (IR) section of the electromagnetic spectrum is used to detect temperatures higher than the threshold in order to decrease the spread of COVID-19. A smart IoT thermal detection solution has been used without needing to be in proximity with infected people through a Wi-Fi or 4G mobile network with-ina range of 1–3m and by scanning 20 people at maximum capacity in an instance. Surveillance network based on unmanned aerial vehicles (UAVs) or drones can be used to determine suspicious activities or crowds, hazardous activities: forest fires, rivers flooding, volcanic activity, and facial recognition. Using robot applications, tracing systems based on IoT include sensors and GPS for location tracking of people and medical equipment, surgical masks and sanitizer shipments, patient monitoring tools to remotely provide their services, body temperature sensors at hospitals, and wearable IoT (WIoT) are some other projects applied in Malaysia (Saadon et al., 2021).

13.4.7 RWANDA

Recording travel info, checking symptoms, risk assessment, contact tracing by phone apps, etc. have been done in some African countries. 3D printing for personal pro-tection equipment (PPE) like nasopharyngeal swabs, valves, face shields, facemasks, respiratory valves, hands-free devices, safety goggles, isolation wards, ear saver, iso-lation chambers, field respiratorsand ventilators, indigenous low-cost ventilators and repurposed phone-based HIV self-care, taking temperature, monitoring the health of patients, and delivering food and medicine are some emerging technological applica-tions in these countries (Maharana et al., 2021; Agarwal, 2022).

The workload of healthcare decreased with efficient diagnosis procedure, quick interventions, and monitoring. Artificial intelligence (AI), robotics for decreasing contact and viral spread to doctors and other medical staff by measuring blood pressure, body temperature and oxygen saturation, data recording in hospitals, drones for broadcasting the appropriate information to the public, geolocalized hotspot map-ping used for contact tracing, and self-testing of unstructured supplementary services data (USSD) have been applied in Rwanda during the pandemic with the help of the

United Nations. Ultraviolet (UV) can be used to disinfect surfaces by robots and the temperature of patients can be measured in isolation rooms by robots. Healthcare workers and patients can be educated by robots about the danger of the virus. AI can give warnings about danger zones. Breathing tests and thermoscan machines are used for temperature measurements of passengers with a high fever. Food and medicine are transported by robots for reducing the risk of exposure to infection (Musanabaganwa et al., 2020).

13.4.8 URUGUAY

All health services of the country, the Ministry of Health, self-monitoring, remote patient monitoring, and telemedicine were integrated in Uruguay to provide a better healthcare system, making full tracking more efficient for following citizens' and patients' situation in the pandemic by software engineering projects done by governments and private organizations where technology can play a key and strategic role in the healthcare system. Preventing contact with possibly infected people by phone calls, preventing the collapse of the health system, being proactive in managing the epidemic, making operative solutions, and providing secure data management are the aims of this new development. To create a better communication, WhatsApp, Facebook Messenger, and WeChat were integrated in the system. Coronavirus.Uy systems including a contact-tracking system for exposure notification were developed. Doctors and nurses were protected by integrating telemedicine into the system to avoid contact with infected people (Milano et al., 2020).

13.4.9 THAILAND

Self-screening for COVID-19 including educational materials about health literacy of the virus and important telephone numbers for self-assessment inputting any signs and symptoms and risk exposures, countries and places visited in the last 14 days and their occupation in Thai, English, and Chinese languages to evaluate the possibility of having a contact with the virus in tabulated data format with a color notice where red identifies as suspected COVID-19, "Medical evaluation is immediately needed"; Self-Health Check at Chiang Mai International Airportfor COVID-19 targeting Thai and foreign visitors coming to Thailand from high-risk countries by downloading that application to report their health condition about COVID-19 daily for 14 days so that travelers can report themselves by that application in case of symptoms of COVID-19 and responsible healthcare staff can contact and guide risky travelers within Thailand and Chiang Mai. COVID-19 (CMC-19) hospital information systems have been developed to manage increasing patients under investigation (PUI) throughout the healthcare system, to assist in allocating resources among hospitals with real-time information systems on suspect cases, and to inform patients that all data from different stakeholders is collected within one platform and applications of technologies are applied based on a bottom-up approach in order to decrease the workload of healthcare staff and tracing people in Thailand as shown in Figure 13.5 (Intawong et al., 2021).

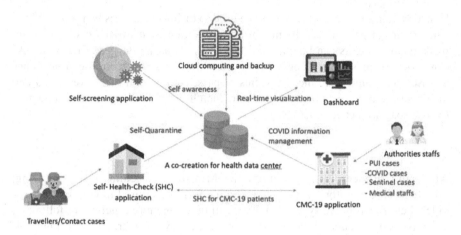

FIGURE 13.5   Overview of a public health platform during COVID-19 in Thailand (Intawong et al., 2021).

13.5   CHALLENGES OF SMART TECHNOLOGIES IN HEALTH

Real-world validation has to be done so that technology has function, accuracy, and robustness. Registration, connecting appropriate clinicians, and data protection are to be validated. High security, high privacy, high reliability, energy efficiency, low response time, transparency for end user, high availability, high performance, high effectiveness, high agility, low cost, and high flexibility were found to be the main advantages of CC while low intelligence, sensor failure, signal interference, difficult access, low security, and low privacy can be disadvantages if suitable precautions are not taken (Pashazadeh & Navimipour, 2018).

Costs of tele-health ophthalmic equipment and additional personnel training are potential barriers to new technology adoption amongst physicians and patients, and heterogeneity in the insurance policy and medico-legal regulations are key challenges. Cases of poor image quality, the legal ramifications of adopting AI for system failure or misdiagnosis from autonomous AI systems, data integrity, protection and cyber security, limited computer literacy, speed of internet, discrimination by AI based on ethnicity, socio-economic status, religion etc., and biases in algorithms are other challenges in new digital technologies in healthcare.

13.5.1   PRIVACY

Protecting personal privacy and the collection, reasonable use, and exchange of personal information are necessary during the pandemic. The consent of the person is required to use personal information. Privacy protection-related laws are required in future for the exchange of epidemiological data among governments, agencies, and communities (Ye et al., 2020). The leakage of important information at the network gateway, malicious codes and information over the network, denial of service attacks, and replay attack are major types of attacks (Tewari & Gupta, 2020). Increased risk of enclosure and subsequent exposure of children to the internet has become a risk for families due to inappropriate content, cyberbullying, and misinformation about health, affecting the mental wellbeing negatively such that strict privacy protection guidelines are required for children (Lampou, 2020).

Preventing the widespread infection of the virus by methods of smart technologies has been done by China's regime so that they have little freedom for expressing their opinions. The techno-driven approach of China and the human-driven approach of Western democracies have different privacy approaches that the majority of interventions have drawn from the central government in China. A techno-driven approach is directly applied in China to identify the most-likely infected individuals, to create awareness for isolation, and to track citizens with their mobile phones, while data is collected anonymously for the protection of privacy and both technology and manual lockdowns are tried since data protection laws in the Western democracies ensure personal data protection. Voluntarily share of data is advised, limiting the government's options to collecting data from citizens directly (Kummitha, 2020).

Facial recognition technologies are banned due to privacy concerns in the USA. Transparency is required in collecting and preserving data, thereby ensuring personal data protection. Although smart cities have adopted the IoT to a different degree in the EU, they collect private data. Mobile operators share anonymous and aggregated data about the concentration and movements of the citizens in areas where COVID-19 is prevalent, and the USA has conducted talks with Facebook and Google to access its anonymized data (Kummitha, 2020). "By default, only personal data which are necessary for each specific purpose of the processing are processed" is applied in the EU for the privacy of patients. Not requiring registration or the divulgation of any private data can be a way of protecting patients' data by registering it as anonymous in a central system (Golinelli et al., 2020).

Healthcare systems using IoT can be protected with chaos-based cryptographic applications for securing patients' privacy. Risk of malicious attacks through internet networks is a threat for privacy (EL-Latif et al., 2020). More sophisticated IoT healthcare services can improve the trust of users (Kim & Kim, 2018). Reliable data mining, qualified services, and enhanced user privacy and security can increase the trust of users (Nord et al., 2019). Blockchain systems can be used to transmit data in a secure way, collecting data from body sensors. Unauthorized access can be prevented in that system by a digital signature algorithm (Islam & Shin, 2020).

13.5.2 BIAS

Systematically biased data can give wrong observations, interpretations, and recommendations, leading to under-representation of some races, ages, genders, etc. Biased from collected data, work flows and environment can lead to wrong outcomes. Training AI from these data resources has the risk of inaccurately generalizing to non-representative populations. Criminal justice sentencing, human resource hiring, education, and other systems are other bad sides of these outcomes, resulting in inequalities and discrimination (Matheny et al., 2019).

13.5.3 INFODEMIC

The spread of "fake news" and rumors, as well as ethical and privacy issues, are challenges of smart technologies and AI and big data technologies are to be designed with these considerations. The spread of COVID-19 misinformation that causes fear and panic, creating COVID-19 stigmatization and anxiety and generation of noisy data, are challenges of social media platforms. Irrational fear due to the infodemic can have long-term results on mental wellbeing like depression, anxiety, insomnia, etc. (Lampou, 2020). Non-reliable sources in the form of videos, articles, interviews, and pictures become information sources when there are not clear answers for something and people believe in false claims through repeated exposure to them (Anzar et al., 2020). The WHO director highlights "we're not just fighting an epidemic (pandemic); we're fighting an infodemic" due to wide use of social media in Western countries (Kummitha, 2020). Furthermore, the WHO general director stated, "we're not just battling the virus, we're also battling the trolls and conspiracy theories that undermine our response" (BBC News, 2020). False information is disseminated knowingly by trolls and conspiracy theorists, political activists, scammers, alternative news media, and hostile governments (Doyle & Conboy, 2020). Increasing the transparency of information on epidemics can prevent rumors and decrease public panic, making them confident in the measures taken to combat epidemics (Ye et al., 2020).

13.5.4 INCONSISTENCIES

China was blamed for concealing information from the public and international community (Kummitha, 2020). Inconsistent and incomplete data can cause failures and incorrect data estimation. Vital signs can be misused due to activities and the surrounding environment. Data analytic techniques such as machine learning are applied on cloud servers, so they can cause inconsistencies. Sensors disunited from the body, losing connection, and running out of battery can cause missing data. Forgetting to use wearable sensor(s) can be another problem of missing data. Missing data can generate biased estimates leading to high error rates in health applications. Personalized models called personalized pooling are used for implying missing values. The last observation carried forward imputation, regression imputation, hot-deck imputation, cold-deck imputation, and K- Nearest- Neighbor imputation are methods to estimate missing data. Moreover, artificial neural networks, support vector machine, and generic algorithms can be used to evaluate missing data (Azimi et al., 2019).

13.5.5 QR CODES CHALLENGES

All data is registered in the servers such as where they go and with whom they talk, categorized as automated social control that governments may use this data in the future for their political agenda and may control people in normal life too. "It divides people up based on where they're from" and is a discrimination, causing regional prejudice. There are also code dilemma: "Even if a yellow code or a red code appears, don't be nervous," is a problem because many people do not know why they have a red code and cannot go work (NY, 2020).

13.5.6 Internet Quality Issues

A good internet speed to work properly for audio, video quality, and transfer of data is necessary. Online traffic is to be prevented for good connection, reducing network congestion. Extra bandwidth is required to recover faster. Furthermore, algorithms can be used to get better video resolution. A good internet speed to work properly for audio, video quality, and transfer of data is required and also reduces network congestion and can increase the recovery speed (Alashhab et al., 2020).

13.5.7 Wearables Devices Challenges

Validation: validation of most consumer wearables such as home blood pressure measurement devices mainly sold in online marketplaces is a challenge in this kind of device market. Validation for accuracy of performance is required to measure the effectiveness of devices.

Accuracy of findings: accuracy of wearable devices is another challenge because a low rate of accuracy can endanger the life of patients.

Clinical implications: lack of robust evaluation of the clinical outcomes is another challenge creating an uncertain impact using those findings, making physicians unsure about thresholds.

Implementation: pre-prototype stage economic and health technology assessments for evaluating their affordability in the healthcare system is another challenge because most devices cannot reach the real-world implementation or are taking a long time (Brahmbhatt et al., 2021).

13.5.8 Other Challenges

Difficulties in construction of IoT skills of small and medium enterprises, high costs of deployments needing supports from governments, medical standards and regulations, talented workers in smart technologies, and collection and management of data are other challenges of informatics in healthcare (Al-Ogaili et al., 2021). Different formats of data cannot be easily integrated in cloud computing. Re-use of older devices can be infected by cybercriminals with malware (Botnets) increasing internet traffic resulting in distributed denial of service attacks that need to be updated (Alashhab et al., 2020). Indistinguishability from other viral pneumonia and the hysteresis of abnormal CT are other problems in diagnosing COVID-19 (Kumar et al., 2020b). Addiction to digital technology/media and dependence on digital technology causing mental problems are other issues (Kaur et al., 2020).

13.6 CONCLUSION

Remote services and digital transactions will increase in the future with digital transformations. Digital technologies have been an effective way of fighting the COVID-19 pandemic in China's case, but privacy concerns are still questionable because these healthcare data and tracks can be used by governments for their political agendas. The infodemic has decreased greatly in China with censorships, and many people do not trust Chinese data because they hide information, while an open data share policy is applied in Western countries. Strong centralization in administration is an advantage in China, while lack of coordination has been in seen in Western countries.

The Spanish Flu pandemic (1918–1920) was discovered after 15 years, but now COVID-19 has been discovered in weeks (Lee & Trimi, 2021). Developed digital technologies and innovations have increased the capability of humanity to fight against outbreaks and make them more resilient by increased tracking and diagnosing features. Remote healthcare management and control, strong collaboration at the national and international level, and developed prediction capabilities for the outbreak to plan resources are the key actions against COVID-19. Agile innovations like producing medical ventilators and production of hand sanitizer in a short time, new solutions of mask shields, collaborating for searching for new vaccines, etc. are other examples of successful attempts. Lack of international standards and validation of wearable devices is another important issue at the international level.

REFERENCES

Agarwal, R. (2022). The personal protective equipment fabricated via 3D printing technology during COVID-19. *Annals of 3D Printed Medicine*, *5*, 100042. https://doi.org/10.1016/j. stlm.2021.100042.

Alashhab, Z. R., Anbar, M., Singh, M. M., Leau, Y. B., Al-Sai, Z. A., & Alhayja'a, S. A. (2021). Impact of coronavirus pandemic crisis on technologies and cloud computing applications. *Journal of Electronic Science and Technology*, *19*(1), 100059. https://doi. org/10.1016/j.jnlest.2020.100059.

Aminullah, E., & Erman, E. (2021). Policy innovation and emergence of innovative health technology: The system dynamics modelling of early COVID-19 handling in Indonesia. *Technology in Society*, *66*, 101682. https://doi.org/10.1016/j.techsoc.2021.101682.

Anzar, W., Baig, Q. A., Afaq, A., Taheer, T. B., & Amar, S. (2020). Impact of infodemics on generalized anxiety disorder, sleep quality and depressive symptoms among Pakistani social media users during epidemics of COVID-19. *Merit Research Journal of Medicine and Medical Sciences*, *69*, DOI: 10.5281/zenodo.3727246.

Aslan, I. (2019). The role of industry 4.0 in occupational health and safety. *International European Congress on Social Sciences*, *IV*, 334–345, Diyarbakır, Turkey.

Aslan, I. (2021). *Technologies and applications internet of things (IoT) in healthcare*. Applications of Big Data in Large- and Small-Scale, IGI Global Systems.

Aslan, I., & Aslan, H. (2019). *Industry 4.0: The role of industry 4.0 in agri-food, process safety & environmental protection*. Chapter I: Industry 4.0: Industry 4.0 Concept and Common Technologies & Systems, Lap Lambert Academic Publishing, ISBN: 978-620-0-49906-6.

Attaran, M. (2020). Blockchain technology in healthcare: Challenges and opportunities. *International Journal of Healthcare Management*, 1–14. DOI: 10.1080/20479700.2020.1843887.

Azimi, I., Pahikkala, T., Rahmani, A.M., Niela-Vilén, H., Axelin, A., & Liljeberg, P. (2019). Missing data resilient decision-making for healthcare IoT through personalization: A case study on maternal health. *Future Generation Computer Systems*, *96*, 297–308.

BBC News (2020). Coronavirus: WHO chief warns against 'trolls and conspiracy theories'. *BBC News*. Accessed Date: 14 June 2020. www.bbc.com/news/world-51429400.

Berchin, I. I., & Andrade Guerra, J. B. S. O. (2020). GAIA 3.0: Effects of the Coronavirus Disease 2019 (Covid-19) outbreak on sustainable development and future perspectives. *Research in Globalization*. http://dx.doi.org/10.1016/j.resglo.2020.100014.

Bhatia, M., & Sood, S.K. (2017). A comprehensive health assessment framework to facilitateIoT-assisted smart workouts: A predictive healthcare perspective. *Computers in Industry*, 92–93, 50–66. http://dx.doi.org/10.1016/j.compind.2017.06.009.

Blomberg, N., & Lauer, K. B. (2020). Connecting data, tools and people across Europe: ELIX-IR's response to the COVID-19 pandemic. *European Journal of Human Genetics*, 28(6), 719–723. https://doi.org/10.1038/s41431-020-0637-5.

Brahmbhatt, D. H., Ross, H. J., & Moayedi, Y. (2021). Digital technology application for improved responses to health care challenges: Lessons learned From COVID-19. *Canadian Journal of Cardiology*, 38(2), 279–291.

Chen, M., Xu, S., Husain, L., & Galea, G. (2021). Digital health interventions for COVID-19 in China: a retrospective analysis. *Intelligent Medicine*, 1(01), 29–36. https://doi.org/10.1016/j.imed.2021.03.001.

Cosmo (2020). *The acceleration of digital health in Japan amid COVID-19*. Accessed Date: 21 December 2020. http://cosmopr.co.jp/en/the-1st-cosmo-innovation-seminar-en/.

Council of Europe (2020). *Digital Solutions to Fight COVID-19*, Accessed Date: 21 December 2020. https://rm.coe.int/prems-120820-gbr-2051-digital-solutions-to-fight-covid-19-text-a4-web-/16809fe49c.

Dhanvijay, M.M., & Patil, S.C. (2019). Internet of Things: A survey of enabling technologies in healthcare and its applications. *Computer Networks*, 153, 113–131. https://doi.org/10.1016/j.comnet.2019.03.006.

Donsimoni, J. R., Glawion, R., Plachter, B., & Wälde, K. (2020). Projektion der COVID-19-Epidemie in Deutschland. *Wirtschaftsdienst*, 100(4), 272–276, DOI: 10.1007/s10273-020-2631-5.

Doyle, R., & Conboy, K. (2020). The role of IS in the covid-19 pandemic: A liquid-modern perspective. *International Journal of Information Management*, 55, 102184. https://doi.org/10.1016/j.ijinfomgt.2020.102184.

Econsultancy (2019). *10 examples of the Internet of Things in healthcare*. Accessed Date: 5 May 2020. https://econsultancy.com/internet-of-things-healthcare/.

EL-Latif, A.A., Abd-El-Atty, B., Abou-Nassar, E., & Venegas-Andraca, S.E. (2020). Controlled alternate quantum walks based privacy preserving healthcare images in Internet of Things. *Optics and Laser Technology*, 124. https://doi.org/10.1016/j.optlastec.2019.105942

Farrugia, G., & Plutowski, R. W. (2020, August). Innovation lessons from the COVID-19 pandemic. In *Mayo clinic proceedings* (Vol. 95, No. 8, pp. 1574–1577). Elsevier. https://doi.org/10.1016/j.mayocp.2020.05.024.

FDA (2020). *Remote or Wearable Patient Monitoring Devices EUAs*. Accessed Date: 21 December 2020. www.fda.gov/medical-devices/coronavirus-disease-2019-covid-19-emergency-use-authorizations-medical-devices/remote-or-wearable-patient-monitoring-devices-euas.

Goel, S., Hawi, S., Goel, G., Thakur, V. K., Agrawal, A., Hoskins, C., . . . & Barber, A. H. (2020). Resilient and agile engineering solutions to address societal challenges such as coronavirus pandemic. *Materials Today Chemistry*, 17, 100300. https://doi.org/10.1016/j.mtchem.2020.100300.

Golinelli, D., Boetto, E., Carullo, G., Nuzzolese, A. G., Landini, M. P., & Fantini, M. P. (2020). How the COVID-19 pandemic is favoring the adoption of digital technologies in healthcare: a literature review. *MedRxiv*, https://doi.org/10.1101/2020.04.26.20080341.

Heo, K., Lee, D., Seo, Y., & Choi, H. (2020). Searching for digital technologies in containment and mitigation strategies: Experience from South Korea COVID-19. *Annals of Global Health*, 86(1), 109. DOI: http://doi.org/10.5334/aogh.2993.

Intawong, K., Olson, D., & Chariyalertsak, S. (2021). Application technology to fight the COVID-19 pandemic: Lessons learned in Thailand. *Biochemical and biophysical research communications*, 534, 830–836. https://doi.org/10.1016/j.bbrc.2020.10.097.

Islam A., & Shin S.Y. (2020). A blockchain-based secure healthcare scheme with the assistance of unmanned aerial vehicle in Internet of Things. *Computers and Electrical Engineering*, 84. https://doi.org/10.1016/j.compeleceng.2020.106627

Kamani, V. (2019). *5 ways cloud computing is impacting healthcare*. Accessed Date: 18 December 2020. www.healthitoutcomes.com/doc/ways-cloud-computing-is-impacting-healthcare-0001.

Kamaraj, S. K. (2020). The perspective on bio-nano interface technology for COVID-19. *Frontiers in Nanotechnology*, *2*, 18, doi: 10.3389/fnano.2020.586250.

Kaplan, B. (2020). Revisiting health information technology ethical, legal, and social issues and evaluation: Telehealth/telemedicine and COVID-19. *International journal of medical informatics*, *143*, 104239. https://doi.org/10.1016/j.ijmedinf.2020.104239.

Kaur, D., Sahdev, S. L., Chaturvedi, V., & Rajawat, D. (2020). Fighting COVID-19 with technology and innovation, evolving and advancing with technological possibilities. *International Journal of Advanced Research in Engineering and Technology (IJARET)*, *11*(7), 395–405, DOI: 10.34218/IJARET.11.7.2020.039.

Khan, A. W., Kader, N., Hammoudeh, S., & Alabdulla, M. (2021). Combating COVID-19 pandemic with technology: perceptions of mental health professionals towards telepsychiatry. *Asian Journal of Psychiatry*, *61*, 102677. https://doi.org/10.1016/j.ajp.2021.102677.

Khan, O. F., Bebb, G., & Alimohamed, N. A. (2017). What oncologists need to know about its potential—and its limitations. *Artificial Intelligence in Medicine*, *16*(4).

Kim, S., & Kim, S. (2018). User preference for an IoT healthcare application for lifestyle disease management. *Telecommunications Policy*, *42*, 304–314. http://dx.doi.org/10.1016/j.telpol.2017.03.006

Kumar, K. K., Ramaraj, E., Geetha, P., & Srikanth, B. (2020a). A Study on Evolving Technologies for COVID-19 contact Tracking. *Solid State Technology*, *63*(5), 4459–4467.

Kumar, R., Nagpal, S., Kaushik, S., & Mendiratta, S. (2020b). COVID-19 diagnostic approaches: Different roads to the same destination. *Virusdisease*, *31*(2), 97–105. https://doi.org/10.1007/s13337-020-00599-7.

Kummitha, R. K. R. (2020). Smart technologies for fighting pandemics: The techno-and human-driven approaches in controlling the virus transmission. *Government Information Quarterly*, *37*(3), 101481. https://doi.org/10.1016/j.giq.2020.101481.

Kutlu, R. (2020). What we have learned about the new coronavirus pandemic, current diagnostic and therapeutic approaches and the situation in Turkey. *TJFMPC*, *14*(2), 329–44, DOI: 10.21763/tjfmpc.729917.

Lampou, R. (2020). Socioeconomic changes, digital technologies and neuro education during the COVID-19 era. *Homo Virtualis*, *3*(2), 28–42. doi:https://doi.org/10.12681/homvir.25447.

Lee, S. M., & Trimi, S. (2021). Convergence innovation in the digital age and in the COVID-19 pandemic crisis. *Journal of Business Research*, *123*, 14–22. https://doi.org/10.1016/j.jbusres.2020.09.041.

Lida, J. (2020). Digital transformation vs COVID-19: The case of Japan. *Digital Law Journal*, *1*(2), 8–16. https://doi.org/10.38044/2686-9136-2020-1-2-8–16.

Lousado, J. P., & Antunes, S. (2020). Monitoring and support for elderly people using lora communication technologies: Iot concepts and applications. *Future Internet*, *12*(11), 206, doi:10.3390/fi12110206.

Maharana, A., Amutorine, M., Sengeh, M. D., & Nsoesie, E. O. (2021). COVID-19 and beyond: Use of digital technology for pandemic response in Africa. *Scientific African*, *14*, e01041. https://doi.org/10.1016/j.sciaf.2021.e01041.

Marbouh, D., Abbasi, T., Maasmi, F. et al. Blockchain for COVID-19: Review, opportunities, and a trusted tracking system. *Arabian Journal for Science and Engineering*, *45*, 9895–9911 (2020). https://doi.org/10.1007/s13369-020-04950-4.

Marbury, D. (2020). *What does the future hold for AR and VR in healthcare?* Accessed Date: 21 December 2020. https://healthtechmagazine.net/article/2020/11/what-does-future-hold-ar-and-vr-healthcare.

Matheny, M., Thadaney Israni, S., Ahmed, M., & Whicher, D. (Eds.) (2019). *Artificial intelligence in health care: The hope, the hype, the promise, the Peril.* NAM Special Publication, National Academy of Medicine.

Mayo (2020). *Remote monitoring of COVID-19 symptoms.* Accessed Date: 21 December 2020. www.mayo.edu/research/remote-monitoring-covid19-symptoms/about.

Medical Technology (2020). *COVID-19 accelerates remote patient monitoring device market growth.* Accessed Date: 21 December 2020. https://medical-technology.nridigital.com/medical_technology_jul20/remote_patient_monitoring_device_market_growth.

Milano, G., Vallespir, D., & Viola, A. (2020). A technological and innovative approach to COVID-19 in Uruguay. *Communications of the ACM, 63*(11), 53–55, DOI:10.1145/3422826.

Mofti, M., A., & Vasudevan, H. (2020). "Motivation" medical team as front liners to fight COVID-19 disease. *International Journal of Social & Scientific Research, 5*(IV), 24–32.

Mukati, N., Namdev, N., Dilip, R., Hemalatha, N., Dhiman, V., & Sahu, B. (2021). Healthcare assistance to COVID-19 patient using internet of things (IoT) enabled technologies. *Materials Today: Proceedings.* https://doi.org/10.1016/j.matpr.2021.07.379.

Musanabaganwa, C., Semakula, M., Mazarati, J. B., Nyamusore, J., Uwimana, A., Kayumba, M., . . . & Nsanzimana, S. (2020). Use of technologies in COVID-19 containment in Rwanda. *Rwanda Public Health Bulletin, 2*(2), 7–12.

Narla, N. P., Surmeli, A., & Kivlehan, S. M. (2020). Agile application of digital health interventions during the COVID-19 refugee response. *Annals of Global Health, 86*(1), DOI: https://doi.org/10.5334/aogh.2995.

Nasrullah, P. (2020). *Internet of things in healthcare: applications, benefits, and challenges.* Accessed Date: 5 May 2020. www.peerbits.com/blog/internet-of-things-healthcare-applications-benefits-and-challenges.html

Nguyen, C. T., Saputra, Y. M., Van Huynh, N., Nguyen, N. T., Khoa, T. V., Tuan, B. M., . . . & Ottersten, B. (2020). A comprehensive survey of enabling and emerging technologies for social distancing—Part I: Fundamentals and enabling technologies. *IEEE Access, 8*, 153479–153507. DOI: 10.1109/ACCESS.2020.3018140.

Nord, J.H., Koohang, A., & Paliszkiewicz, J. (2019). The Internet of Things: Review and theoretical framework, *Expert Systems With Applications, 133*, 97–108. https://doi.org/10.1016/j.eswa.2019.05.014.

Nuffield Council on Bioethics (2018). *Artificial intelligence (AI) in healthcare and research,* Bioethics briefing note, Nuffield Council on Bioethics.

Nugraha, A., Daniel, D. R., & Utama, A. A. G. S. (2021). Improving multi-sport event ticketing accounting information system design through implementing RFID and blockchain technologies within COVID-19 health protocols. *Heliyon, 7*(10), e08167. https://doi.org/10.1016/j.heliyon.2021.e08167.

NY (2020). *In Coronavirus Fight, China Gives Citizens a Color Code, With Red Flags.* Accessed Date: 21 December 2020. www.nytimes.com/2020/03/01/business/china-coronavirus-surveillance.html.

Pashazadeh, A., & Navimipour, N. J. (2018). Big data handling mechanisms in the healthcare applications: A comprehensive and systematic literature review. *Journal of Biomedical Informatics, 82*, 47–62. https://doi.org/10.1016/j.jbi.2018.03.014.

Pillai, A. S., Sunil, R., & Guazzaroni, G. (2022). Leveraging immersive technologies during the COVID-19 pandemic—opportunities and challenges. In *Extended reality usage during COVID 19 pandemic* (pp. 75–87). Springer.

Prada, E.J.A. (2020). The internet of things (IoT) in pain assessment and management: An overview. *Informatics in Medicine Unlocked, 18.* https://doi.org/10.1016/j.imu.2020.100298.

Qi, J., Yang, P., Min, G., Amft, O., Dong, F., & Xu, L. (2017a). Advanced internet of things for personalised healthcare systems: A survey. *Pervasive and Mobile Computing, 41,* 132–149. http://dx.doi.org/10.1016/j.pmcj.2017.06.018.

Qi, X., Liu, C., Song, L., Li, Y., & Li, M. (2017b). PaCYP78A9, a cytochrome P450, regulates fruit size in sweet cherry (*Prunus avium* L.). *Frontiers in Plant Science, 8,* 2076. https://doi.org/10.3389/fpls.2017.02076

Rahmani, A.M., Gia, T.N., Negash, B., Anzanpour, A., Azimi, I., Jiang, M., & Liljeberg, P. (2018). Exploiting smart e-Health gateways at the edge of healthcare Internet-of-Things: A fog computing approach. *Future Generation Computer Systems, 78,* 641–658.

Razvi, S. M. H., & Prasad, N. (2020). COVID-19: a potential threat. *International Journal of Innovative Medicine and Health Science, 12,* 1–4.

Reeves et al. (2020). Rapid response to COVID-19: health informatics support for outbreak management in an academic health system. *Journal of the American Medical Informatics Association,* 0(0), 2020, 1–7, doi: 10.1093/jamia/ocaa037.

Rodin, D., Lovas, M., & Berlin, A. (2020, December). The reality of virtual care: Implications for cancer care beyond the pandemic. In *Healthcare* (Vol. 8, No. 4, p. 100480). Elsevier. https://doi.org/10.1016/j.hjdsi.2020.100480.

Rooney, L., & McNicholas, F. (2020). 'Policing' a pandemic: Garda wellbeing and COVID-19. *Irish Journal of Psychological Medicine, 37*(3), 192–197, doi:10.1017/ipm.2020.70.

Saadon, A., Al-Ogaili, A., Ameer, R., Agileswari, M. B., Marsadek, T. J. T., Hashim, A. S., ... & Aadam, A. (2021). IoT technologies for tackling COVID-19 in Malaysia and worldwide: Challenges, recommendations, and proposed framework. *Computers, Materials, & Continua,* 2141–2164.

Saheb, T., & Izadi, L. (2019). Paradigm of IoT big data analytics in the healthcare industry: A review of scientific literature and mapping of research trends. *Telematics and Informatics, 41,* 70–85.

Sahiner, D. D., & LaçinÖzer, L. (2020). *Digitalisation of the health services in Turkey.* Accessed Date: 21 December 2020. www.lexology.com/library/detail.aspx?g=02970f1d-2f88-4936-a958-d3e8fa8576f5.

Sharma, A., Bahl, S., Bagha, A. K., Javaid, M., Shukla, D. K., & Haleem, A. (2020). Blockchain technology and its applications to combat COVID-19 pandemic. *Research on Biomedical Engineering,* 1–8. https://doi.org/10.1007/s42600-020-00106-3.

Shammi, M., Bodrud-Doza, M., Islam, A. R. M. T., & Rahman, M. M. (2020). COVID-19 pandemic, socioeconomic crisis and human stress in resource-limited settings: A case from Bangladesh. *Heliyon, 6*(5), e04063. https://doi.org/10.1016/j.heliyon.2020.e04063.

Sheikhzadeh, E., Eissa, S., Ismail, A., & Zourob, M. (2020). Diagnostic techniques for COVID-19 and new developments. *Talanta, 220,* 121392. https://doi.org/10.1016/j.talanta.2020.121392.

Shebrain, S., Nava, K., Munene, G., Shattuck, C., Collins, J., & Sawyer, R. (2021). Virtual surgery oral board examinations in the era of COVID-19 pandemic. How I do it! *Journal of Surgical Education, 78*(3), 740–745. https://doi.org/10.1016/j.jsurg.2020.09.012.

Smith, L., Jacob, L., Yakkundi, A., McDermott, D., Armstrong, N. C., Barnett, Y., ... & Tully, M. A. (2020). Correlates of symptoms of anxiety and depression and mental wellbeing associated with COVID-19: a cross-sectional study of UK-based respondents. *Psychiatry Research, 291,* 113138. https://doi.org/10.1016/j.psychres.2020.113138.

Tewari, A., & Gupta, B.B. (2020). Security, privacy and trust of different layers in Internet-of-Things (IoTs) framework. *Future Generation Computer Systems,* 108, 909–920.

Tuli, S., Basumatary, N., Gill, S.S., Kahani, M., Arya, R.C., Wander, G.S., & Buyya, R. (2020). HealthFog: An ensemble deep learning based smart healthcare system for automatic diagnosis of heart diseases in integrated IoT and fog computing environments. *Future Generation Computer Systems,* 104, 187–200.

Uschamber (2020). *Leveraging Digital Technologies to Improve Health Care Outcomes: An opportunity for policy collaboration in the U.S.-Japan bilateral relationship.* Accessed Date: 21 December 2020. www.uschamber.com/report/leveraging-digital-technologies-improve-health-care-outcomes-opportunity-policy-collaboration

Vetrò, A., Santangelo, A., Beretta, E., & Martin, J.C.D. (2019). *AI: From rational agents to socially responsible agents.* Digital Policy, Regulation and Governance, pp. 291–304, ISSN: 2398–5038.

Vieira, C. M., Franco, O. H., Restrepo, C. G., & Abel, T. (2020). COVID-19: The forgotten priorities of the pandemic. *Maturitas*, *136*, 38–41. https://doi.org/10.1016/j.maturitas.2020.04.004.

Ye, Q., Zhou, J., & Wu, H. (2020). Using information technology to manage the COVID-19 pandemic: development of a technical framework based on practical experience in China. *JMIR Medical Informatics*, *8*(6), e19515, doi: 10.2196/19515.

Zeadally, S., & Bello, O. (2021). Harnessing the power of internet of things based connectivity to improve healthcare. *Internet of Things*, *14*, 100074.

14 Two-Layered Security Mechanism for Enhanced Biometric Security of IoT-Based Digital Wallet

Tanya Chauhan and Ajmer Singh

CONTENTS

14.1 INTRODUCTION

Demand for the digital wallet system in the IoT system is growing day to day. But there remains the issue of security. There are some existing solutions to security, but there is need of improvement. The IoT systems used to authenticate persons for digital wallet systems are supposed to be more reliable and fast because if it is slow then the quality of service is influenced, and if it is not reliable then the security issue may arise. In order to improve efficiency while protecting the

DOI: 10.1201/9781003245469-14

digital wallet in the IoT system, a bio meter security system has been included in edge detection. Since we all know that the use of digital wallets grows every day, such an efficient methodology is needed to safeguard different digital wallet products. Many techniques or processes that provide security for digital baggage have been available; however, these current methods or tactics are limited. Some of today's safety approaches give a little safety, while others work poorly. Thus, iris-based biometric processes for the safety of digital wallets were proposed with the inclusion of edge detection. Inclusion of a canny mechanism has been made to reduce biometric model spatial usage. In addition, the time consumption of a comparison procedure is also minimized. This digital wallet security solution, based on the iris, will be used in the future as a safe technique above other biometric techniques. In the course of biometric comparison activities, the current study has concentrated on enhancing performance, further reducing file size in order to preserve space. The study offered saved time, space, and efficiency. In the Matlab environment, comparisons between the projected working time and the use of space were simulated.

Certain IoT devices offer digital wallets that may store money digitally. The security of these digital assets is an important problem. To authenticate these digital wallets, biometric techniques [1–2] are introduced. By using its biometric traits and pattern online, it encourages users to input their properties [3]. Some research focused on biometric binary string [4] while some authors considered multispectral trends [5].

This may include retinal, iris [6–8], voice, fingerprint and other biometric properties. However, the use of one biometric function has remained risky for some time. Multiple biometric inputs are then made to secure the digital assets of a non-authentic individual.

In other words, more than one bodily trait is taken into account for biometric safety. The difficulty with many biometric samples is the space utilization and contrast time. Much physical space is taken up by multi-biometric samples. The period when numerous biometric samples are comparable is, furthermore, greater than one. Therefore, the accuracy is compromised when many biometric samples are utilized to increase the IoT security.

14.1.1 WORKING OF THE TRADITIONAL BIOMETRIC MODEL

This section presented the working of the traditional biometric model to provide security to the IoT-based digital wallet. In the traditional biometric system there remains one single biometric sample for verification. If that sample is verified then the user is authorized to access the digital wallet.

14.1.2 NEED OF RESEARCH

The suggested effort should include biometrics into the digital IoT wallet in order to overcome the problems of present investigations. Edge detection is taken into account to minimize biometric content size and to minimize comparison and validation time, while using an iris-based biometric technique to overcome past research limitations.

Traditional model

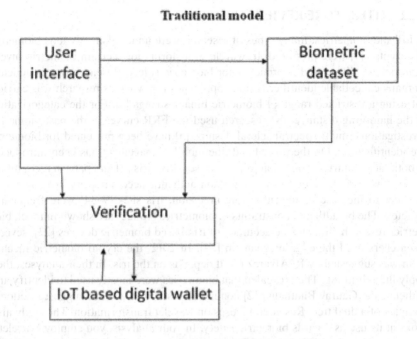

FIGURE 14.1 Traditional security system for IoT-based digital wallet.

The research aims for research to explore and investigate the extent of the biometric safety and the limits of current research on biometric safety in IoT systems. In addition, the study proposes a new technique to reduce time consumption in the deployment of biometric security in IoT. Research, in order to give a more dependable and safe solution, will finally conduct a comparative study of time, safety, storage consumption, and recommended biometric safety.

14.1.3 RESEARCH CHALLENGES

Technological biometrics tests and monitors any physical or inherent behaviors in order to identify individuals. The characteristics of the human body include voice patterns, retinas, fingerprints, irises, and face patterns. In biometric systems, biometrics technologies are employed. It automatically verifies, authenticates, and identifies a human person. On the other hand, there is increasing demand for IoT applications nowadays. Several research items in the IoT and biometrics sector have nonetheless been conducted. The problem that the author works on regarding restrictions on biometrics in IoT systems has been studied before. In addition, current research has time to employ the biometric methods. The need is for biometric safety systems to be improved with precision and performance. A biometric safety method has to be proposed to be able to function with a better-performing IoT environment. The next study aims to increase the security of current biometric safety mechanisms as well as their performance.

14.2 LITERATURE REVIEW

In IoT and biometrics, many types of research were undertaken. Various biometric iris cryptos [1] were proposed. In wavelet transformation, certain biometric investigators have discovered approaches for biometric safety. Research was conducted on many cancellable identifier fusion approaches [3], whereas research was carried out using a restricted range of biometric binary strings [4]. For the categorization of the hamming distance, the research used the FRR curve. In the past biometric investigations, multi-spectral scleral designs [5] have been presented for biometric eye identification. On the other hand, the mobile engagement has been introduced to both face and iris recognition [6]. Safety-sensitive iris [7] and pattern recognition systems were determined to be safe for non-authentic access in prior investigations. In order to increase biometric device precision, iris sites [9–14] were frequently targeted. The breadth and constraints of biometrics [16–22] are shown in much biometrics research. To enhance accuracy of iris-based biometric devices [23], several researchers used the edge mechanism [24]. In 2012, the crypto biometric mechanism was suggested by R. Álvarez [1]. It depends on the iris. In their analyses, they apply fluid thinking. They revealed that biometrics were mostly used to identify and authenticate. Gaurav Bhatnagar [2] conducted a study in 2012 to suggest fractional complex of a dual tree. Research focuses on wavelet transformation. The study also looks at its use as regards biometric safety. In your analysis, you employ Wavelet's transformation. In 2013 Anne M.P. [3] carried out a fusion investigation. A technique is suggested for multi-biometric cancelable identification. Research has offered a technique for biometric safety. Biometric binary strings were extracted in 2011 by C. Chen [4]. A limited area operation was been performed. It used FRR curve to perform hamming distance classification. In 2012, multi-spectral scleral trends were hypothesized by Simona Crihalmeanu [5]. The objective is the identification of eye biometrics. In 2014, Maria de Marisco [6] proposed a technique for the identification of both face and iris. For mobile contact, research is suitable. Research has shown that mobile and cellphones are utilized more and more to affect technology advancement. In 2013 Javier Galbally [7] worked to restore the image using irises. The binary templates operation is carried out. A biometric bimodal technique was proposed by Ujwalla Gawande in 2013. Research is doing fusion in the iris and in fingerprints. In 2014, Marta Gomez-Barrero [9] proposed multimodal biometric systems. As a research approach, the author employed pattern analysis. In 2011, the genetically similar irises of Karen Hollingsworth [10] had the same texture. The author uses image and insight. The placement of the iris was investigated by Farmanullah Jan [11] in the front eye picture in 2012. The objective is to provide a less limited iris identification approach. For the purpose of the study, digital signal processing was applied. The Hough-dependent Iris Location Plan was proposed by Farmanullah Jana [12] in 2013. The Radial Gradient Generator is considered by research. This study was recommended by Hough Transform, a radial gradient operator. The 2013 localization method of non-circular iris was suggested by Farmanullah Jana [13]. Research employs the grey stage statistical and photo projection tool. The exact location of the iris was revealed in 2013 by Farmanullah Jana [14]. The Hough transform is used in research. Histogram bisection and eccentricity were commonly taken into consideration in research. Farmanullah Jana [15] established in 2013 a complicated approach to locate irises that were not circulating. Research involves non-ideal information. The investigation utilizes the iris location. In 2018, the investigation of biometric methods was carried out by Choudhury B. [16]. This research explored the benefits and functions of several biometric technologies. The latest biometric upgrades were verified in 2015 by Sharadha Tiwari [17]. New biometric safety technologies and trends are being tackled. A biometric research was undertaken by Gursimarpreet Kaur [18] in 2014. Research has compared modalities using different biometric methods. Research considers biometrics as an urgent tool for expanding protection problems. The suspect's different biometrics were compared in 2013 with Himanshu Srivastva. The usage of biometrics for safety was considered in this research. A. K. Jain [20] discussed biometrics as a safety instrument in 2006. In forensics and protection systems, the function of biometrics has been defined. The study presents a biometrics overview and has solved several recognized flaws in previous studies. It is necessary to tackle these challenges. A. K. Jain suggested to implement and perform biometric recognition [21] in 2004. Research has employed picture recognition during biometric application. In the past 50 years until 2016, A. K. Jain [21] highlighted biometrics' accomplishments, potential and problems. The relevance and the difficulty of biometric evolution in the research was discussed. Vineet Kumar [23] employs an edge method to increase the accuracy of iris photographs in pupil situations in 2016. In 2016, Charan S. G. [24] created an optimization function supporting biometric iris-dependent identification. Research has created an optimum strategy during biometric processing. R. M. Bolle [25] based this on several biometric features in 2004. This is the biometrics guideline, which illustrates how biometric technologies function and reach.

14.3 PROBLEM STATEMENT

Several biometric researches have been carried out comparing modalities using different biometric approaches. The use of biometrics for security has been discussed in prior studies. Various authors presented work in the field of biometrics as a safety weapon. Research has employed picture recognition during biometric application. Various research studies have also taken into account biometrics for research based on IoT. However, time consumption during decision-making is the problem in the present study. Therefore, study using edge detection must be considered to lower the time needed to make important decisions during the picture processing. Taking into account problems in implementing improved biometric security for IoT digital wallet systems, the suggested system utilizes an edge detection mechanism and IoT to offer a more scalable, flexible and trustworthy solution: IoT approach.

14.4 PROPOSED WORK

Proposed work is considering existing research related to biometric security for the IoT-based digital wallet. Then research is investigating limitations of security in the existing IoT-based digital wallet. These security threats could be attacks from the hacker side. A further objective of proposed work is building a reliable and secure model to enhance the protection of such IoT-based digital wallets. Performance enhancement during decision making is another objective of research that would be fulfilled by making use of an edge detection mechanism. Finally, comparative analysis between proposed and existing research work to ensure the accuracy, performance, and reliability of proposed work would be made.

Research would propose a web-based IoT system to manage the automation for enhancing biometric security. Such security would be provided to IoT-based digital wallets. In order to increase the security, a cryptographic mechanism has been built for an IoT system where collection of biometric samples has been considered for security purposes. Research work is considering existing research related to IoT where security of digital wallets is a major issue. The traditional biometric system that would be used for the IoT system has been simulated using Matlab. Canny edge detection approach would be integrated into the existing biometric system to minimize the duration as well as space. Performance of proposed work would be compared to existing research to assure the reliability, security and flexibility of proposed work.

The proposed research is providing biometric security to improve the protection of the digital wallet in the IoT system. As we all know, the use of digital wallets is growing on a daily basis, therefore the requirement to develop such kinds of efficient models to secure different digital wallets came into existence.

There exist many techniques or mechanisms available that are supposed to enhance the security of digital wallets, but these existing methods or techniques have their limitations.

Therefore, iris-based biometric mechanisms are introduced for the security of digital wallets. The integration of canny-based edge detection is made to decrease the space consumption of the biometric model. In the future, this iris-based digital wallet security system would be used as a more secure system than other biometric techniques.

14.4.1 Working of Proposed Model

In proposed work edge detection has been applied on both images of iris and finger, extracted from the user interface in order to remove the useless data from the image. This helps to reduce storage requirements. Moreover, the time taken to compare the biometric image also gets reduced. Then the biometric dataset is verified for authentication. Finally access to the IoT based digital wallet is provided after successful verification.

Process flow of proposed work is discussed in the following list:

1. Image acquisition: Digital coding is made from biometric samples. Thus more than one biometric image is captured during image acquisition.
2. Pre-processing: This step involves edge detection from biometric samples. The contrast adjustment and multiplier are also considered here.

Proposed model

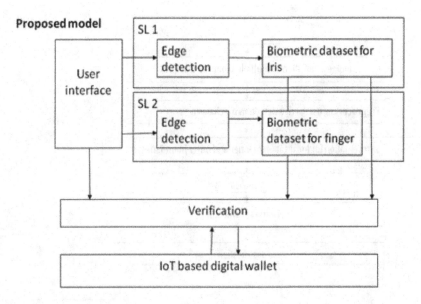

FIGURE 14.2 Proposed work.

3. Segmentation: Localization of inner and outer boundaries is made. Additionally, the localization of the boundary is performed.
4. Normalization: Conversion to Cartesian coordinates from polar is executed in this step. In addition, the normalization related to the biometric image is made.
5. Feature extraction: Noise extraction to the biometric sample is made. After that the biometric sample code is generated.
6. Classification and matching: Comparing biometric code with the codes already saved in the database.

14.4.2 TOOLS AND TECHNIQUES

This section explores the hardware and software used in research work along with the edge detection mechanism. The edge detection mechanism has been used to increase the performance and accuracy during biometric comparison.

1. HARDWARE AND SOFTWARE REQUIREMENTS

Hardware requirement

a) Central processing unit above one gigahertz
b) Random access memory of four gigabytes
c) High-resolution monitor
d) Mouse and keyboard

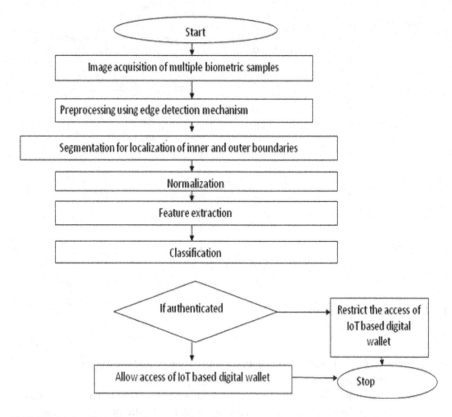

FIGURE 14.3   Flowchart of proposed work.

Software Requirements

  a) Windows operating system
  b) Matlab 2020

 2. Edge detection

The technique of locating and recognizing sharp discontinuities in a picture is known as edge detection. The discontinuities are sudden variations in the grey level value of the pixel intensity. The standard approach of edge detection is convolving the picture with an operator designed to be noise sensitive. The edge detector is a set of crucial local image processing methods for detecting abrupt changes in intensity value. Many image processing applications, such as object identification, motion analysis, pattern identification, and medical image processing, use edge detection. The edge detection methodology is being employed in this study to minimize the size of biometric samples. Matlab was used to complete this task.

There are many sorts of edge variables to consider when selecting a sensitive edge detector, including:

- Edge orientation: A typical direction in which the operator is most responsive to edges is determined by the geometry of the operator. It is possible to optimize the operator to seek horizontal, vertical or diagonal edges.
- Image noise: In noisy pictures, edge detection is different. Attempts to minimize noise result in blurred and distorted edges because both noise and edges include high-frequency information. On noisy pictures, operators are usually wider in scope so that they can average enough data to dismiss localized noisy pixels.
- Edge structure: Objects with borders defined by gradual change in intensity might occur from refraction or poor focus, although not all edges include a step change in intensity. We don't have difficulties with false edge detection, missing genuine edges, edge localization or excessive computing time since the operator must be sensitive to such slow change. One of the most widely used methods in digital image processing is edge detection. Edges are directional localized variations in picture intensity caused by the borders of object surfaces in a scene. Because edge detection is a challenging job, there is no objection to comparing and analyzing the performance of various edge detection approaches under varied situations. In this study, a clever edge detecting approach was used.

14.5 RESULT AND DISCUSSION

During simulation, various iris boundaries in the localized image have been considered from the iris database of University of Bath and ASIA Iris Interval. The extracted image of the iris has been processed with an edge detection mechanism before performing comparison. On the other hand, when the image is passed for comparison then the input image is also processed with an edge detection mechnsim. This processing reduces the comparison time. Moreover, the accuracy of the result also increases.

14.5.1 IRIS RECOGNITION IMPLEMENTATION

Step 1: Picture of the iris acquisition: Scan the image of the eye or capture it with a digital camera.

Before comparing, we crop the eye picture:

We may utilize Photoshop or a photo package manager to trim an eye image.

Step 2: After cropping, the canny method is used to locate eye edges.

Step 3:

Store image as matrix within i

>>i=imread('eye1.jpg')

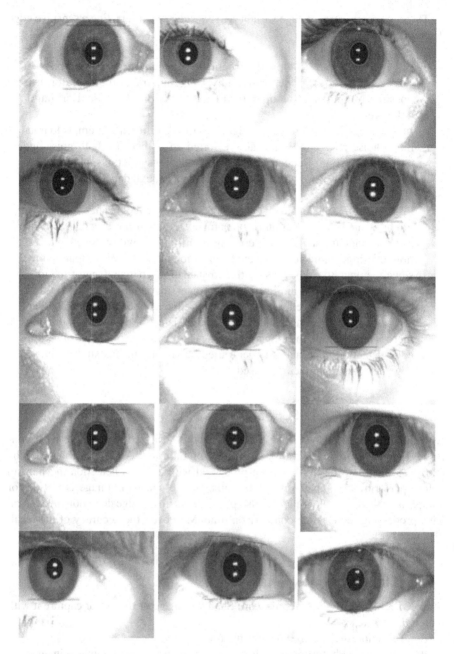

FIGURE 14.4 Localized iris boundaries for a number of eye images (pictures; iris database: University of Bath).

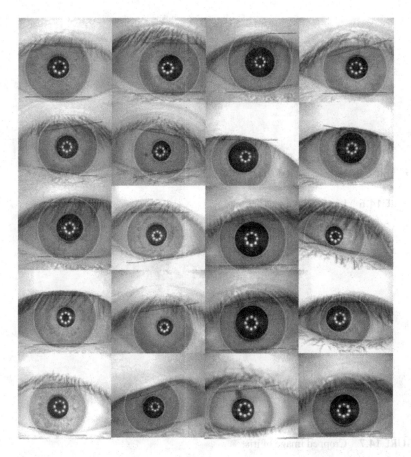

FIGURE 14.5 Localized iris boundaries for a number of eye images (pictures; iris database: ASIA Iris Interval).

Step 4:
Apply canny to i matrix and store within ii
>> ii=canny(i,1,1,1)
Step 5:
Create histogram using surf command
>>surf(ii)
Step 6: In same way we could take a different eye image then crop it and store its matrix within different ranges.
Step 7: Now find edge of cropped eye.
Step 8: Take histogram from matrix of edge-based iris and compare both histograms.

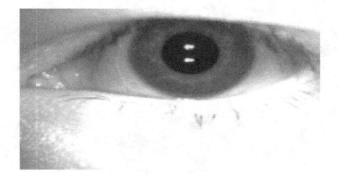

FIGURE 14.6   Image of eye.

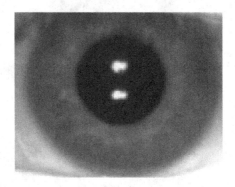

FIGURE 14.7   Cropped image of iris.

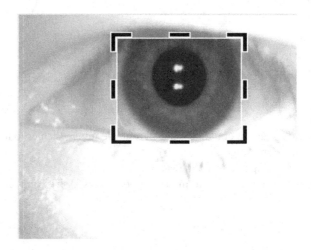

FIGURE 14.8   Image cropping in photo package manager.

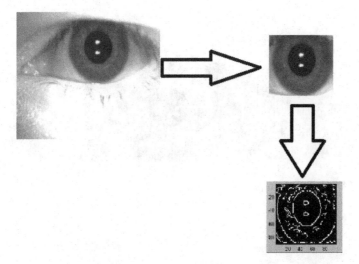

FIGURE 14.9 Cropping and extraction of edges of iris.

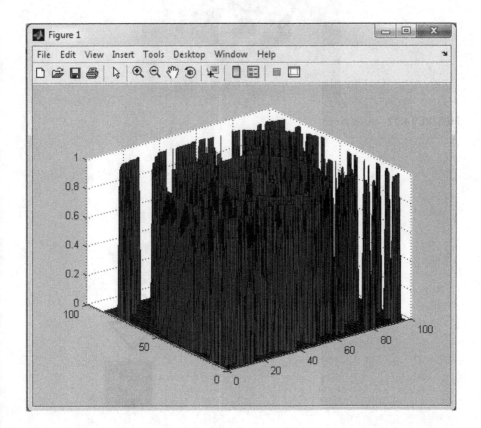

FIGURE 14.10 Histogram of matix of iris image.

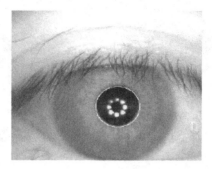

FIGURE 14.11   Before crop.

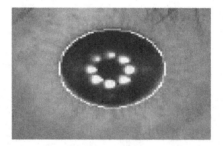

FIGURE 14.12   After crop.

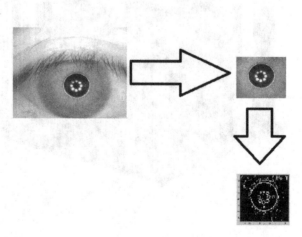

FIGURE 14.13   Process of cropping and extracting different eyes.

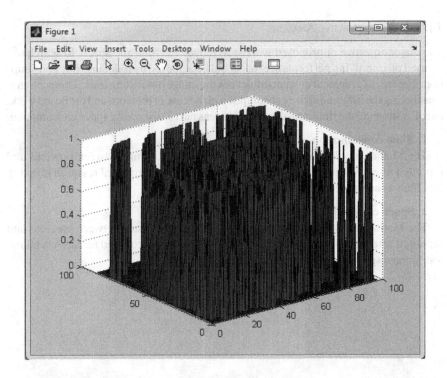

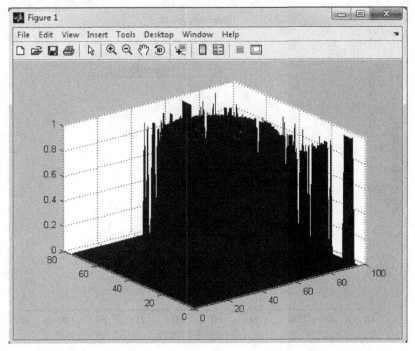

FIGURE 14.14 Histogram from matrix of edge-based iris.

14.5.2 COMPARATIVE ANALYSIS

The major objective of proposed work is to reduce the time and space. Research has focused on improving performance that could be achieved by reducing time consumption. The size of graphical content influences the image processing time. This section is considering the time and space consumption in the case of previous and proposed work. Part 1 is showing the time consumption whereas part 2 is showing space consumption.

1. Time consumption

The Matlab simulation presenting the comparison of time between previous and proposed research has been shown in Figure 14.15 where proposed research is taking less time as compared to previous research.

2. Space consumption

The Matlab simulation presenting the comparison of space between previous and proposed research has been shown in Figure 14.16 where proposed research is taking less storage space as compared to previous research.

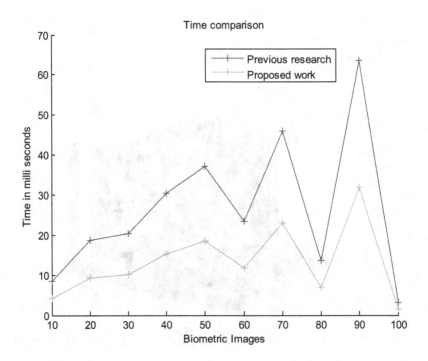

FIGURE 14.15   Time comparison of previous and proposed work.

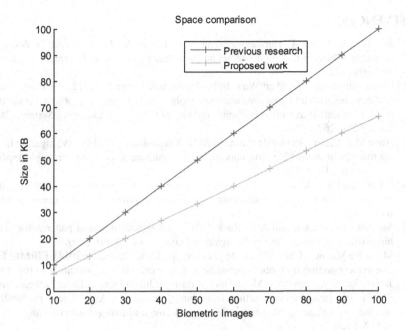

FIGURE 14.16 Comparison of space consumption.

14.6 CONCLUSION

Through trustworthy solutions and increased performance, the planned work has met the goals. The combination of picture processing and the edge detection method has improved biometric security of IoT-based digital wallets. Proposed work has made the decision-making process more accurate, adaptable and scalable. Integration in the biometric system of the edge detection method enhances its accuracy. In addition, the space use of graphical material is also decreasing. In a less accurate way, the proposed system would be able to compare the biometric content. These solutions may improve the security of the digital wallet based on IoT. By using a multicapacitive biometric method that takes into account numerous biometric techniques, the system plays a key role in blocking non-authentic access. In addition, access would be allowed in a shorter time to the privileged individual.

14.7 SCOPE OF RESEARCH

Any other digital payment mechanism might use such a security approach. During the biometric comparison, such a system would safeguard digital wallets using both iris and retina. To lower the space consumption of the biometric model, integration of canny-based edge detection is needed. These techniques would work at minimal consumption of time and space. An iris-based module might often be utilized as a secure biometric system with edge detection mechanism in the near future. For safe banking transactions and other financial relationships it would be quite handy.

REFERENCES

[1] R. Álvarez Mariño, F. Hernández Álvarez, L. Hernández Encinas. (2012). A crypto-biometric scheme based on iris-templates with fuzzy extractorse. *Information Sciences*, 195, p91–102.

[2] GauravBhatnagar, Jonathan Wua, Balasubramanian Ramanb. (2012). Fractional dual-tree complex wavelet transforms and their application to biometric protection at the time of communication as well as transmission. *Future Generation Computer Systems*, 28, 1, 2012, p254–267.

[3] Anne M.P. Canuto, Fernando Pintro, João C. Xavier-Junior. (2013). Investigating fusion approaches in multi-biometric cancellable recognition. *Expert Systems with Applications*, 40, 6, p1971–1980.

[4] C.Chen and R.Veldhuis. (2011). Extracting biometric binary strings with the minimal area in FRR curve to implement hamming distance classifier. *Signal Processing*, 91, 4, p906–918.

[5] Simona Crihalmeanu and Arun Ross. (2012). Multispectral scleral patterns for ocular biometric recognition. *Pattern Recognition Letters*, 33,1, p1860–1869.

[6] Maria De Marsico, Chiara Galdi, Michele Nappi, Daniel Riccioc. (2014). FIRME: Face and iris recognition for mobile engagement. *Image and Vision Computing*. p1161–1172.

[7] Javier Galbally, Arun Ross, Marta Gomez-Barrero, Julian Fierrez, Javier Ortega-Garcia. (2013). Iris image reconstruction from binary templates: An efficient probabilistic mechanism depending on genetic. *Computer Vision and Image Understanding*, 117, 1, p1512–1525.

[8] Ujwalla Gawande, Mukesh Zaveri, AvichalKapur. (2013). bimodal biometric system: feature-level fusion of iris and fingerprint. *Biometric Technology Today*, p7–8.

[9] Marta Gomez-Barrero, Javier Galbally, Julian Fierrez. (2014). Efficient software attack to multimodal biometric systems and its application to face and iris fusion. *Pattern Recognition Letters*, 36, 1, p243–253.

[10] Karen Hollingsworth, Kevin W. Bowyer, Stephen Lagree, Samuel P. Fenker, Patrick J. Flynn. (2011). Genetically identical irises have texture similarity that is not detected by iris biometrics. *Vision and Image Understanding*, 115, 1, p1493–1502.

[11] Farmanullah Jan, Imran Usman, Shahrukh Agha. (2012). Iris localization in frontal eye images for less constrained iris recognition systems. *Digital Signal Processing*, 22, 1, p971–986.

[12] Farmanullah Jana, Imran Usman, Shahid A. Khana, Shahzad A. MalikaaDepartment. (2013). Iris localization is based on the Hough transform, a radial-gradient operator, as well as gray-level intensity. *Optik- International Journal for Light and Electron Optics*, 124, 1, p5976–5985.

[13] Farmanullah Jan, Imran Usman, Shahrukh Agha. (2013). A non-circular iris localization mechanism with help of image projection function as well as gray level statistics. *Optik—International*, 241.

[14] Farmanullah Jan, ImranUsman, ShahrukhAgha. (2013). Reliable iris localization with help of Hough transform, histogram-bisection, and eccentricity. *Signal Processing*, 93, 1, p230.

[15] Farmanullah Jan, Imran Usman, Shahid A. Khan, Shahzad A. Malik. (2014). A dynamic noncircular iris localization technique for non-ideal data. *Computers & Electrical Engineering*, p215–226.

[16] B. Choudhury, P. Then, B. Issac, V. Raman, M. K. Haldar. (2018). A survey on biometrics and cancelable biometrics systems. *International Journal of Image and Graphics*, p1–28.

[17] Shradha Tiwari, J. N. Chourasia, Vijay S. Chourasia. (2015). Reviewing enhancement in case of biometric systems. *International Journal of Innovative Research in Advanced Engineering*, 2, 1, p187–204.

[18] Gursimarpreet Kaur, Chander Kant Varma. (2014). Comparative analysis of biometric modalities. *International Journal of Advanced Research in Computer Science and Software Engineering*, 4, 4, p603–613.

[19] Himanshu Srivastva. (2013). A comparison based study on biometrics for human recognition. *International Journal of Computer Engineering*, 15, p22–29.

[20] A. K. Jain, A. Ross, S. Pankanti. (2006). A tool for information security. *IEEE Transactions on Information Forensics and Security, Biometrics*, 1, 2, p125–144.

[21] Jain K. Anil, Ross Arun, Prabhakar Salil. (2004). An introduction to biometric recognition. *IEEE Transactions on Circuits and Systems for Video Technology*, 14, 1, p4–20.

[22] Anil K. Jain, Karthik Nandakumar, Arun Ross. (2016). 50 years of biometric research, accomplishments, challenges and opportunities. *Pattern Recognition Letters*, 79, p80–105.

[23] Vineet Kumar, Abhijit Asati, Anu Gupta. (2016). A novel edge-map creation approach for highly accurate pupil localization in unconstrained infrared iris images. *Hindawi International Journal of Electrical and Computer Engineering*, p1–10.

[24] S. G. Charan. (2016). Iris recognition using feature optimization. *IEEE International Conference on Applied and Theoretical Computing and Communication Technology*, p726–731.

14.8 AUTHORS' PROFILES

Tanya Chauhan born in Faridabad on December 10, 1996. She has earned her B.Tech degree from GJUST, Hisar, Haryana, India in 2019 and her M.Tech degree in computer science and engineering from DCRUST, Murthal, India in 2021. Her research interests include biometrics and web mining.

Dr. Ajmer Singh earned his B.Tech and M.Tech degrees from Kurukshetra University, Kurukshetra, India in 2004 and 2007 respectively. He did his PhDat at DCRUST, Murthal, Sonepat, India. He is presently working as Assistant Professor in the Department of Computer Science and Engineering at DCRUST, Murthal has more than 13 years' experience of academic and administrative affairs. He has published more than 20 research papers in various international/national journals and conferences of repute as author/co-author. His research interests include software testing, information retrieval, and data sciences.

15 Fused Deep Learning for Distributed Denial of Service Attack Detection in IoT

Sangeeta Lal, Ritu Rani and Ajmer Singh

CONTENTS

15.1 INTRODUCTION

With the advent of new technologies like Internet of Things (IoT), new challenges in connectivity and communication technologies are being noticed. The Internet of Things (IoT) has the potential to gather, quantify, and analyze the surrounding environment in addition to modernizations that boost life capacity [1]. Examples such as this one make it easier for people and things to communicate, which in turn makes it possible to build better cities [2]. This new and exciting field in computer science is known as the Internet of Things (IoT). An estimated 50 billion gadgets will be on the market by the end of 2020 [3–4]. However, IoT technologies play an important part in the development of real-life sensible applications, such as smart healthcare, home automation, and transportation. As a result of the complexity of IoT systems and the number of components engaged in their preparation, new security issues have arisen.

There are numerous ways in which network users profit from the internet's constant expansion and troubled use. The need of keeping a secure network grows as the internet becomes more extensively utilized. Thus, network security is tightly tied to computers, networks, and programs; the purpose of defense is to prevent unauthorized access or change.

With dozens or even hundreds of sensors and actuators (e.g. for important systems like the HVAC), IoT-connected homes also have a wide range of capabilities. It's possible for each gadget to connect to the Internet via a different protocol (e.g. Wi-Fi or Bluetooth), but most of them don't appear to be ready yet. Thus, any device that can collect and analyze sensor data before sending it to web servers is a critical component of the Internet of Things (IoT) [5–6].

Increasingly sophisticated internet-connected systems in the financial, commercial, and military sectors are vulnerable to network attacks, which can cause a wide spectrum of harm and danger. Network security and threat detection are critical for the Internet of Things (IoT). The best approach to characterize network attacks in the IoT is the most difficult topic in network security because of the variety of attacks. Attacks that have never been seen before are particularly dangerous. In recent years, researchers have used several machine learning techniques to classify network risks. Recent advances in machine learning have been made thanks to deep learning algorithms, which mimic the organization of neural networks in the human brain. Deep learning's enhanced original structures and domain-oriented applications may be employed in this research to demonstrate how hybrid deep learning algorithms can be used to check for network security vulnerabilities. Deep learning algorithms have been extensively studied in the past for their potential application in attack detection. [2].

The assault detection method depicted in Figure 15.1 is based on traffic analysis. All network interfaces collect the information packets that change when the Internet of Things (IoT) enters the network. Once these packets have been processed, several packet-level metrics are extracted so that network assaults can be recognized. As input to a pre-trained IoT track classification engine, these parameters indicate how likely it is that a home's IoT-connected environment is now under attack.

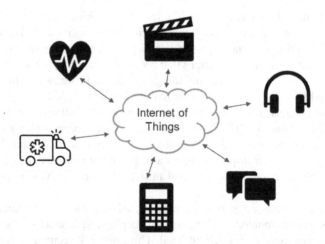

FIGURE 15.1   General IoT security threats.

15.2 LITERATURE REVIEW

Several studies [7–11] involving deep learning concepts form the basis of our chapter.

Deep learning methods and their applications in attack detection can be described by reading a huge volume on the topic, according to Berman and colleagues. While Berman [9] provides an extensive overview of this topic, Apprizes et al. [8] provide a more in-depth look at attack detection approaches connected with intrusion detection, malware analysis, and spam detection. Using deep learning techniques, Wickramasinghe et al. [11] examine numerous sorts of cyber-attacks and their accompanying detection systems in order to secure the Internet of Things.

For the review of the intrusion detection system using deep learning technology, Aleesa and colleagues [7] have published their findings. A scientific literature review employing keywords "deep learning," "invasion," and "attack" provides a wide selection of materials for researchers. Seven types of network datasets are described by Ferrag and colleagues, which include 35 well-known network datasets [10].

As a way to provide a path for future IoT security research, a number of researchers have been conducting studies on IoT security issues. Few studies have focused on the role of machine learning and deep learning in the security of the Internet of Things (IoT).

Many researchers [12–18] have conducted surveys to investigate research and categorize IoT system security concerns. Further, many solutions to this issue have been proposed. IoT communication security was underlined by Granjal, Monteiro, and Timber [19] in their analysis of IoT security concerns and solutions. A survey of IoT intrusion detection was undertaken by Zarpelo et al. [20]. Using legal and regulatory methodologies, Weber [21] sought to determine whether or not IoT frameworks met privacy and security requirements. It was brought forward by Roman, Chou Dynasty, and Lopez [22–23] in the context of distributed IoT. The distributed IoT strategy can assist some of the security and privacy concerns that need to be addressed. According to a recent survey, ransomware attacks and other dangers to IoT systems are growing more prevalent. Metric capacity unit approaches can help protect IoT privacy and security, according to Xiao et al. [24]. Metric capacity unit implementation in IoT systems faces three key challenges in the future, according to the experts (i.e. computation and communication overhead, backup security solutions, and partial state observation). Alternative survey studies like [25–28] examine how knowledge mining and machine learning can be used in cyber security to help detect intrusions. Data processing and machine learning techniques, as well as the detection of Internet abuse and abnormalities, were the key topics covered in the surveys.

15.3 RESEARCH PROBLEM

The Internet of Things (IoT) relies on network attack detection since the number of attacks is increasing as the network grows. Based on deep learning technology, hybrid deep learning is a form of deep learning. There are many different deep learning approaches that are used by everyone in the field of deep learning. On the other hand, the demand for IoT applications and deep learning is expanding every day. IoT and deep learning, on the other hand, are the subject of a number of studies. In prior studies, researchers focused on network threat detection for IoT using deep learning, although they did it with minimal effort. Deep learning is the only technique currently being employed in scientific studies. The difficulty of using only one deep learning technique has to be improved. Fusion deep learning is expected to improve network attack detection for IoT in upcoming research projects.

15.4   NEED OF RESEARCH

Hybrid deep learning based solution is proposed on the basis review of previous studies. Network attack detection using hybrid deep learning and implementation of hybrid deep learning to overcome the limitations of earlier research would be examined in this study. A hybrid deep learning model for attack prediction, implementation, and performance evaluation of our model are the goals of this research. IoT attack detection utilizing hybrid deep learning will be used to ensure the network's security.

15.5   IMPLEMENTATION AND RESULTS

MATLAB methodology is used. MATLAB is a programming language that is created by Math Works. Matlab has functions like parceling of data, which helps in MATLAB to perform calculations. Graphics functions are used as 2D and 3D, which helps in visualizing the data. MATLAB provides interactive features that are useful in many areas like iterating, exploring, and designing. It is capable of being used in technical computing.

Figure 15.2 depicts the proposed methodology in a flowchart. The flowchart shows that initially we will use the training dataset for the implementation. After

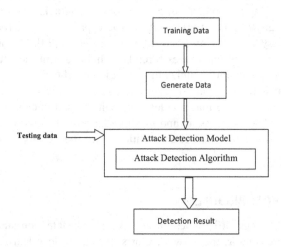

FIGURE 15.2   Flow chart to show various steps of proposed methodology.

that with the help of neighborhood component analysis (NCA) we filter the training dataset and find out that our necessary data means that we will generate data from training data with the help of NCA, in which we are using the Decision Tree, deep learning and KNN algorithm for final result. All the inputs are sent to the attack detection models and attack detection algorithm where again with the help of NCA it will choose best classifier from the previous three classifiers, and it will give us the detection result in terms of accuracy, false alarm rate, precision, and F-measure.

Step 1: Step 1 is primarily utilized for the filter the data. We obtain the NSL-KDD dataset from the Canadian institution.

Pre-Processing: In which we use different types of attacks.

Attacks	Numeric Conversion
Normal	- 0
DoS	- 1
Probe	- 2
R2L	- 3
U2R	- 4

1. DoS—Denial of service
2. U2R—User to Root
3. R2L—Remote to Local
4. Probe—It target detects and reports it with a recognizable "fingerprint" in the report

At least four types of probes are used in the first step. These are the four varieties of normal. We ignore Choice's U2R attack. As indicated earlier, we assign a numeric value to each of the attacks. The size of the data is represented by n and m where n is used for the row and m is used for the column. When we talk about class imbalance, we use the phrase "count"; when the amount of data is small, the count ignores it. We have the option of manually altering the count value. Everything rests on our shoulders. $[n,m]$ = size(features)

The result indicates that from all data it select the only restricted size of data $n = 22{,}543$, $m = 41$, and $Dos = 1$; from all the previous values the label selected is the label 1 of DoS.

Step 2: Multi-classification without fusion

A confusion matrix, also known as an error matrix, is a special table structure that permits visualization of the performance of an algorithm, typically a supervised learning one, in the field of machine learning and specifically the problem of statistical classification (in unsupervised learning it is usually called a matching matrix).

Three classifiers are: KNN = 1; Decision Tree = 2; Ensemble = 3.

Figure 15.3 shows training process without fusion.

Step 3: Apply fusion

We applied the LSTM model with the following configurations:

layers = 5 x 1 layer array with layers:

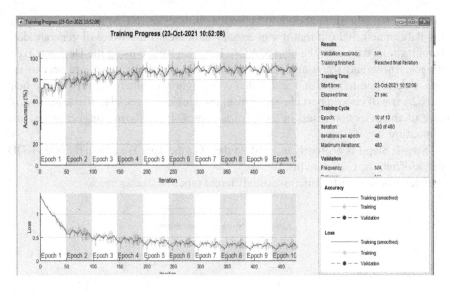

FIGURE 15.3   Training process without fusion.

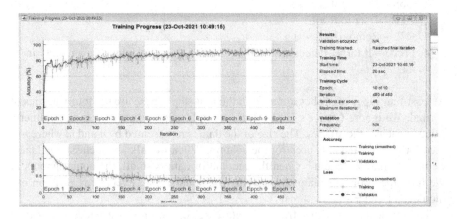

FIGURE 15.4   Training process with fusion.

1	-	Sequence Input	-	Sequence input with 18 dimensions
2	-	BiLSTM	-	BiLSTM with 100 hidden units
3	-	Fully Connected	-	4 fully connected layer
4	-	Softmax	-	softmax
5	-	Classification Output	-	crossentropyex

Figure 15.4 depicts the process with fusion.

Figure 15.5 and Figure 15.6 show the false alarm rate and attack detection rate of different algorithms.

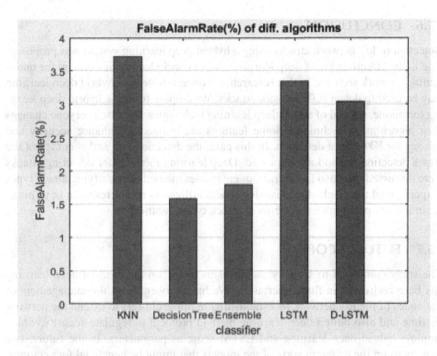

FIGURE 15.5   False alarm rate% of different algorithms.

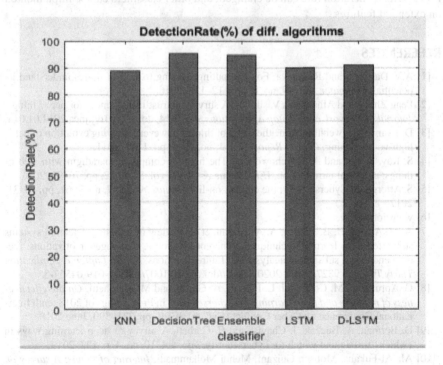

FIGURE 15.6   Attack detection rate of different algorithms.

15.6   CONCLUSION

Detection of IoT network attacks using a hybrid deep learning system was proposed. Due to the effectiveness of deep learning, we've concluded that it's crucial for maintaining network security. In this research, we demonstrate how hybrid deep learning may be used to detect IOT network attacks. We employ fusion, a hybrid deep learning technique, instead of single deep learning techniques, because everyone changes their algorithm or technique. Some features are utilized to enhance security and privacy for IOT threat detection. In this case, the detection rate and accuracy of the attack detection method are improved. Deep learning methods for detecting attacks were discussed. We also found that our ensembles method to classifying attack types outperformed the single deep learning model utilized in recent research in terms of accuracy and performance in terms of attack type classification.

15.7   FUTURE SCOPE

The study connected to security mechanisms in IoT on the basis of deep learning has been addressed in this dissertation. We have spoken about the management of the nodes of the IoT network by establishing multiple clusters to extend the network lifetime and also built a deep learning method protocol to regulate security, which employs hybrid deep learning and LSTM score as parameters. In the future, one may focus on the different sorts of the models that might be beneficial for accuracy, F-measure. Generation rate can be changed, and other classifiers can be implemented on extended features.

REFERENCES

[1] A.V. Dastjerdi and R. Buyya, Fog computing: serving to web of things understand its potential, *Computer*, vol. 49, no. 8, pp. 112–116, 2016.
[2] Peng Zhang and Athanasios Vasilakos, A survey on trust management for net of things, *Journal of Network and Laptop Applications*, 42, 2014, doi:10.1016/j.jnca.2014.01.014.
[3] D. Evans, {the web|the online|the web} of things: however following evolution of net is high-vo everything, *CISCO Report*, vol. 1, no. 2011, pp. 1–11, 2011.
[4] S. Ray, Y. Jin, and A. Raychowdhury, The high-vo computing paradigm with web of things: a tutorial introduction, *IEEE Vogue & Check*, vol. 33, no. 2, pp. 76–96, 2016.
[5] S. Aftergood, Cybersecurity: the conflict on-line, *Nature*, vol. 547, no. 7661, pp. 30–31, 2017.
[6] www.iotthreats.
[7] A. M. Aleesa, B. B. Zaidan, A. A. Zaidan, et al. Review of intrusion detection systems supported deep learning techniques: coherent taxonomy, challenges, motivations, recommendations, substantial analysis and future directions. *Neural Laptop & Application Thirty Two*, pp. 9827–9858, 2020, https://doi.org/10.1007/s00521-019-04557-3
[8] G. Apruzzese, M. Colajanni, L. Ferretti, A. Guido, and M. Marchetti, *On the effectiveness of machine and deep learning for cybersecurity*, in Proceedings of 2018 tenth International Conference on Cyber Conflict (CyCon), IEEE, pp. 371–390, June 2018.
[9] D. Berman, A. Buczak, J. Chavis, and C. Corbett, A survey of deep learning ways in {which} throughout which for cybersecurity, *info*, vol. 10, no. 4, p. 122, 2019.
[10] Ala Al-Fukaha, Mohsen Guizani, Mehdi Mohammadi, *Internet of things: A survey on enabling technologies, protocols, and applications*, IEEECommunication, 2015.

[11] C. S. Wickramasinghe, D. L. Marino, K. Amarasinghe, and M. Manic, *Generalization of deep learning for cyber-physical system security: a survey*, Proceedings of IECON 2018–44th Annual Conference of the IEEE Industrial science Society, IEEE, USA, pp. 745–751, October 2018.

[12] A. R. Sfar, E. Natalizio, Y. Challah, and Z. Chtourou, *A roadmap for security challenges among internet of things*, Digital Communications and Networks, 2017.

[13] S. Sicari, A. Rizzardi, L. A. Grieco, and A. Coen-Porisini, Security, privacy in internet of Things: 'the road ahead', *Computer Networks*, vol. 76, pp. 146–164, 2015.

[14] F. A. Alaba, M. Othman, I. A. T. Hashem, and F. Alotaibi, Net of things security: a survey, *Journal of Network and laptop Applications*, vol. 88, pp. 10–28, 2017.

[15] K. Zhao and L. Ge, *A survey on web of things security*, in procedure Intelligence and Security (CIS), 2013 ninth International Conference on, 2013, pp. 663–667, IEEE.

[16] J. S. Kumar, and D. R. Patel, A survey on internet of things: security and privacy problems, *International Journal of Laptop Applications*, vol. 90, no. 11, 2014.

[17] H. Suo, J. Wan, C. Zou, and J. Liu, *Security among internet of things: a review*, computing and science Engineering (ICCSEE), 2012 international conference on, 2012, vol. 3, pp. 648–651, IEEE.

[18] D. E. Kouicem, A. Bouabdallah, and H. Lakhlef, Net of things security: a top-down survey, *Laptop Networks*, vol. 1, 2018.

[19] J. Granjal, E. Monteiro, and J. S. Silva, Security for web of things: a survey of existing protocols and open analysis problems, *IEEE Communications Surveys & Tutorials*, vol. 17, no. 3, pp. 1294–1312, 2015.

[20] B. B. Zarpelão, R. S. Miani, C. T. Kawakami, and S. C. state Alvarenga, A survey of intrusion detection in internet of things, *Journal of Network and Laptop Applications*, vol. 84, pp. 25–37, 2017.

[21] R. H. Weber, Net of things—new security and privacy challenges, *Laptop Law & Intelligence*, vol. 26, no. 1, pp. 23–30, 2010.

[22] R. Roman, J. Zhou, and J. Lopez, On the alternatives and challenges of security and privacy in distributed internet of things, *Laptop Networks*, vol. 57, no. 10, pp. 2266–2279, 2013.

[23] I. Yaqoob et al., The increase of ransomware and rising security challenges among internet of Things, *Laptop Networks*, vol. 129, pp. 444–458, 2017.

[24] L. Xiao, X. Wan, X. Lu, Y. Zhang, and D. Wu, IoT security techniques supported machine learning, *arXiv preprint arXiv:1801.06275*, 2018.

[25] A. L. Buczak and E. Guven, A survey of data mining and machine learning ways in {which} throughout which for cybersecurity intrusion detection, *IEEE Communications Surveys & Tutorials*, vol. 18, no. 2, pp. 1153–1176, 2016.

[26] P. Mishra, V. Varadharajan, U. Tupakula, and E. S. Pilli, *A careful investigation and analysis of follow machine learning techniques for intrusion detection*. IEEE Communications Surveys & Tutorials, 2018.

[27] A really distinctive amount DDoS Attack Detection Mechanism supported MDRA formula in massive Data-Scientific Figure on ResearchGate. *On the market from.* www. researchgate.net/figure/Flowchart-of-attack-detection_fig3_308572010   (accessed twenty seven solar calendar month, 2021).

[28] M. Brown_eld, Y. Gupta, and N. Davis. *Wireless device network denial of sleep attack.* In Proceedings 2005 IEEE workshop on data assurance and security, U. S. Academy, West Point, NY, 2005.

16 Analyzing the Suitability of Deep Learning Algorithms in Edge Computing

Dr. Nidhi Srivastava

CONTENTS

16.1 INTRODUCTION

IoT computing has moved ahead to a new level now, that is way ahead of the traditional computing. IoT is becoming a part of everyone's day to day life. The majority of the physical devices like sensors, smartphones, actuators, cameras, game controllers, music players, wearables, etc. are interconnected. These devices produce massive amounts of data and communicate and interchange information with data centers. The number of devices is in fact increasing day by day. Connecting the devices that themselves are capable of sensing and responding to a stimulus in the real world without any human intervention is the main consideration in IoT [1].

DOI: 10.1201/9781003245469-16

IoT devices are crucial and in near future will become source for generation of enormous amount of data. Handling and managing this massive data can be easily done by cloud computing technology.

Cloud computing is an extended version of distributed computing. It provides its users with a range of computation services. It is an on-demand, pay-as-you-go model. It basically frees the users of requirements of storage, network, services, applications, etc. The users can on demand acquire the infrastructure and resources as per their need and can release the unused resources once no longer required. Organizations today are preferring cloud computing due to the large-scale benefits provided by it as compared to the traditional computing. It offers three types of services—Software as a Service (SaaS), Platform as a Service (PaaS) and Infrastructure as a Service (IaaS) [2]. Cloud computing provides many benefits, but with the introduction of numerous interconnected devices that lie at a significant physical distance, certain problems are faced. A few of these include greater latency time, traffic congestion, more computation cost, etc. This introduces the concept of edge computing.

Sustainable development is becoming an imperative aspect of our lives today. It is needed to ensure solutions to the problems of stabilizing quality of services and their costs. Therefore, extensive research is being carried out to develop procedures, equipment, and technology to help support the vision of attaining sustainability of firms.

Application of edge computing is seen in countless fields like smart city, transportation, autonomous driving, smart home, etc. Use of deep learning (DL) in edge computing can significantly increase the power of the delivery of services in the aforementioned applications and areas [3].

16.2   EDGE COMPUTING

Edge computing is a technology that performs computation and all services at the edge, which helps in faster computation and speedier execution. Any network resource or computing resource that lies between the cloud and the data source can be termed an edge device. This technology basically helps in computation of the data both while uploading on the cloud, called upstream—and while downloading from the cloud, called downstream [4]. The basic intention or target is to keep the processing of data nearer to the IoT device. More and more devices like microwaves, refrigerators, and ACs are interconnected nowadays and generate a sizeable amount of data. Devices like smart phones etc. in earlier cloud computing paradigms were only consuming the data but now they are also producing the data. For example, mobiles are used to upload videos and photos on the cloud. Thus, now it has become a two-way communication—from just being consumers now these devices have changed to being producers of data as well [1].

16.2.1   ADVANTAGES OF EDGE COMPUTING

Cloud computing has gained a lot of importance in previous years due to its numerous benefits. Edge computing adds to these benefits and assists cloud computing. Edge and cloud computing are complementary to each other. Some of the advantages of edge computing can be listed as follows:

a. Processing of data is fast in cloud computing, but if for every small computation access to the cloud is to be made, the response time will be too high, and a lot of network bandwidth will be required at the same. Real-time computing should be very fast for critical applications and requires a millisecond response. For example, a Boeing 787 generates a massive amount of data in one second, and so the bandwidth between satellite and airplane needs to be good enough for transmission of this voluminous data. Another example being that, in case of videos, the edge device can reduce the resolution and the size of the video before uploading, which will now consume less bandwidth and increase the speed.

b. In a very short span of time, IoT has gained a lot of momentum, and several devices are connected to each other and with the cloud. Each of these devices produces data. Processing of this data at the edge will reduce the load on the cloud as in the future more devices will join the Internet of Things.

c. Since the users have now become used to these services, a 24/7 availability of these services has become a necessity for the users rather than a luxury. This is a challenge for cloud providers, and it can be catered to some extent by edge computing.

d. Several devices generate raw data, which needs processing even before it can be uploaded in the cloud. Uploading of this raw data is time consuming, utilizes lots of bandwidth, and security of this data becomes a major issue. Processing of this raw data even before it is uploaded or sent to the cloud can increase the security and privacy of the data as most processing will be done at the edge and closer to the source of data. This is especially useful in the case of sensitive data like health, traffic management, etc.

e. Edge computing devices are closer to the end devices and thus give better quality of service (QoS) to the users along with reduced latency.

f. Edge computing can help several IoT connected devices by reducing their load of performing power-consuming tasks and thus saving energy and increasing their performance [3, 5, 6].

16.2.2 CHALLENGES OF EDGE COMPUTING

a. The number of devices in edge computing are large and are also heterogeneous in nature. Identifying the exact device for data communication is a great challenge for edge computing. The edge experts need to figure out the various network protocols for smooth communication with these systems. To ease this problem, a naming scheme is to be devised. As every system is given a name, similarly a name is given to all these edge devices. Until now, there has been no standardized naming scheme for edge computing, and so identifying these devices becomes difficult. The naming scheme is required to help in the mobility of the device, network topology, scalability, and privacy to name a few. The naming in the current network is satisfied by the traditional schemes like Domain Name Server (DNS) and uniform resource identifier. But these networks are not able to satisfy the wide range of edge network due to mobility of edge devices, their complexity, and overhead. Named data network (NDN) and Mobility First are some of the new naming schemes that can be used in edge computing [4].

b. In edge computing, as already said, heterogeneous devices are connected. This increases the load on programmers as they face difficulty in writing an application that can be run on these heterogeneous devices. This is easy to handle in cloud computing as the responsibility to handle the computing is that of the cloud provider and the programmer is not responsible for how the application is handled by end-users. Thus, there is a need for tools and frameworks that are easy to use and are easily manageable.

c. Securing the data is an important challenge in edge computing. Since data from different systems are captured by the edge, it has a lot of privacy issues. This data can be easily accessed by anyone if proper security protection is not provided. E.g., if a smart city is taken into consideration, then it contains lots of personal information of the public. This data is processed at the edge, and while uploading or downloading the data it should be properly secured or else hackers can misuse it. One way of securing the data can be that private information can be removed from the data.

d. The edge computing should be able to manage the services based on four features. It should be able to differentiate between the critical and not-so-critical applications e.g. health of a person should be preferred over games. Edge devices should provide an extensibility feature. New devices or systems can be easily added to the existing edge system. This calls for design of the service management layer. Isolation and reliability also play an important part in the edge. Failure of one application or device should not crash the whole system. Also, in case of such failure, proper action should be taken for recovery immediately [5].

16.2.3 ARCHITECTURE OF EDGE COMPUTING

In edge computing, we place computer nodes in a mesh that is situated very close to the devices that need it. The benefit of using edge computing is that computations are performed at a very high speed and the latency requirement is very low. Thus, the upload bandwidth to the clouds is drastically reduced.

Figure 16.1 shows a very basic architecture of edge computing. The edge here is in the middle of the end devices like mobile phone, etc. and the cloud data center that contains enormous data.

All the devices are connected via Internet. The end devices are connected to the cloud through the edge devices. Thus, for every small piece of information the end device need not connect to the cloud. The edge device provides the information as far as it can and in turn connects to the cloud for data where large computation is required. It provides hierarchical architecture of end devices, edge compute nodes, and cloud data centers that can provide computing resources and scale with the number of clients, avoiding network bottlenecks at a central location [7].

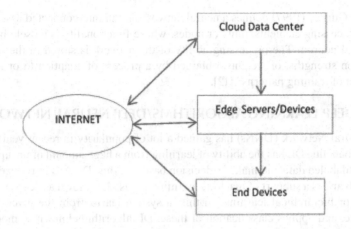

FIGURE 16.1   Architecture of edge computing.

16.3   MACHINE LEARNING ALGORITHMS

Machine learning (ML) plays an important part in edge computing. Before we describe the importance of ML and Deep Neural Network (DNN) in edge computing, let us talk briefly about these technologies. ML is a technology or a part of AI that uses learning algorithms to learn from the data provided to it. Given a dataset, ML is used to work on that data and find out the hidden patterns in it. ML algorithms can be subdivided into supervised learning, unsupervised learning, or reinforcement learning. ML works well on data that is not too large.

For a given large dataset it is difficult to extract the non-linear patterns or complex patterns using the traditional ML and feature engineering algorithms. However, deep learning (DL) is helpful in this and is also very effective and efficient in classification and prediction [8].

Due to its large-scale computational facilities, DL is able to build more precise models in relatively less time, which tends to give good results as compared to traditional or ML algorithms. DL uses a neural network and works in a way similar to the human brain. DL algorithms analyze the data in a similar fashion as the human brain does and accordingly draw a conclusion based on their learning. For achieving the aforementioned goal, it uses neural network. It is a type of Artificial Neural Network (ANN) with lots of layers. ANN helps the DL models to reach the goal that ML is incapable of. ANN is a mathematical structure that solves problems by self-learning as our brain does. ANN is said to be a computational model because it processes information by using artificial neurons. Certain inputs are given to the ANN, and certain weights are applied on these inputs, which finally gives the output. The weights are adaptive in nature and adjust on their own. This process of adjusting the weights is known as learning [9, 10].

Dr. Robert Hecht-Nielsen (1989) defines a neural network as "a computing system made up of a number of simple, highly interconnected processing elements, which process information by their dynamic state response to external inputs" [11].

Kevin Gurney (1997) defines a neural network as "an interconnected assembly of simple processing elements, units or nodes, whose functionality is loosely based on the animal neuron. The processing ability of the network is stored in the interunit connection strengths, or weights, obtained by a process of adaption to or learning from a set of training patterns" [12].

16.4 DEEP LEARNING ALGORITHMS/DEEP NEURAL NETWORKS

Deep Neural Network (DNNs) has gained a lot of popularity in recent years due to its many benefits. DL has the ability of learning from a huge amount of unsupervised data or unlabeled data and makes a decision based on this. DL was introduced in the year 2006 and is a subpart of machine learning. DL is also known as deep structured learning or hierarchical learning. In this, a system learns from the given images, sound, etc. and applies classification of these. DL algorithms known as models are preferred as they can give a very high level of accuracy, far exceeding what humans can perform on it. Given the raw data, DL algorithms can extract meaningful information from it by using a hierarchical multi-level learning approach. The learning at the lower level forms the basis for a higher level that has more abstract and complex representations. DL can be applied to both labeled and unlabeled data.

DNN can work on a very large dataset, and the neural network used has many layers. Given the input data in deep learning, this data passes through several layers one after the other. The output of the first layer is input to the second layer; the output from the second layer is input to the third layer, and so on and so forth. Each layer works on the given input, and the last layer gives us the output in the form of a feature or a classification output. The reason this is known as DNN is because the model works on multiple layers that are present one after the other, and each layer performs matrix multiplication on the data. The more the layers in the model, the better it is able to learn the complex pattern and recognize or characterize the features. However, the drawback is that training this model is difficult, and it also consumes a lot of time. There are different types of DNN: Convolutional Neural Networks (CNN), Recurrent Neural Networks (RNN), and Recursive Neural Networks. If convolutional filter operations are applied on matrix multiplication, then we call it CNN. RNN is designed to deal with time series prediction [7, 13].

DNN is helpful when the dataset is huge. The accessibility of the hardware resources like GPU and the DL libraries helps in using DL in many applications.

Figure 16.2 is an example of DL that has three hidden layers between the input and output layer.

It is a little difficult to decide which DNN model will be useful for which application as it is based on several parameters. But a few of the key considerations that can be kept in mind while deciding on a model can be memory, speed, accuracy, or any other system resource [7].

16.4.1 Types of Deep Learning Algorithms

There are various types of deep learning algorithms given by various researchers that are used and are helpful in prediction or classification. Some of these are mentioned in the following sections.

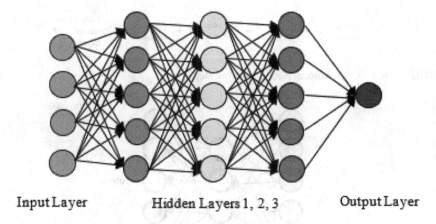

Input Layer   Hidden Layers 1, 2, 3   Output Layer

FIGURE 16.2 Deep learning with three hidden layers.

16.4.1.1 Convolutional Neural Network

One of the most popular DL algorithms is CNN. It is specially designed to work on two-dimensional image data. CNN is fed with an input image. Through various weights and biases, CNN gives importance to various objects visible in the image, thus differentiating one object from the other. The advantage of using ConvNet is that the preprocessing required is way less than that of other classification algorithms. CNN can learn the filters on its own that had to be hand-engineered with enough training in earlier methods. CNN works in the same way as the visual cortex does in the human brain. CNN is composed of a convolutional layer, pooling, a Rectified Linear Unit for a non-linear operation (ReLU), and fully connected and loss layers.

In the convolution layer a filter is used that is applied to the input repeatedly. This results in a feature map that shows the location of the detected feature in the input image. The pooling layer reduces the size of the convoluted feature. This helps in extracting the most dominant features. Pooling is of different types like max, average, etc. As the name suggests, max gives us the maximum value from the portion of the image to which kernel is applied, while average calculates the average of the values from the portion of the image to which kernel is applied. Pooling generally performs downsampling or upsampling to reduce the dimensionality. ReLU is the activation function $f(x) = \max(0, x)$ that is used on layers. Other activation functions like tanh or sigmoid can also be used. However, ReLU performs better than these activation functions. In the full connection (FC) layer the matrix is converted to a vector and features in each layer are combined to form a model. Finally, in the loss layer softmax or sigmoid activation function is applied, which helps to classify the outputs like tree, sun, cat, etc. [14, 15].

Figure 16.3 shows the steps involved in CNN.

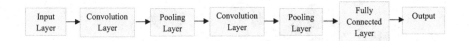

FIGURE 16.3   Convolutional Neural Network steps.

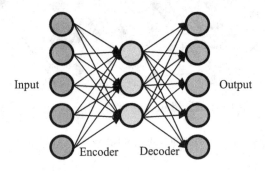

FIGURE 16.4   Autoencoder with encoder and decoder.

16.4.1.2   Autoencoder

It is an unsupervised learning algorithm. In this, the dimension of the input and the output is the same. Autoencoder contains encoder and decoder. It has three layers—input layer, hidden layer, and output layer. Autoencoders take unlabeled data as input, but they tend to produce their own label data for training. So, it can be called a self-supervised algorithm. An encoder encodes the given input data, and the decoder decodes the output of the encoder and reconstructs it to give the output. For both these phases, the same weights are used. A back propagation algorithm is used by the autoencoders. This is a lossy method as decoders cannot reconstruct exactly the same input, but it is very close to it. To achieve the encoding, the number of hidden layers used are fewer so that the dimensionality of the input layer is reduced. This calls for extracting the most discriminative features. For a high dimension input data, more than one layer will be needed to represent all data. An alternative to get a deep autoencoder is to stack one encoder after the other. There are different types of encoders like sparse encoders, denoising encoders, contractive autoen-coders, convolutional encoders, etc. [14, 16]. Figure 16.4 depicts an autoencoder.

16.4.1.3   Recursive Neural Network

In Recursive Neural Network the computation performed in one layer is used by the other layer. So, each layer is dependent on the last computation. For this RNN uses inbuilt memory. The samples given to RNN have interdependencies, and it remem-bers all the previous steps.

Thus, for a particular output it uses both the present and the recent past, i.e., a series of data is used. It is a recursive approach. One of the problems of this is that when there are long input sequences, a vanishing gradient is generated. This problem is dealt with by Long Short-Term Memory Units (LSTM), a variant of RNN. Like any other DNN, RNN also uses weights in all steps. The only difference is that the weights used in this are same throughout the layers. This reduces the complexity,

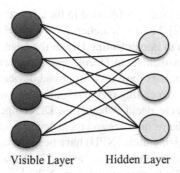

Visible Layer   Hidden Layer

FIGURE 16.5   Restricted Boltzmann machine.

and it is easy for the network to learn. Recurrent Neural Network is a special type of recursive neural network that has a particular structure [3, 17].

16.4.1.4   Restricted Boltzmann Machine

It is a two-layer network known as visible layer and hidden layer. In this, each node in visible layer is mapped to each node in the hidden layer. In a Boltzmann Machine the nodes in one layer are connected to the nodes in the other layer. Along with that, all nodes in a layer are also interconnected. RBM is a restricted version of a Boltzmann machine as there is restriction in the second condition. So, in RBM only nodes of the visible layer are connected with each node of the hidden layer. RBM follows both the forward pass and the backward pass. Given the input, it is encoded in the form of a number. To reconstruct the input, backward translation is applied [3, 14, 16]. Figure 16.5 depicts the working of it.

16.5   APPLICATIONS OF CONVERGENCE OF DL WITH EDGE COMPUTING

DL is making an impact in almost all fields. A few of the areas where it is widely used are speech recognition, cancer cell detection, content-based search, drug discovery, face recognition, video surveillance, cyber security, Natural Language processing, etc. It is now being used extensively in edge computing as well [18]. Some of the applications where DL is being used on the edge are mentioned in the following sections:

16.5.1   Video Analytics

Shifting from traditional video to real-time video analytics, the application of it is seen in innumerable fields like augmented and virtual reality (AR/VR), automatic pilot, person authentication through facial recognition, etc. [19]. With edge computing the videos can be processed locally instead of being processed in the cloud. This helps in reducing the latency as the data is processed near the device [20]. Deep learning can contribute a lot in this field. But only few of the applications make use of it due to constraints like unsuitable infrastructure support. Such applications are also computationally intensive and so cannot be run on the edge devices [21]. Offloading is a solution for this where the video is offloaded to the edge server/device so that the DL models can be run on it [22]. Many researchers have worked in this area and proposed a solution that would help in implementing DL on the edge. The authors of [21] have used DL and proposed mathematical DeepDecison framework, which is measure-ment driven. In this the authors have used front-end devices that are weak in nature and back-end helpers that, depending on the resources like network bandwidth, video resolutions, battery usage, etc. decide whether the DL computation is to be performed locally or remotely. This helps in faster and more accurate object detection.

Recently, neural processing units (NPU) have been introduced and are being used by many companies in new mobile phones as NPU is capable of processing AI fea-tures. [22] have proposed a framework wherein NPU has been used to process the DL in video analytics on mobiles. They implemented and tested this on smartphones and found two problems of maximizing the utility and accuracy. Depending on the network bandwidth the framework takes the decision of whether to use the NPU or to offload the computation. The system has proven to be very effective.

16.5.2 HEALTHCARE

Improvement in healthcare and the medicinal field is the need of the hour especially after the pandemic. Recently much work has been done in healthcare in both the domains of edge computing and deep learning separately. But, to unleash and avail the full opportunities of edge computing, it is essential that more processing power should be provided at the edge. The convergence of edge and DL can help in creating new opportunities and in exploiting the two latest research areas to their full potential [3]. In fact, the capability of DL can be exploited in edge computing to give an effec-tive healthcare system, which can be of help to users. [23] in their study of COVID-19 have found that the lack of a large dataset and optimization lead to a problem where DL techniques cannot be effectively used in edge computing. In this paper they have introduced Deep Transfer learning as a solution where, in the cloud using DL, a large dataset can be trained. The pretrained system would then be used in the edge where it would be retrained with limited data and used. This would help in mitigation of COVID-19 and any other such pandemic if it arises in the future. [24] have used DL in healthcare and proposed an automatic bi-modal (speech and video) emotion recog-nition system. The authors have used a CNN model in this, but for an efficient system with low latency they have used edge caching. The speech and video are processed using CNN, and for fusion ELM is used. The training of the data in CNN and ELM is done on the cloud but before passing to the cloud some of the tasks of processing of audio and video are done using edge computing to reduce the data to be transmitted to cloud. Also, after the processing by cloud, an edge cache was used to store the trained parameters for testing to ease the load of transmitting the data to the cloud.

16.5.3 INTELLIGENT TRANSPORTATION SYSTEM (ITS)

The population is increasing day by day and so is the number of people who are driving cars. This has led to huge traffic on the roads and consequently an increase in traffic jams and pollution level. Self-driving cars/vehicles have also increased on roads. These vehicles have several sensors for making calculations and different measurements like choosing the optimal path, controlling the speed and direction on their own, etc. To ease this situation, an intelligent transportation system has become a necessity. Computers, internet, edge computing, etc. form the components of ITS. Combining deep learning algorithms in edge devices has paved the way for better traffic flow detection by quicker detection and tracking of the objects moving around. This decision-making was time consuming with cloud computing due to accessing of cloud time and again for uploading of data generated by the vehicles and sensors. Also, once the processing and computation has been done by the cloud, downloading takes a long time. Edge algorithms have helped in this [25]. In [26] the authors have proposed a new architecture for ITS that uses deep learning on edge analytics for better and effective computation.

In [25] the authors have proposed an algorithm for vehicle detection based on the You Only Look Once (YOLO) V3 model. For multiple object vehicle training they have used a Deep Simple Online and Realtime Tracking (DeepSORT) algorithm. To detect the traffic flow, a vehicle detection and a tracking algorithm have been used. This has finally been implemented on the edge device. The results are promising and implementing the deep learning on the edge device in this case gave an average accuracy of about 92%.

Authors of [27] have developed a lightweight deep learning model that is useful and supports mobile edge computing applications in the transportation cyber physical system. The new algorithm CNN-FC shows a good result by achieving higher accuracy as compared to the conventional CNN.

16.5.4 SMART INDUSTRY

Industry 4.0 is the new buzzword that is much talked about nowadays. This lies on two principles of production automation and data analysis [20].

The advanced technologies like AI, IoT, edge, big data, etc. can help the industry in making decisions based on complex and real-time manufacturing. The companies for security reasons tend to maintain their closed networks on cloud, but that leads to the same problem as seen in the previous fields i.e., high response time, protection and safety of data, and high network bandwidth. The edge along with DL can help in this sector. In this chapter the authors have used deep Q network (DQN), which is a combination of deep learning and reinforcement learning, on edge computing to solve the job of shop scheduling problems [28].

The authors of [29] have designed a system through which they can optimize the deep learning model so that it can work on edge devices with low computational facility. In Industrial Internet of Things (IIoT), the manufacturing components were classified using CNN. The CNN was optimized and implemented on an edge node for which an edge computing testbed was created on Google cloud. The results were very optimistic, and accuracy was high.

In the manufacturing process of the lenses for mobile cameras there are many parameters that affect its quality. [30] have proposed a method wherein they have used deep learning for early prediction of the defective lens so as to avoid sending it to further stages. Further, to make sure the prediction is real-time and low-cost, it has been implemented on an edge node.

16.6 FRAMEWORKS OR TOOLKITS FOR USING DL ON THE EDGE

Many frameworks and libraries have been developed that can help in deep learning. These, if well optimized, can be used on the edge. Some of these toolkits are Keras, PyTorch, TensorFlow, Caffe, Theano, etc. A brief description of some of these are as follows:

TensorFlow was initially created by Google and is the most common library used in deep learning. It uses dataflow graphs and is used for numeric computations. TensorFlow has various APIs in it that are written in different languages. It helps users in training and testing the required model. A program in this has two sections—the construction phase and the execution phase. Building of a computation graph using edges and nodes is done in the construction phase, and in the execution phase these previously created graphs are executed or run. The parameters in these are improved to train the model well [31].

TensorFlow was further refined to create TensorFlow Lite in 2017 for mobile devices. It is a lightweight deep learning framework. TensorFlow Lite is used because the size of its binary is very small and it can implement machine learning on embedded devices. For improving the performance, it also uses hardware acceleration and Android Neural Network API, etc. [32].

For building deep learning models PyTorch, a library of Python, is widely used. It supports both usability and speed. It is supposed to be better than Tensor as it does not need to perform the compilation phase. In this the mathematical expression to calculate gradient of an expression is easily evoked using an operator [33].

Berkeley Vision and Learning Center developed convolutional architecture for fast feature embedding (Caffe). It is written in C++. It is helpful in image classification due to its high speed and easy transition between CPU and GPU. It is very fast and is modular in nature [34, 35].

16.7 CONSTRAINTS IN USING DL IN EDGE COMPUTING

a. DNN generates weights that are required to be stored in memory. Also, for computation of billions of multiplications, more memory is required. Edge devices have low power and do not have a large memory to support DL. Even if the edge devices can process the entire data, then also the entire purpose of hiring edge computing fails as then in that case it will not be able to deliver the results in real time. High computing power is a necessity in the DL model as it trains the model on a large dataset.

b. If the training set available with the DNN is very small, then it leads to overfitting. In some cases, the data is now being stored electronically but still large amounts of data need to be fed in the system that are not easily available like data for rare diseases.

c. The raw data cannot be used in DNN, so preprocessing of data is required. This preprocessing is also not easy as, depending on the data, different procedures must be applied. This increases the training time of the model. The healthcare data is very extensive, so applying different preprocessing techniques will be time consuming.

d. Edge computing demands a specialized hardware accelerator so that the processing of DNN is faster and reduces latency, keeping in mind the security and privacy issues.

e. Training a large dataset in DL requires special hardware accelerators like GPUs, Tensor Processing Units (TPUs), or FPGAs, otherwise it would take months to train a model. However, the classical ML algorithms give good results and work very fast on CPU as the model is trained on a relatively smaller dataset. For a highly accurate DNN there is the requirement of large memory. The researchers are working toward developing a DNN that requires less memory. Many DNN accelerators are being designed that will compress the DNN by parameter pruning, sharing, and quantization.

f. Many of the computations in DNN are redundant in nature so it is not necessary to do these computations at each and every stage. A method can be designed to reduce these redundancies and remove unnecessary computations.

g. Since a huge amount of data is dealt with in DNN, security of data is of great concern. Security vulnerabilities like data poisoning etc. raise the need for safe implementation of DNN [16].

Thus, we see that there are many constraints in using DL in edge computing. However, as can been seen from Section 16.6, researchers are making an effort to minimize these restrictions, and many new models are being developed to overcome these limitations.

16.8 FUTURE TRENDS AND CONCLUSION

With advancement in the IT industry and introduction of new technologies, the facilities have improved. People are moving toward smart city, smart health, smart transportation, etc. Cloud computing has played a significant role in this by delivering the services at the doorstep. However, due to certain shortcomings in cloud like increased response time and decreased network bandwidth, edge technology is used nowadays. The edge technology provides support to cloud and IoT devices in delivering these services efficiently and effectively. The previously listed areas in the chapter need faster processing of data to be delivered in real-time for better decision-making. The edge along with DL provides us with lots of advantages. If implemented in hospitals, they can monitor the patient's data locally and easily. In transportation or in autonomous vehicles like self-driving vehicles it tends to reduce the computation time and takes quick decisions required on roads. This chapter explains the edge computing technology, highlights its benefits and challenges, and discusses how the edge can be used so that the amenities are substantially improved. Deep learning tends to learn from the data provided to it, which helps it in making decisions. Deep learning in edge computing can further enhance these facilities. The ongoing research work in this area has been highlighted and discussed in the chapter. The researchers are experimenting and providing lighter variations of DL that can be used in the edge and are computationally inexpensive. So, although edge computing has proved to be quite beneficial, it still faces the challenge of overlooking the data. In an effort to reduce the latency and bandwidth, it is possible that it may overlook the important data as insignificant. Further, such algorithms need to be developed that can process on edge devices with low computation power. Thus, there is a need to maintain the balance in edge computing so as to maximize the performance.

REFERENCES

[1] Yu W. et al. (2018): A Survey on the Edge Computing for the Internet of Things. *IEEE Access*, 6, 6900–6919.

[2] Meher N. K., Lokhande S.P. (2013): Cloud Computing: An Architecture, its Security Issues & Attacks. *International Journal of Advanced Research in Computer Engineering & Technology (IJARCET)*, 2(3).

[3] Wang F., Zhang M., Wang X., Ma X., Liu Y. (2020): Deep Learning for Edge Computing Applications: A State-of-the-Art Survey. *IEEE Access*, 8, 58322–58336. doi: 10.1109/ ACCESS.2020.2982411

[4] Shi W., Dustdar S. (2016): The Promise of Edge Computing. *Computer*, 4(5), 78–81.

[5] Shi W., Cao J., Zhang Q., Li Y., Xu L. (2016): Edge Computing: Vision and Challenges. *IEEE Internet of Things Journal*, 3(5), 637–646.

[6] Hamdan S., Ayyash M., Almajali S. (2020): Edge-Computing Architectures for Internet of Things Applications: A Survey. *Sensors (Basel)*, 20(22), 6441. https://doi. org/10.3390/s20226441

[7] Chen J., Ran X. (2019): Deep Learning with Edge Computing: A Review. *Proceedings of the IEEE*, 107(8), 1655–1674. doi: 10.1109/JPROC.2019.2921977

[8] Najafabadi M. M., Villanustre F., Khoshgoftaar T.M. et al. (2015): Deep Learning Applications and Challenges in Big Data Analytics. *Journal of Big Data*, 2, 1.

[9] Firoz S. A., Raji A. S., Babu P.A. (2009): Automatic Emotion Recognition from Speech, Using Artificial Neural Networks with Gender-Dependent Databases. *International Conference on Advances in Computing, Control, and Telecommunication Technologies, IEEE*, 162–164, doi: 10.1109/ACT.2009.49

[10] Kumar P., Sharma P. (2014): Artificial Neural Networks—A Study. *International Journal of Emerging Engineering Research and Technology*, 2(2), 143–148.

[11] Caudil M. (1987): Neural Network Primer: Part I. *AI Expert*, 2(12), 46–52.

[12] Gurney K. (1997): *An Introduction to Neural Networks*, CRC Press.

[13] Muniasamy A., Tabassam S., Hussain M.A., Sultana H., Muniasamy V., Bhatnagar R. (2020) Deep Learning for Predictive Analytics in Healthcare. In: Hassanien A., Azar A., Gaber T., Bhatnagar R., F. Tolba M. (eds) The International Conference on Advanced Machine Learning Technologies and Applications (AMLTA2019). AMLTA 2019. Advances in Intelligent Systems and Computing, vol 921. Springer, Cham.

[14] Faust O., Hagiwara Y., Hong J. T., Lih S. O., Acharya R. U. (2018): Deep Learning for Healthcare Applications Based on Physiological Signals: A Review. *Computer Methods and Programs in Biomedicine*, 161, 1–13.

[15] Pinto, C., Furukawa, J., Fukai, H., & Tamura, S. (2017): Classification of Green coffee bean images basec on defect types using convolutional neural network (CNN). *International Conference on Advanced Informatics, Concepts, Theory, and Applications (ICA-ICTA)*, 1–5. doi: 10.1109/ICAICTA.2017.8090980.

[16] Ravi D. et al. (2017): Deep Learning for Health Informatics. *IEEE Journal of Biomedical and Health Informatics*, 21(1), 4–21.

[17] Pham T., Tran T., Phung D., Venkatesh S. (2017): Predicting Healthcare Trajectories from Medical Records: A Deep Learning Approach. *Journal of Biomedical Informatics*, 69, 218–229. doi: 10.1016/j.jbi.2017.04.001.

[18] Casalino F.W., L. P., & Khullar, D. (2018): Deep Learning in Medicine- Promise, Progress, and Challenges. *JAMA Internal Medicine*, 179(3):293–294. doi:10.1001/jamainternmed

[19] Wang X., Han Y., Leung M.C.V., Niyato D., Yan D., Chen X. (2020): Convergence of Edge Computing and Deep Learning: A Comprehensive Survey. *IEEE Communications Surveys & Tutorials*, 22(2), 869–904, doi: 10.1109/COMST.2020.2970550.

[20] Ananthanarayanan G., Bahl P., Bodík P., Chintalapudi K., Philipose M., Ravindranath L., Sinha S. (2017): Real-Time Video Analytics: The Killer App for Edge Computing. *Computer*, 50(10), 58–67.

[21] Ran X., Chen H., Zhu X., Liu Z., Chen J. (2018): DeepDecision: A mobile deep learning framework for edge video analytics. *IEEE INFOCOM 2018—IEEE Conference on Computer Communications*, 1421–1429.

[22] Tan, T., Cao G. (2020): *FastVA: Deep learning video analytics through edge processing and NPU in mobile*. https://ieeexplore.ieee.org/document/9155476.

[23] Sufian A., Ghosh A., Sadiq S.A., Smarandache F. (2020): *A survey on deep transfer learning and edge computing for mitigating the COVID-19 pandemic*. Preprint submitted to Journal of Systems Architecture, Elsevier.

[24] Hossain S.M., Muhammad G. (2019): An Audio-Visual Emotion Recognition System Using Deep Learning Fusion for a Cognitive Wireless Framework. *IEEE Wireless Communications*, 26(3), 62–68.

[25] Chen C., Liu B., Wan S., Qiao P., Pei Q. (2021): An Edge Traffic Flow Detection Scheme Based on Deep Learning in an Intelligent Transportation System. *IEEE Transactions on Intelligent Transportation Systems*, 22(3), 840–1852.

[26] Ferdowsi A., Challita U., Saad W. (2017): Deep learning for reliable mobile edge analytics in intelligent transportation systems. *IEEE Vehicular Technology Magazine*, 99. DOI: 10.1109/MVT.2018.2883777.

[27] Zhou J., Dai, HN., Wang H. (2019): Lightweight Convolution Neural Networks for Mobile Edge Computing in Transportation Cyber Physical Systems. *ACM Transactions on Intelligent Systems and Technology*, 10(6), 1–20, https://doi.org/10.1145/3339308.

[28] Lin C., Deng D., Chih Y., Chiu H. (2019): Smart Manufacturing Scheduling with Edge Computing Using Multiclass Deep q Network. *IEEE Transactions on Industrial Informatics*, 15(7), 4276–4284.

[29] Liang F., Yu W., Liu X., Griffith D., Golmie N. (2020): Toward Edge-Based Deep Learning in Industrial Internet of Things. *IEEE Internet of Things Journal*, 7(5), 4329–4341. doi: 10.1109/JIOT.2019.2963635.

[30] Lee T.K., Lee S.Y., Yoon H. (2019): Development of edge-based deep learning prediction model for defect prediction in manufacturing process. *International Conference on Information and Communication Technology Convergence (ICTC)*, 248–250.

[31] Pang B., Nijkamp E., Wu N.Y. (2019): Deep learning with tensorflow: A review. *A Journal of Educational and Behavioral Statistics*, 45(2), 227–248.

[32] Louis S.M., Azad Z., Delshadtehrani L., Gupta S., Warden P., Reddi J.V., Joshi A. (2019): *Towards Deep Learning Using Tensorflow Lite on RISC-V*, CARRV.

[33] Ketkar N. (2017): Introduction to PyTorch. *Deep Learning with Python*, 195–208. doi:10.1007/978-1-4842-2766-4_12.

[34] Cengil E., Çınar A., Özbay E. (2017): Image classification with caffe deep learning framework. *International Conference on Computer Science and Engineering (UBMK)*, 440–444.

[35] Erickson, B.J., Korfiatis, P., Akkus, Z. et al. (2017): Toolkits and Libraries for Deep Learning. *Journal of Digital Imaging*, 30, 400–405.

17 Effectiveness of Machine and Deep Learning for Blockchain Technology in Fraud Detection and Prevention

Yogesh Kumar and Surbhi Gupta

CONTENTS

17.1 INTRODUCTION

The emergence of machine learning or deep learning algorithms in various domains is evident. These algorithms have provided immense power in analyzing the collected data and making decisions using that data. But, for the model to be efficient and accurate requires training with an enormous amount of data

[1]. Data sharing within organizations can lead to various threats like tampering, attack by various hackers, and decentralization of data. Another technology, blockchain, possesses the capabilities to overcome all the described shortcomings. The amalgamation of these technologies results in systems resistant to the tinkering, robust, decentralized, secure [2]. One such application of such systems has been studied in this chapter, fraud detection and prevention using blockchain and machine learning or deep learning algorithms. An unknown group, Satoshi Nakamoto, behind Bitcoin, described the importance of blockchain technology in solving the problem of maintaining the order of transactions. Bitcoins consist of blocks that are constrained-size structures containing transactions. These blocks are connected with the help of hash values [8], present in the previous blocks. As described, blockchain was formerly originated to prevent fraud in digital currency exchanges. Blockchain being unaffected by tinkering gives the confirmed contributors access to store, view, and share the digital material in a situation rich in safety, which in turn supports development of trust, liability, and transparency in business relations. To capitalize on this specified assistance, companies have now started exploring ways in which blockchain technology could prevent fraud in numerous industry verticals [31].

Figure 17.1 describes the structure of blockchain technology. It is visible from Figure 17.1 that each block is a collection of transactions occurring at the same time-stamp. It also contains the hash value of the next block in a chain [21].

Protection from identity theft and fraud is an endless challenge for everyone to elaborate on buying and selling. Merchants, consumers, issuers, and acquirers know there are susceptibilities in how payments and data are secured. With each novelty in security technology, hackers and fraudsters learn how to outsmart the technology and breach these networks. Blockchain seems to possess capabilities that help to overcome the described problems [22].

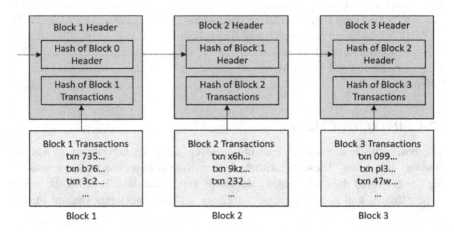

FIGURE 17.1   Structure of blockchain [22].

Figure 17.2 displays the various properties of blockchain:

- **Programmable:** Blockchain uses the concept of smart contracts (SC). SCs can be defined as: "a computerized transaction protocol which helps in executing the terms of a contract," which minimizes external risks. In the context of blockchain, SCs can be defined as scripts that run in a decentralized manner [9] without relying on third parties.
- **Distributed:** This is one of the biggest strengths of blockchain. In this, all the computers (nodes) in a network have a copy of the full ledger to ensure transparency [10].
- **Immutable:** The data present in the ledger cannot be tampered with, hence making the blockchain robust.
- **Consensus:** The data present in the ledger is transparent and distributed so all the nodes in a network agree to the validity of the records.
- **Secure:** All the records present in the blockchain are encrypted individually.
- **Time-stamped:** All the records present in the ledger are time-stamped. These time-stamps are also stored in the ledger with their corresponding records.
- **Anonymous:** Blockchain doesn't reveal the information of the user, hence keeping them anonymous.

Sensing devices like IoT devices, web applications, etc. have led to the massive production of data [26]. The data produced can be effectively analyzed with the help of various algorithms of machine learning and deep learning [27]. But, such algorithms in application work on centralized models where various organization servers can access this model on their specific data without sharing it [28]. But, this centralized nature of models leads to the problem of tampering with data [29]. Moreover, data authenticity cannot be verified [30]. So, the outcomes of these models cannot be trusted completely. We can say that decentralized AI is a blend of blockchain and machine learning/deep learning algorithms. Such systems allow securely shared data and allow better personal analysis of the data. This allows authentication of the data source and stores data in a highly secured manner.

The chapter's primary concern is to highlight the various features in the blockchain that can prevent the different types of fraud from occurring. Another focus is to highlight the different types of frauds, identify theft, and supply chain, which can be prevented and detected using blockchain technology. This chapter will also cover the various implementation challenges to prevent fraud and the importance of machine learning and deep learning for fraud detection in blockchain technology [41]. The machine learning models for blockchain include the anomaly transaction detection system, Bitcoin fraud detection, and other types of fraud detections. The structure of the chapter will comprise a framework for automatic fraud detection systems using blockchain, the motivation of the work, different benefits of fraud prevention, and detection using blockchain technology, applications, and role of learning models in blockchain for fraud prevention and

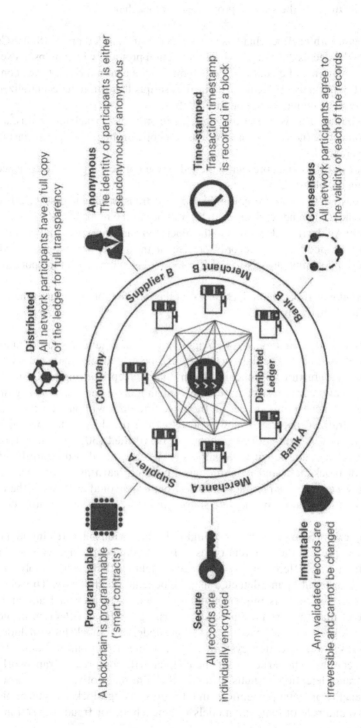

FIGURE 17.2 Properties of blockchain [28].

detection. The chapter will also cover the work that different researchers in the analysis section have done.

17.2 BACKGROUND STUDY

This section of the chapter gives information about the framework of blockchain, the different types of frauds, the importance of machine and deep learning in blockchain, benefits associated with the work, and most importantly the research challenges associated with it.

17.2.1 AUTOMATIC FRAMEWORK OF BLOCKCHAIN

This section of the chapter describes the framework of the blockchain for an automated transaction.

Figure 17.3 explains the flow of transactions in a blockchain-based framework. Suppose A requests B for the transaction; this request is broadcasted via peer to peer. The algorithms are used for the validation of transactions and users by the nodes; nodes represent the computer systems in a network of multiple systems. Further, it awaits the verification of the transaction. Once the transaction is verified, it can involve crypto currencies, records, or confidential information. A new block is created in a ledger that is being added to the existing blockchain, which is immutable and permanent. This addition of the block to the blockchain marks the completion of the transaction.

There are many industry-specific frameworks of blockchain. Some of them are [10,34,51]:

- **Exonum:** It's an open-source project given by the Bitfury group and developed in Rust. It has applications in fintech and legal tech.
- **Hyperledger:** It's an open-source project given by the Linux Foundation and developed in python. It has applications in cross-industry.
- **Openchain:** It's an open-source project given by Coinprism and developed in Javascript. It has applications in cross-industry.
- **Graphene:** It's an open-source project given by Cryptonomex and developed in C++. It has applications in cross-industry.
- **Corda**: It's an open-source project given by R3 Consortium and developed in Kotlin, Java. It has applications in the financial services industry.
- **Multichain**: It's an open-source project given by Coin Science and developed in C++. It has applications in the financial services industry.

All of the previously mentioned frameworks allow smart contracts and are generally private. Only hyperledger and graphene are public. Most of them allow regular updates, and only a few are not active.

17.2.2 TYPES OF FRAUD

The technological advancements have also posed immense threats like phishing, malware, and many more. Many threats that have been caused due to misuse of

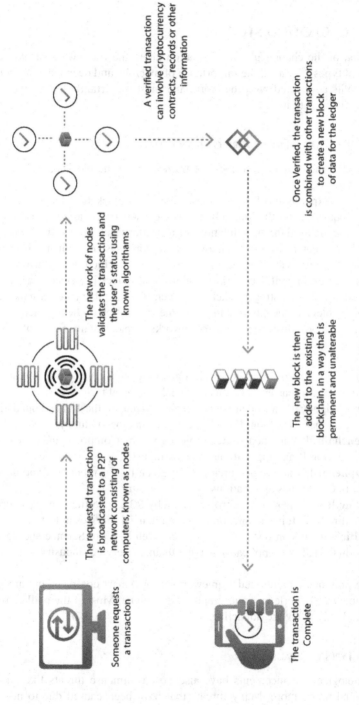

FIGURE 17.3   Blockchain framework for automated transaction [39].

technology are categorized in to multiple types. Nine types of frauds are discussed here [21, 25, 36]:

- **Mail Fraud:** It is a kind of fraudulent activity that revolves around postage mail. It involves a letter sent with the intent of scamming money or stealing the personal information of a person. In simple words, mail fraud involves the use of mail.
- **Driver's License Fraud:** At first glance, this fraud doesn't seem to cause a lot of harm to anyone, but it has an immense capability to ruin someone's life. A driving license can help in fraudulent activities like opening a bank account in someone's name, boarding a plane, etc.
- **Health Care Fraud:** In this type of fraud, insurance information is used by any other individual for personal gain. To avoid such frauds, it is necessary to keep updated on all medical bills, personal information, and insurance claims.
- **Debit and Credit Card Fraud:** It is one of the most common fraudulent activities we all come across. It happens when a number present on the card or the card itself is stolen. To prevent this fraud, continuous monitoring of bank accounts every week is crucial. One should keep their cards safe and not display the number present on the card. These days, hackers are also using "skimmers" to tamper with information. Never save card numbers online, instead use e-wallets for the purpose.
- **Bank Account Takeover Fraud:** Getting account information is comparatively easy. Personal information required to access an account can be obtained by getting access to the cheque in the cheque box or through an email scam, or maybe with the help of malware. It is necessary to keep a check on account statements. Never login to the bank account using unsecured Wi-Fi.
- **Stolen Tax Refund Fraud:** It is one of the top scams encountered by the IRS. It happens when someone steals a Social Security number and files the taxes themselves; this leads to receiving of a refund by a thief. It is necessary to be vigilant about to whom we provide our information.
- **Voter Fraud:** This broad term includes various malfunctions like voting under a false identity, buying votes, voting twice, and voting of a felon.
- **Internet Fraud:** As it sounds, it is the type of fraud that occurs via the Internet. It includes breaching of data, email account compromise (EAC), phishing, and malware. Every year, millions of dollars are stolen through the internet. It is necessary to keep an anti-virus update for your system to prevent malware and viruses from entering your system. Always read links to ensure that you are being directed to an official site.
- **Elder Fraud:** Senescence makes the person more susceptible, kind-hearted, and trusting. All these qualities of an elderly person are extorted by scammers and lead to illegal activities like phone scams offering lottery winnings or healthcare services [46–47]. These activities help them to gain access to their personal information.

17.2.3 Importance of Machine Learning and Deep Learning-Based Models for Blockchain in Fraud Prevention and Detection

Machine learning is a paradigm that automatically learns from the data feed. It gains knowledge by observing the data provided to it. Various patterns are analyzed, and inferences are made based on those observations. Machine learning-based models facilitate automatic learning in several fields. The model's efficiency is dependent on large volumes of data. The machine learning process has two phases, namely, the training and validation phases. These models learn through the training cases in the training phase; labels of a large portion of the data are fed to the model as input (also known as the training set). Further, the constructed model trained using the training set is evaluated using the validation data. The validation phase involves testing trained models and is validated by cross-checking the predicted labels of the data. It helps the system understand complex perception tasks, trying to achieve maximum performance. Learning can be classified into the following broad types:

- **Supervised learning** is learning in which machines are taught or trained using data that is well labeled, which means some data is already tagged with the correct answer. Majorly used supervised classification algorithms include Decision Tree, support-vector machines, linear regression, logistic regression, and Naive Bayes.
- **Unsupervised learning:** Unsupervised learning is the training of a machine that uses information that is neither classified nor labeled and allows the algorithm to act on that information without guidance. Unsupervised classification algorithms like the K-means algorithm, Neural Networks, and K-Nearest Neighbor algorithm have been widely employed.
- **Semi-supervised algorithm:** Semi-supervised learning proves to be very useful when the cost associated with labeling the dataset is too high to allow a fully labeled dataset. Semi-supervised learning methods include regression algorithms such as ordinary least, cluster assumption, manifold assumption, generative models, low-density separation, graph-based methods, and heuristic approaches.
- **Reinforcement learning:** This is reward-based learning, i.e., for each correct prediction, the learning approach gets a positive reward. Popular reinforcement learning approaches are State—action—reward—state—action (SARSA-Lamda) and Deep Q Network (DQN).
- Deep learning is a subsection of machine learning in artificial intelligence. *Deep learning approaches* are the networks that can learn from complex/intrinsic data. Deep learning is also recognized as deep neural learning and deep neural networks. As the name indicates, deep neural networks imitate the functioning of the human brain, i.e., the neurons in a neural network and those in the human brain work similarly. The deep learning model consists of various layers that include a non-linear processing unit for feature extraction. Also, deep learning models exhibit a hidden layer processing that works like a human brain and performs classification with more accuracy. Popular deep learning models are Artificial Neural Networks, Convolutional Neural Networks, and Stacked Auto-encoders.

Applications of deep learning include biological image classification (diagnosing images like x-ray), computed tomography (CT), magnetic resonance imaging, deep vision systems, healthcare (disease identification, robotic surgery), parking systems, object detection (moving object detection in videos), medical applications (drug targeting, personalized medicine, predictive analytics, personal care), document analysis and recognition, data flow graph, remote sensing, stock market analysis (predicting the rise and downfall of shares in the stock market based on the patterns), semantic image segmentation, synthetic aperture radar, person re-identification, and many more.

Machine learning applications are image recognition, speech recognition, medical diagnosis, automatic language translation, online fraud detection, stock market trading, email spam and malware filtering, virtual personal assistant, self-driving cars, traffic prediction, product recommendation, and many more [28].

Some of the work of the various researchers using machine learning or deep learning for fraud detection has been discussed in the following list:

- It helps to control corruption in governmental bodies.
- It provides automatic digital signature during the transaction with the help of machine learning.
- It helps solve the access control problem of IoT.
- It is useful to protect customers from online fraud ratings.
- It helps remove the anonymity in the blockchain which leads to illicit activities. A supervised machine learning technique is useful by providing 80.2% accuracy.
- It is useful in detecting suspicious activities by using various algorithms like XGBoosting, support vector machines, Random Forest, and many more.
- It helps detect frauds with the help of various graphs with considerably high results.
- Various clustering techniques are also helpful in detecting suspicious activities and frauds.
- By detecting fraud machine learning also helps an organization to increase profits. Frauds in telecommunication can be precisely detected with the help of fuzzy logic with an accuracy of 97.6%. It also helps organizations to divide customers into various categories.
- By analyzing a series of transactions, a problem of high false-positive rates can be successfully addressed. Algorithms like SVM, LSTM-RNN are helpful by giving AUC as 89.322% and 95% respectively.
- Generative Adversarial Networks help increase the data available for the minority class. This helps to reduce the problem of overfitting and increase the performance of the model.
- Considering the previously provided results, we can considerably check that the results of deep learning and machine learning are significant in fraud detection. They help detect fraud and also the problems associated with it. So, it is a no-brainer to merge these two emergent techniques with blockchain [26].
- Bringing AI and blockchain together can result in a revolutionized paradigm. On the one hand, blockchain is a distributed ledger that is transparent and immutable, whereas AI (machine learning and deep learning) can result in

the study of the enormous data gathered in the blockchain. Machine learning and deep learning can help analyze the data gathered more efficiently. They can be fruitful in enhancing the security feature of the chain. Since blockchain works on a decentralized ledger, enormous data can be gathered to train the machine learning model, creating a robust model. Their amalgamation has resulted in various frameworks for fraud detection like credit cards, insurance, and much more. The proposed work of many researchers in this field is demonstrated in the reported work section of the chapter, where we will come across advantages their merger has posed [16].

17.2.4 BENEFITS OF BLOCKCHAIN

Blockchain possesses three main features that increase its robustness and make it suitable for fraud detection. Those three features are Distributed Networks, Immutable, and Permissible. As a result, the relevance of blockchain technology increases exponentially [10]. According to a survey conducted by IBM in 2017, almost 33% of executives are engaged with blockchain. With the advancements, three generations of blockchain have been witnessed [11]. Blockchain has many advantages:

- It provides more efficient performance for security and transactions [12–17]. Since blockchain is transparent and transactions are digitally signed and encrypted, they are highly secured.
- It inhibits the involvement of third parties, so no involvement of hidden fees. Blockchains have made the involvement of third parties or settlers zero because all the users in a network can check the ledger, all the activities are transparent, and blockchain also involves the usage of smart contracts.
- It provides transparency. It means that all the associated nodes (computers) in a network can see the ledger instantly. All the transactions are known to everyone connected in a network.
- Since it is based on a distributed ledger, it creates a sense of trust among its participants or connected nodes in a network.
- Highly cost-effective. Since it expedited the involvement of third parties.
- Transactions are cryptographically signed and verified. All the transactions made in blockchains require users to sign them digitally, and all the transactions are encrypted after that.
- It provides security against hackers as the data present in the ledger cannot be tampered with.
- It provides high speed. Since blocks in the blockchain are connected via hash values, it is easy and faster to navigate between them.
- It helps to check fraudulent activities like corruption because of its transparent and distributed characteristics.

17.2.5 RESEARCH CHALLENGES

This section discusses the challenges faced to bringing blockchain and AI together. Some of the main issues are listed in the following list:

- **Privacy:** The data stored in ledgers of blockchain are publicly available. It can lead to harm to privacy. However, making the data private can lead to the inefficient performance of AI models.
- **Blockchain Security:** Blockchain is susceptible to cyber-attack by 51% [21]. Decentralized characteristics of blockchain make it more prone to attacks. Issues related to security are more evident in public blockchains like Bitcoin and Ethereum.
- **Trusted Oracle:** Smart contacts cannot pull data from the outside world. It cannot trigger an event automatically. Trusted oracles, commonly known as third parties, come into play [22].
- **Scalability and Side Chains:** Bitcoin can complete four transactions in a second, whereas Ethereum can complete 12 transactions. Sidechains are useful in enhancing scalability [23]. These days researchers are working significantly on improving the scalability of the blockchain [24].
- **Fog Computing:** It allows localized computation of the gathered data. It helps to prevent long delays that can be faced in cloud computing.
- **Lack of Standards, Interoperability, and Regulations:** No standards have been devised. Many bodies are working toward it [25].
- **Smart Contracts Vulnerabilities and Deterministic Execution:** Smart contracts should be error-free and secured against various vulnerabilities. For instance, DAO was hacked in 2016, resulting in a loss of 3.6 million Ethers [26].
- **Governance:** It is crucial to manage, deploy, and construct blockchain effectively among its participants.
- **Quantum Computing:** Blockchain requires a digital signature that is cryptographically stored. It is estimated that by the year 2027, 10% of the world's GDP will be handled through the blockchain network [28]. Subsequently, sound research that can solve blockchain's scalability, security, and breakability issues using quantum computing should be worked on.

17.2.6 VARIOUS FEATURES OF FRAUD DETECTION/PREVENTION SYSTEMS

Multiple systems detect the fraud in advance and those that prevent the fraud. A brief description of such systems is given in the following list:

- **Real-time transaction screening and automated review:** Machine learnings help to monitor incoming data. This real-time processing of data helps employees in reviewing data [27].
- **Deep insights into user behavior:** Machine learning or deep learning-based systems help understand user behavior, and analysts work accordingly.
- **Reduction in false-positives:** It means the decline of legitimate transactions. Accuracy can be maintained by going beyond the transaction amount and the user's location.
- **Real-time operations tracking and reporting:** Customers can track their performance in real-time. They can also check information like location, channel, payment method, etc. Analysts use various fraud patterns using visualization software to find user behavior [28].

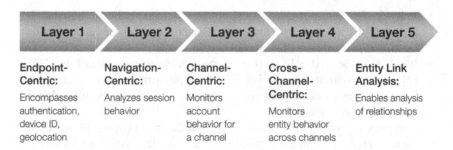

| Layer 1 | Layer 2 | Layer 3 | Layer 4 | Layer 5 |

Endpoint-Centric:
Encompasses authentication, device ID, geolocation

Navigation-Centric:
Analyzes session behavior

Channel-Centric:
Monitors account behavior for a channel

Cross-Channel-Centric:
Monitors entity behavior across channels

Entity Link Analysis:
Enables analysis of relationships

FIGURE 17.4   Five-layer architecture for fraud detection or prevention [30]

- **Automation level:** Organizations have two choices: either completely rely on automated fraud detection models or consider taking an additional service of analysts to comply with the model's results [29].
- **Comprehensiveness and self-learning capability:** Systems used for fraud detection must be versatile. The systems used by organizations must be comprehensive to cover all the information systems. These systems should be able to automatically learn from the data gathered in the organization so that not only known but also yet to be discovered threats can be discovered [30].
- **Multiple protection layers:** The Gartner group suggested a five-layer approach to detect fraud and prevent it. Figure 17.4 describes five layers.
- **Integration and deployment:** Organizations should encourage discussions and reviews on their products so that they can work on the pitfalls. They should also be focused on integrating current facilities in their software.
- **Compliance with security standards:** Organizations should keep the document of their requirements handy. They must ensure that every customer complies with these pre-settled requirements of the organization.
- **Cost:** The cost of fraud detection software varies according to the expectations of the organization. The choice is between having fixed subscription plans and flexible pricing according to the requirements. For example, Fraud screening tools are comparatively less expensive as compared to full-service fraud tools.
- **Customer Support:** Organizations should provide robust and easily accessible customer support services. They should work on a document that describes the technical support customers can get from the service provider in case of difficulty, the response time, and the contact authority.
- **Approval rates and false-positive handling:** It is necessary to evaluate how the software handles cases of false declines that are originally legitimate.

17.3   REPORTED WORK

In this section, we explicate the finding of different researchers for fraud detection and prevention using blockchain technology. Many significant research works have been reported in this section that enhance the domain-knowledge along with the issues resolved by multiple advanced techniques. This section also highlights how deep and machine learning techniques have been used in blockchain technology for fraud detection and prevention.

17.3.1 FRAUD DETECTION USING BLOCKCHAIN

Many researchers have documented significant achievements in detecting frauds using blockchain concepts. In research by Joshi et al. [1], a framework was worked out to prevent fraud among various governmental organizations. The work was generic and developed a layered architecture for disbursement through various government bodies. They addressed the problems of corruption, which occur at various levels, leading to economic losses to the country. They employed blockchain because of its transparent, decentralized, and immutable characteristics. They used a complete graph to model cash flow across various levels in government. The information generated at any level is broadcasted on a network. The proposed model has four levels: sub-departments, district government, state government, and central government. Their model can be deployed on Ethereum. In their model, central authorities could deploy the smart contracts, whereas other levels could only update them. Their objectives were transparency, time complexity improvement, complete utilization of funds, immutability, security, promptness, and eco-friendliness. Podgorelec et al. [2] carried out a noteworthy research study that proposed an automated blockchain transaction signing using machine learning. They also proposed personalized anomaly detection in their work. Data storage in a blockchain ledger can be done only with the help of a digital signature, but this technique is time-consuming. To address this issue, they proposed a machine learning-based method for the automated signing of blockchain transactions. An analysis has been done using Ethereum. Customers are susceptible to fraud ratings in online shopping. Cai et al. [3] carried a substantial research work, addressed fraud ratings, and discussed blockchain effectiveness in objective rating and its leggings in subjective ratings. In their study, they checked the performance of blockchain under the two types of attacks: ballot-stuffing and bad-mouthing. They found that blockchain is more robust for bad-mouthing as compared to ballot-stuffing. Outchakoucht et al. [4] amalgamated machine learning and blockchain to address the problems associated with access control in IoT (Internet of Things). IoT is associated with the problems of decentralization and dynamicity, which have been addressed in their work with the help of blockchain and machine learning, respectively. Their proposed framework was based on smart contract and reinforcement learning. They used reinforcement learning to make agents learn from the environment and update the smart contract as per the results. Yin et al. [5] proposed supervised machine learning to remove anonymity from the bitcoin blockchain. Bitcoin has enabled widespread illicit activities due to its high degree of anonymity. They used 385 million transactions to train the model and built a classifier to categorize it into 12 categories. Their proposed model had the capability of detecting yet-unidentified entities. They used a Gradient Boosting algorithm with parameters set with default values. They achieved a mean accuracy of 80.42% and F1-score of 79.64%. In their article, they predicted a set of 22 clusters that might be related to cyber-crime and also classified 153,293 clusters to estimate the bitcoin ecosystem.

Another study attempted to address the needs of the ever-growing insurance market Dhieb et al. [6] and proposed a blockchain-based framework to detect frauds and increase monetary profits. Since the framework was based on blockchain, it allowed secure transmissions among various agents. They employed extreme gradient boosting to provide insurance services and compared the performance with other algorithms. They used the auto-insurance dataset and found an increment of 7% invalidation compared to the robust decision trees. To implement a blockchain-based framework, they combined the proposed machine learning model with the hyperledger fabric composer. More substantial research by Piquet et al. [49] detected suspicious users and transactions with the help of graphs that were generated with the help of the blockchain network. In graphs, nodes were represented as users or transactions. They used many algorithms like support vector machine, K-means clustering, and Mahalanobis distance. They also applied ensemble classification approaches like Random Forests and Extreme Gradient Boosting classifier (also popularly known as XGBoost classifier) to detect fraudulent activities in the famous Ethereum blockchain network.

For detecting an anomaly in the records, Monamo et al. [7] used a K-means clustering technique. It validated the model's performance on the Ethereum network, gained better results, and detected more fraudulent transactions. Arjoune et al. (2020) [52] employed multiple classification models to detect jamming attacks. The classification models used to achieve the desired objectives are Random Forest, support vector machine, and neural network. The classification approach projected in this study can process large-sized data in less time with greater efficiency. The transmission link state is monitored to verify if it is attacked or not.

17.3.2 FRAUD PREVENTION USING BLOCKCHAIN

Estevez et al. [40] designed a system to prevent fraud in fixed telecommunication. They addressed the problem associated with the long-distance impact factor. They designed both classification and prediction modules. They classified subscribers into four categories: fraudulent, normal, insolvent, and subscription fraudulent based on their history, whereas the prediction module was proposed to find a fraudulent customer. They employed fuzzy logic on a database with more than 10,000 subscribers residing in Chile. The database considered for the study had a fraud prevalence of 2.2%. Multilayer perceptron was used to predict potential fraudulent customers. The model was robust enough, which identified 56.2% using only 3.5% of the test data. For the total dataset, it gave a tremendous result of 97.6%.

Wiese et al. [41] addressed the problem of high false-positive rates using dynamic machine learning algorithms. They suggested that, rather than analyzing single transactions, a series of transactions must be analyzed for the detection purpose. The biased machine learning methods were used to achieve better results. They used support vector machines and Long Short-Term Memory Recurrent Neural Network. The authors considered the European cardholder dataset for their study, with 30,876 transactions collected over 12 months. PERF Release 5.10 was used to determine performance metrics in each case. They used 41 features out of the total feature set to classify. The AUC valued 95% confidence using 30 trials. The maximum value of AUC of 99.216% was recorded, and the minimum of 91.903%. With the help of SVM, they achieved an AUC of 89.322%; it was also found that SVM took a shorter duration for classification.

Generative Adversarial Networks can mimic minority classes as they are flexible, general, and powerful generative deep learning models. Fiore et al. [42] exploited Generative Adversarial Networks for increasing the dataset for training. As a result, the performance of the classifier considerably improves. The generator and discriminator were both of three layers each. They successfully addressed the class imbalance problem using Generative Adversarial Network by producing great results—specificity as 99.998% and sensitivity as 70.229%. Rushin et al. [43] used a dataset with 80 million transactions gathered over 8 months to detect credit card frauds. The considered dataset has 69 attributes containing account, transactions, and account holder information. The dataset also suffered from imbalance class problems as it consists of 99.864% legitimate credit card holders and only 0.136% frauds. They used autoencoder and domain expertise to obtain a feature set for building the model. They created six different feature sets using original features, autoencoder features, and the ones generated by domain expertise. They tested the performance of all six datasets with the help of the following classifiers: deep learning models, logistic regression, and Gradient Boosted Trees. Deep learning models outperformed most of the cases, and the other two classifiers gave presentable results. Recently Dhir et al. [51] analyzed the application of machine learning and deep learning-based approaches in cyber security. The study explained the threat models that embrace different attacks faced in the network. Another significant research work in healthcare was carried out in 2020; Chen et al. [52] proposed a new approach based on fuzzy entropy and projected a fuzzy entropy weighted natural nearest neighbor method (FEW-NNN) for detecting the attack in the network traffic. The Fisher score and deep graph feature selection methods achieved the dimension reduction objective. The proposed model was validated in the publicly available datasets like KDD99 and CICIDS-2017 datasets and achieved prediction accuracy of 80% on the KDD99 dataset. Gorla et al. [54] explored the issues encountered in blockchain for fraud prevention and detection in forensics. Blockchains can generate a decentralized situation where any third party does not control transactions and data. It also can analyze frauds in the digital setting. This study proposed the inclusion of artificial intelligence to address the issues in fraud prevention. Another prominent research work carried out in 2020 by Nguyen [55] applied a blockchain model intending to prevent roaming fraud and maximize the benefits. This study investigated a profitable model based on Stackelberg game, intending to enhance the profits and encourage users to participate in the blockchain system.

17.4 COMPARATIVE ANALYSIS

This section of the chapter presents the comparison of previous works based on the techniques used and prediction accurateness achieved. The studies used for the comparative analysis are those that employed machine learning and deep learning algorithms in blockchain technology for fraud detection. Multiple frauds like Credit Card Fraud, Financial Fraud, Energy Fraud, Medicare Fraud, Fixed Telecommunication

Fraud, Network Traffic Attack, Jamming Attacks, Technical Loss Fraud, and Auto-Claim Fraud have been detected. Most of the studies that have been analyzed worked on Credit Card Fraud detection using automated learning techniques. A comparative analysis of all the studies is presented in Table 17.1.

TABLE 17.1
Comparative Analysis for Fraud Detections Using Blockchain Technology

Author	Fraud type	Dataset	Technique	Result
Awoyemi et al. [31]	Financial Fraud	ULB Machine Learning Group	Naïve Bayes, K-nearest neighbor, and Logistic Regression	Accuracy= 97.92%, 97.69%, and 54.86% respectively.
Maes et al. [32]	Credit Card Fraud	Serge Waterschool	ANN, Bayesian Networks	Accuracy= 70%, 74% respectively.
Yee et al. [33]	Credit Card Fraud	WEKA	K2, Tree Augmented Naïve Bayes, and Naïve Bayes, logistics and J48.	Accuracy= 95.8%,96.7%, 99.7%,100.0%, 100.0% respectively.
Ford et al. [34]	Energy Fraud	WEKA	ECB neural network	TN: 75.00% TP: 93.75% FN: 6.25% FP: 25.00%
Bahnsen et al. [35]	Credit Card	From a European card processing company.	Logistic Regression (LR) and Random Forest (RF)	Bayes gave savings of 23% more as compared to Random Forest.
Bauder et al. [36]	Medicare Fraud	2015 Medicare PUF data	DNN, GBM, RF	Confidence= 95%
Randhawa et al. [37]	Credit Card Fraud	European cardholders	AdaBoost	Accuracy=99%
Khare et al. [38]	Finance Fraud	European cardholders	Logistic Regression, Decision Tree, Random Forest, and SVM	Accuracy= 97.7%, 95.5%, 98.6%, 97.5% respectively
Estévez et al. [40]	Fixed Telecommunication Fraud	DICOM	A multilayer perceptron neural network, fuzzy logic	Accuracy= 97.6%
Wiese et al. [41]	Credit Card Fraud	European cardholders	Support vector machines, long short term memory	AUC as 89.322%, 99.126% respectively
Thennakoon et al. [44]	Credit Card Fraud	Fraud transactions log file and all transactions log file	Support vector machine, Naive Bayes, K-Nearest Neighbor, and Logistic Regression	74%, 83%, 72%, and 91% accuracy respectively.

Author	Fraud type	Dataset	Technique	Result
Yap et al. [54]	Technical loss fraud	Pre-Populated Database (PPD)	Support vector machines, Genetic Algorithm	Accuracy = 94%
Viaene et al. [46]	Auto-claim fraud	PIP claim data	Bayesian Neural Network	N/A
Dornadula et al. [47]	Credit Card Fraud	European credit card dataset	Isolation Forest, local outlier factor, SVM, LR, DT, RF	Accuracy= 90.11%, 89.90%, 99.87%, 99.90%, 99.94%, 99.94% respectively.
Viaene et al. [48]	Credit Card Fraud	Claims during 2000	Logistic regression	Accuracy= 99.42%
Pinquet et al. [49]	Credit Card fraud	Claims during 2000	Back propagation	N/A
Panigrahi et al. [50]	Credit Card Fraud	NA	Decision Tree, Naïve Bayes	Avg. Accuracy= 98%
Chen et al. [52]	Network Traffic Attack	KDD99	Fuzzy Entropy Weighted Natural Nearest Neighbor Method	Training Accuracy=80% Testing Accuracy= 80%
Arjoune et al. [53]	Jamming Attacks	Customized dataset	Random forest, support vector machine, and neural network	Accuracy =96.6%

As stated in the previous table, we can see that multiple learning techniques such as supervised (Decision Trees, support vector machine, Naive Bayes, K-Nearest Neighbor), ensemble techniques (AdaBoost, Random Forest), multiple hybrid approaches (Bayesian Neural Network, Isolation Forest, local outlier factor), and neural learning approaches (Artificial Neural Networks, ECB neural network, Deep Neural Networks, multilayer perceptron neural network, long short term memory) have been explored in the field of different types of fraud detections. Different machine and deep learning-based algorithms [56–59] have successfully achieved significant prediction accuracy. The ensemble models have proved their supremacy [61–62]. Different studies have used different evaluation parameters to test the performance of the classification models. The parameters used to assess the model's performance are accuracy, the area under the curve, confidence rate, true-positive, and false negative.

17.5 CONCLUSION AND FUTURE WORK

In this chapter, we have walked through the combination of machine learning or deep learning algorithms with blockchain to prevent and detect fraud. We have studied the blockchain framework, the various types of frauds, the benefits associated with the blockchain, and the research challenges. We also checked the contributions of different researchers to detect fraud using various algorithms.

This study shows that bringing AI (machine learning/deep learning) and block-chain together results in tremendous systems that have the following capabilities: secure, transparent, decentralized, robust, encrypted, analytical, accurate, precise, and many more. We can see from the presented related work that machine learning and deep learning algorithms can precisely catch fraud. We have also seen that these technologies counter each other. The disadvantages associated with one can be remedied by the other. For example, a model requires a large amount of data to create better results, and the limitation of data can be rectified using blockchain. Since the model uses personal and private information, there is a grave fear of tampering with data. However, since blockchain ledgers are tampering free and decentralized, the task is easy and suitable. There are many such described limitations of these technologies that can be rectified using the other.

The researchers should work to bring down the associated cost so that the majority can access this technology. Untouched domains like genetics and evolution should also be explored using these technologies, as these domains suffer from the disadvantage of limited and private data. Blockchain and AI show the signs of being a new emergent field that can help establish systems that people did not even think of.

REFERENCES

[1] Joshi, P., Kumar, S., Kumar, D., & Singh, A. K. (2019). *A Blockchain-Based Framework for Fraud Detection*. Conference on Next Generation. IEEE. pp:1–5.
[2] Podgorelec, B., Turkanović, M., & Karakatič, S. (2019). A Machine Learning-Based Method for Automated Blockchain Transaction Signing Including. *Sensors*, Vol:20, Issue:1, pp:147–165.
[3] Cai, Y., & Zhu, D. (2016). Fraud Detections for Online Businesses: A Perspective from Blockchain Technology. *Financ Innov 2*, Article No.:20, pp:1–16.
[4] Outchakoucht, A., Es-Samaali, H., & Leprosy, J.P. (2017). Dynamic Access Control Policy Based on Blockchain and Machine Learning for the Internet of Things. *International Journal of Advanced Computer Science and Applications*, Vol:8, Issue:7, pp:417–424.
[5] Yin, H., Langenheldt, K., Harlev, M., Mukkamla, R., & Vatrapu, R. (2019). Regulating Cryptocurrencies: A Supervised Machine Learning Approach to De-Anonymizing the Bitcoin Blockchain. *Journal of Management Information Systems*, Vol:36, Issue:01, pp:1–24.
[6] Dhieb, N., Ghazzai, H., Besbes, H., & Massoud, Y. (2020). A Secure AI-Driven Architecture for Automated Insurance Systems: Fraud Detection and Risk. *IEEE Access*, Vol:8, pp:58546–58558.
[7] Monamo, P., Marivate, V., & Twala, B. Unsupervised learning for robust Bitcoin fraud detection; *Proceedings of the 2016 Information Security for South Africa (ISSA); Johannesburg*, South Africa. 17–18 August 2016; pp.129–134.
[8] Crosby, M., Pattanayak, P., Verma, S., & Kalyanaraman, V. (2016). Blockchain technology: beyond bitcoin. *Appl. Innovation*, Vol:2, pp:6–10.
[9] Christidis, K., & Devetsikiotis, M. (2016). Blockchains and Smart Contracts for the Internet of Things. *IEEE Access*, Vol:4, pp:2292–2303.
[10] Zhao JL., Fan S., & Yan J. (2016). Overview of Business Innovations and Research Opportunities in Blockchain and Introduction to the Special Issue. *Financial Innovation*, Vol:2, pp:1–28.

[11] Zyskind, G., Nathan, O., & Pentland, A.S. (2015). Decentralizing privacy: using block-chain to protect personal data. *Proceedings—2015 IEEE Security and Privacy Workshops, SPW*, pp. 180–184.

[12] Velde, V., Scott, J., Sartorius, A., Dalton, K., Shepherd, B., Allchin, C., Dougherty, M., Ryan, P., & Rennick, E. (2016). *Blockchain in Capital Markets—The Prize and the Journey*, Oliver Wyman.

[13] Wu, L., Meng, K., Xu, S., Li, S.Q., Ding, M., & Suo, Y. (2017). Democratic Centralism: A Hybrid Blockchain Architecture and Its Applications in the Energy Internet. *Proceedings—1st IEEE International Conference on Energy Internet 2017*, pp. 176–181.

[14] Papadopoulos, G. (2015). Blockchain and Digital Payments. *An Institutionalist Analysis of Cryptocurrencies*, pp:153–172.

[15] Beck, R., Stenum Czepluch, J., Lollike, N., & Malone, S. (2016). *Blockchain—The gateway to trust-free cryptographic transactions*. 24th European Conference on Information Systems, ECIS 2016.

[16] Min, X., Li, Q., Liu, L., & Cui, L. (2016). *A permissioned blockchain framework for supporting instant transaction and dynamic block size*. Trustcom/BigDataSE/ISPA, IEEE, pp. 90–96.

[17] Yamada, Y., Nakajima, T., & Sakamoto, M. (2017). Blockchain-LI: a study on implementing activity-based micro-pricing using cryptocurrency technologies. *ACM International Conference Proceeding Series*, pp: 203–207.

[18] Koch, M. (2018). Artificial Intelligence Is Becoming Natural. *Cell*, Vol:173, Issue:3, pp:531–533.

[19] Schmidhuber, J. (2015). Deep Learning in Neural Networks: An Overview. *Neural Networks*, Vol:61, pp:85–117.

[20] Dinh, T., & Thai, M. (2018). Ai and Blockchain: A Disruptive Integration. *Computer*, Vol:51, Issue:9, pp:48–53.

[21] Qi, Y., & Xiao, J. (2018). Fintech: AI Powers Financial Services to Improve People's Lives. *Communications of the ACM*, Vol:61, Issue:11, pp:65–69.

[22] Li, X., Jiang, P., Chen, T., Luo, X., & Wen, Q. (2017). A Survey on the Security of Blockchain Systems. *Future Generation Computer Systems*, pp:1–13.

[23] Stradling, A., & Voorhees, E. (2018). *System and method of providing a multivalidator oracle*. US Patent App. 15/715,770.

[24] Hwang, G., Chen, P., Lu, P., Chiu, C., Lin, H., & Jheng, A. (2018). Infinitechain: A multi-chain architecture with distributed auditing of side chains for public blockchains. *International Conference on Blockchain*. Springer, pp. 47–60.

[25] Boyen, X., Carr, C., & Haines, T. (2018). Graphchain: a blockchain-free scalable decentralised ledger. *Proceedings of the 2nd ACM Workshop on Blockchains, Cryptocurrencies, and Contracts. ACM*, pp. 21–33.

[26] Gilad, Y., Hemo, R., Micali, S., Vlachos, G., & Zeldovich, N. (2017). Algorand: Scaling byzantine agreements for cryptocurrencies. *IACR Cryptology ePrint Archive*, pp. 454.

[27] Anjum, A., Sporny, M., & Sill, A. (2017). Blockchain Standards for Compliance and Trust. *IEEE Cloud Computing*, Vol:4, Issue:4, pp:84–90.

[28] Kakavand, H., Sevres, N., & Chilton, B. (2017). *The blockchain revolution: An analysis of regulation and technology related to distributed ledger technologies*. SSRN.

[29] Kiktenko, E., Pozhar, N., Anufriev, M., Trushechkin, A., Yunusov, R., Kurochkin. Y. V., Lvovsky. A., & Fedorov, A. (2018). Quantum-Secured Blockchain. *Quantum Science and Technology*. Vol:3, Issue:3, pp:035004–035013.

[30] Rodenburg B., & Pappas S. P. (2017). Blockchain and quantum computing. *MITR*. pp:1–8.

[31] Awoyemi, J. O., Adetunmbi, A. O., & Oluwadare, S. A. (2017). Credit card fraud detection using machine learning techniques: A comparative analysis. *International Conference on Computing Networking and Informatics (ICCNI), IEEE*, pp:1–7.

[32] Sam, M., Tuyls, K., Vanschoenwinkel, B., & Manderick, B. *Credit Card Fraud Detection using Bayesian and Neural Networks*. pp:1–7.

[33] Yee, O.S., Sagadevan, S., & Malim, N. Credit Card Fraud Detection Using Machine Learning As Data Mining Technique. *Journal of Telecommunication, Electronic and Computer Engineering*. Vol:10, pp:23–27.

[34] Mehdi, M., Zair, S., Anou, A., & Bensebti, M. (2007). A Bayesian Network in Intrusion Detection Systems. *International Journal of Computational Intelligence Research*, Vol:3, Issue:1, pp:0973–1873.

[35] Ford, V., Siraj, A., & Eberle, W. (2014). Smart grid energy fraud detection using artificial neural networks. IEEE Symposium on Computational Intelligence. pp:1–6.

[36] Bahnsen, A. C., Stojanovic, A., Aouada, D., & Ottersten, B. (2013). Cost Sensitive Credit Card Fraud Detection Using Bayes Minimum Risk. *12th International Conference on Machine Learning and Applications*, pp:333–338.

[37] Bauder, R. A., & Khoshgoftaar, T. M. (2017). Medicare Fraud Detection Using Machine Learning Methods. *16th IEEE International Conference on Machine Learning and Applications*, pp:858–864.

[38] Randhawa, K., Loo, C. K., Seera, M., Lim, C. P., & Nandi, A. K. (2018). Credit Card Fraud Detection Using AdaBoost and Majority Voting. *IEEE Access*, Vol:6, pp:14277–14284.

[39] Khare, N., & Sait, S. (2018). Credit Card Fraud Detection Using Machine Learning Models and Collating Machine Learning Models. *International Journal of Pure and Applied Mathematics*. Vol:118, Issue:20, pp:825–838.

[40] Estevez, P. A., Held, C. M., & Perez, C. A. (2006). Subscription Fraud Prevention in Telecommunications Using Fuzzy Rules and Neural Networks. *Expert Systems with Applications*. Vol.:31, pp:337–344.

[41] Wiese, B., & Omlin, C. (2009). Credit Card Transactions, Fraud Detection, and Machine Learning: Modelling Time with LSTM Recurrent Neural Networks. *Innovations in Neural Infor. Paradigms & Appli. Sci*, Vol:247, pp. 231–268.

[42] Fiore, U., De Santis, A., Perla, F., Zanetti, P., & Palmieri, F. (2017). Using Generative Adversarial Networks for Improving Classification Effectiveness in Credit Card Fraud Detection. *Information Sciences*, pp:1–20.

[43] Rushin, G., Stancil, C., Sun, M., Adams, S., & Beling, P. (2017). Horse race analysis in credit card fraud—deep learning, logistic regression, and Gradient Boosted Tree. *Systems and Information Engineering Design Symposium (SIEDS)*. Pp:117–121.

[44] Thennakoon, A., Bhagyani, C., Premadasa, S., Mihiranga, S., & Kuruwitaarachchi, N. (2019). Real-time Credit Card Fraud Detection Using Machine Learning. *9th International Conference on Cloud Computing, Data Science & Engineering (Confluence)*, pp:488–493.

[45] Piramuthu, S. (1999). Financial Credit-Risk Evaluation with Neural and Neurofuzzy Systems. *European Journal of Operational Research*, Vol:112, Issue:2, pp:310–321.

[46] Viaenese, S., Dedene, G., & Derrig, R. (2005). Auto Claim Fraud Detection Using Bayesian Learning Neural Networks. *Expert Systems with Applications*, Vol:29, Issue:3, pp:653–666.

[47] Gupta, S., & Gupta, M. (2021, October). Deep learning for brain tumor segmentation using magnetic resonance images. In *2021 IEEE Conference on Computational Intelligence in Bioinformatics and Computational Biology (CIBCB)* (pp. 1–6). IEEE.

[48] Gupta, S., & Gupta, M. K. (2021). A Comparative Analysis of Deep Learning Approaches for Predicting Breast Cancer Survivability. *Archives of Computational Methods in Engineering*, pp:1–17.

[49] Pinquet, J., Ayuso, M., & Guillén, M. (2007). Selection Bias and Auditing Policies for Insurance Claims. *The Journal of Risk and Insurance*, Vol:74, Issue:2, pp:425–440.

[50] Panigrahi, S., Kundu, A., Sural, S., & Majumdar, A.K. (2009). *Credit card fraud detection: A fusion approach using Dempster—Shafer theory and Bayesian learning*. Elsevier, Information Fusion, Volume 10, Issue 4, pp: 354–363.

[51] Dhir, S., & Kumar, Y. (2020). Study of Machine and Deep Learning Classifications in Cyber Physical System. *Third International Conference on Smart Systems and Inventive Technology (ICSSIT)*, pp: 333–338, doi: 10.1109/ICSSIT48917.2020.9214237.

[52] Chen, L., Gao, S., Liu, B., Lu, Z., & Jiang, Z. (2020). FEW-NNN: A fuzzy entropy weighted natural nearest neighbor method for flow-based network traffic attack detection. *China Communications*, Vol:17, Issue:5, pp:151–167, doi: 10.23919/JCC.2020.05.013.

[53] Arjoune, Y., Salahdine, F., Islam, M.S., Ghribi, E., & Kaabouch, N. (2020). A novel jamming attacks detection approach based on machine learning for wireless communication. International Conference on Information Networking (ICOIN), pp.459–464, doi: 10.1109/ICOIN48656.2020.9016462.

[54] Oladejo, M. T., & Jack, L. (2020). Fraud prevention and detection in a blockchain technology environment: challenges posed to forensic accountants. *International Journal of Economics and Accounting*. Vol:9, Issue:4, pp:315–335, doi: 10.1109/MNET.001.2000005.

[55] Nguyen, C. T., Nguyen, D. N., Hoang, D. T., Pham, H. A., Tuong, N. H., & Dutkiewicz, E. (2020, May). Blockchain and stackelberg game model for roaming fraud prevention and profit maximization. IEEE Wireless Communications and Networking Conference (WCNC), pp. 1–6.

[56] Kumar, Y., Sood, K., Kaul, S., & Vasuja, R. (2020). Big Data Analytics and Its Benefits in Healthcare. In *Big Data Analytics in Healthcare* (pp. 3–21). Springer.

[57] Kumar, Y., & Mahajan, M. (2019). Intelligent Behavior of Fog Computing with IOT for Healthcare System. *International Journal of Scientific & Technology Research*, 8(7), 674–679.

[58] Kumar, Y. (2020). Recent Advancement of Machine Learning and Deep Learning in the Field of Healthcare System. In *Computational Intelligence for Machine Learning and Healthcare Informatics* (pp. 7–98). De Gruyter.

[59] Kumar, Y., Kaur, K., & Singh, G. (2020). Machine Learning Aspects and its Applications Towards Different Research Areas. *2020 International Conference on Computation, Automation and Knowledge Management (ICCAKM)*, 150–156.

[60] Gupta, S., & Gupta, M. K. (2021). Computational Prediction of Cervical Cancer Diagnosis Using Ensemble-Based Classification Algorithm. *The Computer Journal*.

[61] Gupta, S., & Gupta, M. K. (2022). A Comprehensive Data-Level Investigation of Cancer Diagnosis on Imbalanced Data. *Computational Intelligence*, Vol:38, Issue:1, pp:156–186.

[62] Gupta, S., & Gupta, M. K. (2021). Computational Model for Prediction of Malignant Mesothelioma Diagnosis. *The Computer Journal*.

18 Advancements in Quantum-PSO and Its Application in Sustainable Development

Anita Sahoo and Pratibha Singh

CONTENTS

18.1 INTRODUCTION

Sustainability is about meeting today's needs without putting the future generation at risk. The agenda of sustainable development is to enhance everybody's lives and opportunities whether by sustainable building design, energy efficiency, cost-effective and competitive mining, sustainable resource management, etc. An exhaustive review of literature (Nishant et al., 2020) indicates that artificial intelligence (AI)-based methods are solving issues and challenges of sustainability and can play a major role in sustainable development. Optimal utilization of available resources is crucial to maintain sustainability. Optimization plays an important role in achieving sustainability. It employs search agents to discover the best feasible solution that

DOI: 10.1201/9781003245469-18

309

satisfies the objectives of sustainable development in the field of interest. Getting an exact solution to complex engineering or a scientific or real-life problem is very challenging and time consuming. Metaheuristic optimization is most suitable for such problems, which are multimodal, highly non-linear and have complex constraints. They provide quality solutions to complex optimization problems in reasonable time.

Kennedy and Eberhart (1995) have proposed a metaheuristic algorithm named Particle Swarm Optimization (PSO) algorithm for solving non-linear continuous optimization problems. The swarm behavior is inspired by birds flocking in nature. The PSO attracts a large group of researchers for its application in various domains with a certain level of enhancements. It is due to its simple implementation, fewer memory requirements, and good computation speed. The PSO follows five main principles: proximity, quality, diverse response, stability and adaptability discussed in this chapter (Millonas, 1994). PSO has a wide range of applications but is less efficient in solving discrete optimization problems. In the PSO, the particles communicate and coordinate to move in groups toward the optimum solution. It may reduce the diversity of the population very fast and may take the solution to be trapped into local minima. It faces the problem of slower convergence and may trap local optima.

The Quantum PSO (QPSO), a PSO combined with quantum theory, was proposed to solve discrete problems (Yang and & Wang, 2004). The quantum mechanism decides the trajectory of the particle's movement. It uses wave function to represent the particle's state and provides the feasible search space to reach the solution. The QPSO steps reduce the holding of velocity, confine the search space, reduce the parameters and simplifies the computation. Quantum mechanics extends the search space and various variants of QPSO proposed in the literature, improves the algorithm, and avoids premature convergence. In QPSO, quantum physics adds stochasticity to the movement of a particle. The Monte Carlo method is used to update the particle's position, and the contraction expansion coefficient is used to control the speed of a particle.

The quantum system is a complex nonlinear system based on state superposition principle; it provides more states than a linear system. Quantum mechanics provides a system of uncertainty letting the particle appear anywhere in the feasible space with a probability before the measurement. QPSO has only one parameter known as the contraction-expansion parameter that controls creativity and imagination of particles during evolution. This makes it easy to control.

Various improvements in QPSO are proposed in the literature, which we have categorized in five types in our paper as parameter control, diversity control, cooperative mechanism, and movement control and fusion mechanism. These variants of QPSO have a wide range of applications; some of these applications reviewed in this paper are solving design problems, networking and communication systems, control systems, planning and scheduling, machine learning and data analytics, real-life/engineering optimization problems, system biology, and image processing.

The conduct of evolutionary algorithms mostly depends on the fitness landscape, whose characteristics rely on the nature of the problem. Discussion of different methods to guide and improve the performance of the QPSO algorithm provides the researchers a platform for development of efficient QPSO-based algorithms suitable for the optimization problem at hand. The aim of this study is to provide an integrated, synthesized overview of the advancements in the QPSO algorithm and drive the inter-disciplinary researchers to utilize the QPSO variants in achieving sustainability in their respective fields of application. It presents a state-of-the-art snapshot of different strategies utilized for the improvement in the performance of a quantum-inspired particle swarm since its advent. This chapter provides the brief overview of QPSO, its representation and algorithm steps in Section 2. Section 3 accumulates the different strategies adapted by researchers for the improvements in basic characteristics of QPSO. Section 4 reviews applications in sustainability and sustainable development in various domains where QPSO and its variants have shown remarkable contributions, and Section 5 concludes with the importance of this literature review.

18.1.1 QUANTUM PARTICLE SWARM OPTIMIZATION

The state of a quantum particle is represented by a wave function $\varphi(\bar{x},t)$ in a quantum time-space domain, instead of velocity and position vectors. Each quantum particle travels in the solution space (in a potential well), the center of which is the convergence point p. The point p is known as the learning inclination point of the particle.

The quantum state of a particle provides the probability density function (PDF) $Q(x) = |\varphi(\bar{x},t)|^2$ that depicts the particle's appearance at position x relative to p. The PDF form depends on the potential field that it lies in. If a quantum harmonic oscillator potential well is used, then the distribution will be a normal distribution. If a Delta potential well is used, then the distribution is exponential. The QPSO algorithm proposed in 2004 (Sun et al., 2004) suggested the use of Delta potential well instead of the harmonic oscillator potential, since the QPSO with normal distribution has a tendency to converge prematurely as compared to exponential distribution. In a Delta potential well, the state of quantum particles is represented by the wave function as

$$\varphi(\bar{x},t) = \frac{1}{\sqrt{L}} e^{-\|p-x\|/L} \tag{1}$$

L is the characteristic length of potential function. The exponential PDF derived is of the form

$$Q(x) = |\varphi(\bar{x},t)|^2 = \frac{1}{L} e^{-\|p-x\|/L} \tag{2}$$

We need to collapse from the quantum state, where the particle position is in a quantized search space to the classical state in order to compute the particle position in the solution space. Particle position in the classical state at time t can be computed using Monte Carlo simulation as

$$x(t) = p \pm \frac{L}{2} \ln\left(\frac{1}{u}\right) \tag{3}$$

Where u is a random number uniformly distributed over $(0, 1)$. L is also known as creativity and imagination parameter of the particle. The method adapted to control the component L is crucial to QPSO algorithm performance and convergence. The length of potential function and quantum particle position are iteratively updated as

$$L(t+1) = 2 * \beta * |p - x(t)| \tag{4}$$

$$x(t+1) = p \pm \beta * |p - x(t)| * \ln\left(\frac{1}{u}\right) \tag{5}$$

Here, β is the contraction-expansion coefficient that influences the convergence behavior of a particle. In QPSO, this is the only variable/parameter that influences creativity and imagination of quantum particles during evolution.

The steps of the basic QPSO algorithm are mentioned later:

Step 1: Initialize the swarm of particles Q(t): The initial population is generated randomly in feasible search space.

$$Q(t) = \left[q^1(t), q^2(t), \dots, q^n(t)\right] \text{ where } q^j(t) = [q_1^j(t), q_2^j(t), \dots, q_m^j(t)] \tag{6}$$

$0 \le q_i^j(t) \le 1$: is the minimum unit to store value for $I = 1$ to m,
m: particle's length and $j = 1$ to n,
n: population size, t: current iteration

Step 2: Initialize the particle's personal best solution ($p_{personalbest}$) and global best solution ($p_{globalbest}$).

Step 3: Evaluate the objective function: Set the objective function to optimize and evaluate fitness value of each individual.

$$P(p_i^{k+1}(t) = 1) = f(p_i^k, v_i^k, p_{personalbest}, p_{globalbest}) \quad (for\, i = 1\, to\, m) \tag{7}$$

where, $-\infty \le x \le \infty,\, 0 \le f(x) \le 1,\, p_i^k \in [0,1]$

Step 4: Repeat steps 4 to 6 until the termination criterion is satisfied
Step 5: Update the particle's position and its fitness value

$$q_{personalbest}(t) = \alpha\, p_{personalbest} + \beta\left(1 - p_{personalbest}\right) \tag{8}$$

$$q_{globalbest}(t) = \alpha\, p_{globalbest} + \beta\left(1 - p_{globalbest}\right) \tag{9}$$

$$q(t+1) = c_1 q(t) + c_2 q_{personalbest}(t) + c_3 q_{globalbest}(t) \tag{10}$$

Here, c_1, c_2, c_3 represents the coefficient of experiences of particle of its own, its personal best experience and global best experience respectively such that $c_1 + c_2 + c_3$ =1, $0 < c_1, c_2, c_3 < 1$. α, β are control parameters with the condition $\alpha + \beta = 1$ and $0 < \alpha, \beta < 1$.

Step 6: Save the best solution.

18.1.2 QPSO Improvements

Individual particle in traditional PSO moves in an attraction potential field until it converges at a point p. From the dynamics point of view, particles must be bounded in order to avoid explosion and guarantee convergence. Jun Sun et al. (2004) used Delta potential as well as the potential field that ensures the bound state of swarm particles with quantum behavior.

Figure 18.1 depicts taxonomy of the QPSO algorithm. The exhaustive survey to understand the scope of improvements in the QPSO algorithm directed us to proceed further with five different types of strategies for controlling its parameters: Parameter control, Diversity control, Movement control, Cooperative mechanism and Fusion mechanism as shown in Figure 18.1. The categorization doesn't show exclusive partitioning; i.e., the methods discussed here in this chapter to control the parameters for the improvements of the characteristics of QPSO may belong to more than one type of strategy.

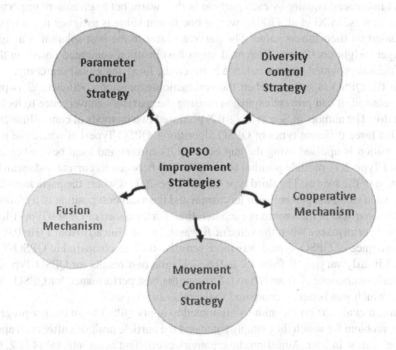

FIGURE 18.1 Taxonomy of QPSO improvement strategies.

18.1.3 PARAMETER CONTROL STRATEGY

In parameter control strategies, a method is devised to vary the parameter's value in a way that improves the efficiency of the algorithm providing better convergence rate and best performance. Researchers have presented many such strategies for the improvement of the searching of QPSO.

The convergence and performance of the algorithm depends on the control method of the only parameter L (the creativity or imagination of a particle) in QPSO algorithm. Traditionally it is computed from the distance between the learning inclination point and position of the particle. In their paper Sun et al. (2004) proposed a novel method of parameter control. Instead of using the local inclination point they used the mean best position or center-of-gravity position of the particle swarm to update the parameter.

Authors in Sun et al. (2005) used adaptive methods to control the QPSO convergence. In one approach, they assigned smaller values to the contraction-expansion coefficient β when particles are far away from the *gbest* and larger values when particles are close to the *gbest*. This improves the local searching ability of QPSO. In another approach, they set a threshold for each particle's L values, and if L is smaller than the threshold, then there is a random selection. This restricts early convergence and improves the global searching ability.

While updating the particle position, authors in Sun et al. (2005) used the mean best position where each particle in the swarm was given equal weight. The search scope (creativity) of the particle is governed by the mean best position. It is not ideal to get influenced equally by each particle in the swarm but to give more importance to the elitists. In Xi et al. (2008), weight coefficient value is assigned to each particle based on their fitness value. The particle nearer to the best solution is assigned a larger weight coefficient. Weighted mean best position computed based on these coefficients provided a better balance between the local and global searching.

In the QPSO algorithm, when the contraction-expansion coefficient β is properly selected, it can prevent explosion causing the particle's convergence to its local attractor. The authors in Sun et al. (2012) proposed two methods of controlling β and applied three different types of QPSO algorithms. QPSO-Type-1 is where the particle position is updated using the gap between its current and local best, whereas in QPSO-Type-2(I), particle position is using the gap between its current and mean best position of the swarm. The third type QPSO-Type-2(II) is where the particle position is updated using the gap between its current and the local best position of a randomly selected particle in the swarm at each iteration. Performance of QPSO-Type-1 led to the best performance when the value of β was decreased linearly from 1.0 to 0.9. The performance of QPSO-Type-1 was more sensitive to β as compared to QPSO-Type-2(I). Linearly varying β from 1.0 to 0.5 yields the best results for QPSO-Type-2(I). Linearly decreasing β from 0.6 to 0.5 shows the best performance for QPSO-Type-2(II), which was better as compared to the other two types.

In Liu et al. (2006) the authors attempted to apply QPSO to an integer programming problem for which they simply truncated the particle positions after each update to the nearest integer. Adjusting the creativity coefficient in an interval of [1.2, 0.4], the results were promising.

In Wang et al. (2020) an enhancement of QPSO algorithm, the Improved Quantum Particle Swarm Optimization (IQPSO) algorithm, is proposed to solve the inverse problems in heat transferring. It applies logistic mapping to initialize chaotic sequence. It uses normal distribution to generate the local attractor to spread the swarm and also applies a mutation operation to avoid premature convergence and escape from local optima. It increases the swarm's diversity and improves the performance.

In Liu et al. (2019) Teamwork Evolutionary strategy with QPSO (TEQPSO) is used to solve multi-peak optimization functions and single-peak cost optimization functions from the IEEECEC 2014. TEQPSO uses cross-sequential quadratic programming to implement the local search procedure, and Gaussian chaotic mutation operators generate new regions of interest (ROI) in the search space. TEQPSO significantly performs better than other PSO variants.

In Xin-gang et al. (2020) Differential Evolution Crossover QPSO (DE-CQPSO) is used for providing optimized solutions for environmental economic dispatch problems. The objective is minimization of both fuel cost and pollution emission subject to the constraints. The penalty factor transforms a multi-objective function into a single objective function. The DE-CQPSO algorithm performs better than QPSO, Multi Objective PSO and New Global PSO.

In Radha and Gopalakrishnan (2020), Improved QPSO (IQPSO) with Intelligent Fuzzy Level Set Method (IFLSM) is proposed to identify the cluster heads, optimize the opening contours and segment the brain tissues from MRI images. It improves the results by 15% as compared to the basic FLSM algorithm.

In Gu and Ma (2021) United Least Squares Support Vector Machine (ULSSVM) with QPSO is used to predict fault in the operating status of wind turbines. QPSO optimizes the penalty and kernel parameters of United LSSVM. United LSSVM combines the prediction model determined for Space LSSVM and Time LSSVM and performs better than them individually.

In Xiong et al. (2022) the Directed Acyclic Graph (DAG) with QPSO is used for scheduling of embedded processors subject to three constraints: partial ordering relations, reliability and the time limit. A QPSO-based approach shows better convergence than Simulated Annealing (SA) and PSO-based solutions. Although selecting the best parameters is still a challenge as in other heuristic-based algorithms, DAG_QPSO_I is efficient and gives better quality of solution. DAG_QPSO_II is used to optimize the execution time and also improves the reliability.

18.1.4 DIVERSITY CONTROL STRATEGY

Diversity control strategies mainly focus on stopping premature convergence of the algorithm and increase the chance of reaching global optima. Many researchers have proposed such diversity control strategies to improve the exploration ability of QPSO.

Diversity must be preserved during the searching process. The authors in Xu and Sun, (2005) used the diversity of population as a component to provide an adaptive control of the creativity coefficient β during position update. If the diversity of the population is low, then the creativity coefficient is assigned a low value and vice versa.

The low diversity with fitness stagnation leads to premature convergence of QPSO. Authors in Sun et al. (2006) presented two different strategies to improve the exploration ability of particles during evolution. They measure the population diversity during iteration; if it is less than a threshold value then (1) the contraction-expansion coefficient is adjusted, and (2) the mean best position of the swarm is reinitialized. This helps in eliminating premature convergence allowing searching for a better solution.

Adaptive IIR filter design is a very challenging global optimization problem, where in a multimodal error surface we need to get the global optima. In QPSO, even if particles' search space is the complete feasible space, due to collectiveness the population might lose diversity, because of which the evolution process becomes stagnant. In Fang et al. (2009) authors suggested a muted version of QPSO, where a random vector is computed based on the difference between the two positional vectors used in the position update. If the random number is less than the control parameter value, then instead of the difference between the two positional vectors, the generated random number will be used for position update. This ensures an increase in exploration ability of particles in the search space, making it quite efficient for multimodal optimization problems such as adaptive IIR filter design.

In Sun et al. (2007), authors proposed a binary QPSO algorithm, where a binary string is used to represent the particle position. The distance between two positions is computed using hamming distance, where the dimension of each binary string shows the number of decision variables. While computing the mean best position (*mbest*) of the swarm, each bit position is assigned a value 1 with probability 0.5 if at least half of the particles in the population has 1 at the corresponding bit position. Further, to maintain diversity, they exert crossover operation on the local best and *mbest* positions to generate two binary strings and then randomly choose one offspring position. This helps in stopping early stagnation. Finally, the binary position is mutated to complete the update operation.

In Su et al. (2009), authors introduced a crossover operator to combine quantum particle's personal best and global best to generate candidate solutions. The best fit particle out of the generated candidates is selected. During evolution, the crossover operator establishes a communication between the particles with the global best in the swarm, thereby improving its ability to reach a better solution.

In Sun et al. (2010), authors presented a diversity preserving QPSO for 2-D IIR filter design. The diversity of the population is computed from the variance and the longest diagonal distance between particles in the population. During evolution, if the particles cluster around a local optima and the diversity goes below a lower bound, then it exerts mutation on the global best position. The guided mutation based on the presented diversity measure maintains the diversity of the population at a favorable level for the search. This makes the modified QPSO a more adaptive and effective approach for complex problems such as 2-D IIR filters design.

In Xin-gang et al. (2020) improved QPSO is used for optimizing the dispatch plan in microgrid to meet the thermal load requirement and to reduce the release of polluting gases.

QPSO is improved using Differential Evolution by updating the position equation. It improves the diversity and enhances the potential of global search with better convergence.

In Jin and Jin (2015a), improved QPSO is used for visual features selection (VFS). It maintains the population diversity and prevents premature convergence. The proposed method has used ensemble strategy for better image annotation. IQPSO is used for VFS, and the selected features are used for automatic image annotation with the most relevant semantics.

In Shi et al. (2019), QPSO, an iterative optimization process, is used for feature subset selection for a Random Forest classifier maximizing the objective function i.e., minimizing the sensitivity and specificity. It is used to predict to disease progression in six months to one year from high resolution computed tomography (HRCT) images. The QPSO-RF method performs better with a smaller subset of features than other wrapper methods. It also shows more balanced sensitivity and specificity.

In Rehman et al. (2019), QPSO is modified to avoid the local optima and to increase the diversity of the population. It is done by randomly selecting the best particle in search space of the current domain, uses gamma probability function for mutation, and then enhancement factor is used with average fitness value for better convergence. Modified QPSO is used to optimize standard benchmark functions and also to solve the TEAM workshop problem. It shows better global solution search ability and faster convergence.

In Zhang and Hu (2019), an Improved Hybrid QPSO is used for solving flexible and effective job shop scheduling problems. It solves the sub-problem of job scheduling by mapping processes with machines using quantum probability maintaining the diversity and aiming the minimum execution time of jobs on machines. Local, global and probabilistic random selections are mixed for machine selection reducing the total execution time avoiding the jumping out the local extremes for better solution.

In Lai et al. (2020), Diversity preserving QPSO (DQPSO) with local optimization method i.e. a variable neighborhood descent method is used to solve the 0–1 knapsack problem. The proposed method controls the diversity of particles in swarm and avoids premature convergence. The local optimization method in a probabilistic way focuses on exploiting the solution search space.

In Li et al. (2020) Stability QPSO and Establishment QPSO are used for multi-tasks set allocation. SQPSO assures a better coalition for every task and shows faster convergence in the multi-tasks allocation. EQPSO reduces the total execution time with optimal scheduling of multi-tasks. The hybrid approach accelerates the convergence speed avoiding local optima and reaching the global optimal solution.

In Islam et al. (2021) Modified QPSO (MQPSO) is used to optimize the sequence of switching angles for producing the Pulse Width Modulation (PWM) signal in real time. The results of proposed MQPSO demonstrate reduced Total Harmonic Distance (THD) in PWM in comparison to other modulation methods. It produces the high quality efficient power.

In Li et al. (2021) diversity-controlled Lamarckian (DCL) QPSO improves the search ability maintaining balance between local and global search ability even in the larger search space. This method uses swarm's diversity and cognitive diversity for enhancing the experiences of swarm. It is used for automated docking software packages.

In Wang et al. (2022) Chaotic QPSO optimized Derived Extreme Learning Machine (DELM) algorithm is used to predict the different tread coordinates and wheel wear value. The proposed approach has higher ability of generalization in prediction and shows better results than other five models discussed in this paper. The results have been generated for four regression datasets available in database of University of California and have also been confirmed on actual sites.

18.1.5 Cooperative Mechanism

Cooperative strategies focus on a way to establish better communication between particles in the swarm so that together they can converge to global optima. Various cooperative mechanisms to improve the global searching ability of QPSO have been presented in past few years.

In traditional QPSO, when a complete d-dimensional particle is updated each time, there is a possibility that some components in respective dimensions move closer to the solution, but a few might move away from the solution. Due to overall improvement, an updated particle might get considered, but the deteriorated components get neglected. So, generally for high dimensional problems, it becomes difficult to achieve global optima. To overcome the curse of dimensionality, H. Gao et al. (2010) presented a modified QPSO algorithm with cooperative method. A high dimensional quantum particle is split into a number of 1-D subparts. Now the particle contributes to each dimension rather than to the population as a whole. The presented QPSO when applied to solve the multilevel thresholding-based image segmentation method, yielded stable solutions in a limited time. This shows an improvement in global searching ability of the particles.

Wang X. et al. (2009) used a variable neighborhood topology model to introduce the parallel mechanism into QPSO. At the beginning, the parallel QPSO maintains the multiple centers of gravitation while exploring the search space and in later stage allows the population to converge to a number of different local bests. Exploitation of more promising regions is thus possible. This improves the searching ability of QPSO reducing the chance of trapping into local optima.

Traditional QPSO has a tendency to converge prematurely. To improve its exploration ability, authors in Zhou et al. (2010) proposed a parallel QPSO with cooperative search strategy. During evolution, swarm is partitioned into a number of sub-swarms, and then each sub-swarm explores a part of the complete search space and finally is recombined to a single swarm. The performance of such a search strategy depends on the frequency of partition and recombination of sub-swarms. Authors studied two different strategies to increase or decrease the frequency with respect to iteration. The study shows that the parallel QPSO performs well when increased with respect to iteration for unimodal functions. But it doesn't have any effect on multi-modal functions. For multi-modal functions, both the strategies work well, where the parallel QPSO has improved global searching leading to better solutions.

In Zhao et al. (2013), authors presented a binary QPSO algorithm with cooperative approach. The quantum particle's personal best and global best positions are binary encoded. The particle positions are not updated as a single vector but in each dimension of the vector position separately. During updates, a particle might not appear fit if considered as whole, but a few of its components might have reached close to optima in their dimensions. In the presented process, since each component evolves in their own dimension, it avoids the loss of such fit components. This increases the speed of convergence and obtaining the chance of global optima.

In Wu et al. (2019) Hybrid Improved Binary QPSO (HI-BQPSO) is used for optimal features' subset selection. This method consists of two steps: The first step uses maximum information coefficient for determining the class-feature correlation. The second step is to identify the feature subset and to input it to BQPSO for optimal feature selection. BQPSO has the disadvantage of premature convergence, which is solved by using the mutation and crossover operators in random particles generation to improve diversity and prevent particles stuck in local extreme points. It has shown the overall good classification accuracy for 36 benchmark UCI datasets and 9 gene expression datasets.

In Ghorpade et al. (2021) Enhanced Differential Crossover QPSO enhances the ability of global search. Crossover operators promote the exchange of information among individuals that accelerates the convergence and maintains the diversity of the population. A differential evolutionary operator increases the diversity and thus global search ability. The suggested approach is used for optimization of 30 benchmark functions and to optimize the problem of locating sensor nodes with minimum energy expenditure in smart car parking application.

18.1.6 MOVEMENT CONTROL STRATEGY

The purpose of movement control strategies is to guide/control the movement of particles in the search space during evolution in such a way that improves the performance of the search algorithm. In the past, many researchers have contributed toward improving the QPSO search by introducing movement control strategies.

A Gaussian disturbance term is added by the authors Sun et al. (2007) during the calculation of the population's mean best position. During iteration, this regular interruption in the mean best position avoids the premature convergence by maintaining the diversity in searching; thereby improving the global searching ability of QPSO.

Liu et al. (2005) had shown that the normal cloud model has characteristics of randomness and stable tendency. Randomness prevents the search from being locked into local optimum, whereas the stable tendency can direct the search toward the global optimum. Inspired by this, authors Zhao et al. (2009) presented a mutation operator based on the normal cloud model to guide QPSO searching. Experiments on benchmark test functions confirm improvement in performance in terms of robustness and efficiency.

In Tian et al. (2011), an improved QPSO with a perturbation was presented. The perturbation applied on each quantum particle during evolution has a nonlinear coefficient that decreases monotonically with iteration. These regular perturbations on particles improve the diversity of the swarm particularly at the later stages. Due to improved diversity, the global searching ability of the QPSO is improved.

In QPSO, during iteration, if the global best position converges to a local optimum or sub-optimum, the quantum particles will be misguided to a bad region in the search space and might stop further exploration. To overcome this, authors in Sun et al. (2012) suggested a multi-elitist strategy. A growth rate is associated with each quantum particle. This value is increased if the particle in iteration has a better fitness than its previous iteration. During position update, the global best position is randomly selected; based on the generated random number it is either the global best of the whole population, or it is the personal best of the particle having maximum growth rate.

In Jin et al. (2014), authors used a logistic chaotic mapping theory to improve searching ability of QPSO. Due to the even distribution and randomness property of the logistic chaotic mapping, the quantum particles in the initial population are evenly distributed and well scattered throughout the search space. This enables a global searching improving the exploration ability of QPSO. Further remolding of locally optimized swarms generates better solutions.

In Jiang and Yang (2014), authors presented a model based on interpolation for improving the exploitation ability of quantum particles in the QPSO algorithm. The linear interpolation method was used to approximate the fitness values around a fit particle selected earlier in the search space. Around the selected fit particle a candidate particle is selected using local search, which is used to update the particle having the worst personal best in the swarm. This enhances the local search yielding an improvement in convergence and precision.

In Sheng et al. (2015) the authors presented four different methods to update quantum particle positions. These involve computation of mean best position and corresponding contraction-expansion parameter control. The first method considers the global best position to be the mean best position of the swarm. The mean of global best and the mean best of the swarm is used instead of mean best position in the second method. While calculating the weighted mean of personal best positions in the swarm, each particle is associated with a coefficient based on its fitness. For assigning the coefficient values, authors used a non-linear parabolic function. In the third strategy, they used an open-side-down parabolic function to compute weighted mean best position, whereas in the third strategy they used open-side-up parabolic function. The use of non-linear coefficients for weighted mean best position calculation improves the convergence speed as well as global searching ability. QPSO based on this strategy showed a stable performance compared to other QPSO algorithms.

In Haddar et al. (2016), three variants of QPSO are used for solving the 0–1 Knapsack problem. First is QPSO alone, the second is QPSO with RO that uses drop/add operator and the third variant uses QPSO+RO with Local Search method to improve the quality of solutions and to accelerate the convergence rate to the optimal solutions.

In Prawin et al. (2016), the Hybrid Dynamic QPSO (HDQPSO) algorithm is used to solve two practical complex structural nonlinear engineering problems. This algorithm with a reverse path can identify multiple non-linear elements even with noisy data. This approach is robust, efficient and provides solutions with minimal error.

In Wei and Wang (2020) Annealing Krill QPSO (AKQPSO) is used to solve the 100-Digit Challenge launched in IEEE Congress on Evolutionary Computation (CEC, 2019). QPSO performs better exploitation, and the Annealing Krill Herd (AKH) algorithm performs better in exploration of the solution. AKQPSO proposed here improves the diversity of the population and demonstrates better than other algorithms as stated in this chapter due to better performance in both properties of exploitation and exploration.

In Alvarez-Alvarado et al. (2021), QPSO in Lorentz potential (LR) field, Rosen–Morse (RM) field and Coulomb-like square root (CS) field determine the potential region for the movement of particle. Probability density function here exhibits the capability of high exploration and exploitation for finding the best solution for 24-benchmark unimodal and multimodal functions.

18.1.7 FUSION MECHANISM

In QPSO, the highly fit particles are always expected to be retained during the process of position update. Now, once such particles are greatly concentrated, the algorithm gets trapped in the local optimum. So, if the particles with low fitness values but with favorable evolutionary tendencies are selected during particle updates, then this limitation may be overcome. Based on the concepts of immune memory and vaccination, authors Liu et al. (2006) presented an immune operator to increase the diversity and precision of QPSO searching. The development of the immune operator is inspired by the biological immune system mechanism, integration of which improved the performance of QPSO.

Based on the Monte Carlo importance-sampling, Simulated Annealing (SA) (Davis, 1987) is a stochastic optimization method. During the annealing process, it accepts not only the evolved but also the deteriorated solutions with the Metropolis acceptance criterion, due to which it has the potential to overcome early convergence and find global minima. Authors in Liu et al. (2006) used this process in the selection phase of QPSO in the latter period of the search to control the premature convergence. The SA complements the evolutionary approach of QPSO.

In mean best position-based QPSO, the position distribution of the particle for the next iteration is decided by the distance between the particle's current position and the mean best position. If several particles in the swarm are far from the global best position, then the mean best position may be dragged away from the global best. This decreases the chance of getting the global optima and resulting in early convergence. To overcome this shortcoming of QPSO, authors in (Long et al., 2009) presented the selection operator to obtain global best particle. During iteration, a particle moved toward the personal best of a randomly selected particle and its own personal best using the selection operator. The random selection is done using the tournament method or roulette method. During iterations, if particles collect closely around the global best of the swarm, then the randomly selected particle will drag the particles away, reducing the chance for the swarm to converge prematurely. Additionally, if the randomly selected particle has a personal best that is closer to the global optimum, then it improves the chance of reaching the global optima.

Guided by the notion of species, authors in Zhao et al. (2012) proposed a species-based QPSO to solve dynamic problems with multiple peaks. Based on similarities, the population of quantum particles is divided into a number of subpopulations known as species. During the evolution, all the species simultaneously optimize toward multiple optima using QPSO. The proposed QPSO is able to track multiple optima simultaneously, and it is quite adaptive to the dynamic environments.

The authors in Xi and Sun (2012) proposed a binary QPSO to address discrete binary optimization problems with a new crossover operator. The quantum particles are binary encoded. For crossover operation, a random vector of particle length is generated. Using the random number, P value for the particle is calculated from its personal best and global best position. Then, for each bit position, if the value is more than 0.5, then the bit position is set to 1; otherwise it is set to 0. With this crossover operation, the diversity during evolution is increased, thereby improving the global searching ability.

In Sun et al. (2013), authors proposed a hybrid approach in which initially the parameters of a fermentation process model are obtained by Genetic Programming (GP), which are then tuned by QPSO algorithm. The limitation of GP is that it doesn't provide guarantee of finding an exact solution or even an acceptable solution, whereas QPSO is a global convergence guarantee algorithm. But, the search space for QPSO is the complete feasible space. Creating a hybrid of both where both complement each other yielded better fitting accuracy for the problem in hand.

Yao (2014) used crossover, mutation and inverse operators to improve the searching ability of QPSO. The proposed method is based on the concepts of swap sequence and swap operator that lets the QPSO overcome the premature convergence.

In Sheng et al. (2014) authors proposed a Q-learning-based strategy to control the contraction-expansion parameter of QPSO. Q-learning is a reinforcement method that provides agents (quantum particles in QPSO) to learn optimally by experiencing the consequences (change in fitness value) of actions (tuning of parameter). The QPSO with Q-learning is found to be more stable with higher convergence speed and global searching ability.

In Yumin and Li (2014) authors presented an improved QPSO based on Artificial Fish Swarm. The proposed QPSO adopts the rear-end activities and swarming concepts as the base for best position computation. The step length of particle movement is dynamically updated, an adaptive method. Further, an adaptive shrinkage factor is used in position update. All these modifications result in an improved QPSO that has better control on diversity and ability to avoid local optima.

In Jin and Jin (2015b) hybrid QPSO with Artificial Neural Network (ANN) is used for prediction of software susceptibility for faults. QPSO is used for selection of software metrics that proved to be more significant than other non-selected ones. ANN used these selected software metrics to predict its class as fault-prone or not fault-prone. QPSO with ANN will reduce the cost of maintaining software.

In Huang (2016) hybrid Support Vector Regression-based Chaotic QPSO (SVRCQPSO) is proposed to solve electric load forecasting problem. Quantum mechanics increases the search space, and cat chaotic function maintains the diversity and avoids premature convergence. This model achieves the best results as compared to other SVR based methods.

In Yu et al. (2017) QPSO with particle filter (PF) is proposed to predict the life remaining for lithium-ion batteries. PF is a Monte Carlo-based tool for random particle generation. Quantum behavior of the particles in swarm increases diversity and converges faster to the optimal solution. PF+QPSO approach is more robust and provides better prediction accuracy than PF and PF+PSO.

In Latchoumi et al. (2019) weighted QPSO with Smooth-Support Vector Machine (SSVM) is used to increase the accuracy in retrieving the useful information from medical data. It maintains the balance between local and global search. The

WQPSO-SSVM approach reduces the complexity of datasets for Wisconsin Breast Cancer Diagnosis (WBCD) and liver disorders, speeds up the computation, increases accuracy and reduces the total execution time.

In El Dahshan et al. (2020) Deep Belief Networks (DBNs) with QPSO are used for identifying emotions of faces. DBN is used to automatically select the parameters used in QPSO. It performs in four stages: selection of the Region of Interest (ROI) by discarding non-significant parts in images; second, determining the best block from many blocks of selected ROI; third, downsampling is used to reduce the size of the subpart of the image for better performance; fourth is the classification of emotions on faces. The results are shown on two popular datasets of Japanese female facial expression (JAFFE) with seven types of facial expressions and Face Emotion Recognition (FER-2013). It reduces the computation time and significantly improves the accuracy.

In Zhang et al. (2021) modified PSO optimized Radial Basis Function (RBF) is used to predict more accurate traffic flow in an intelligent transportation system. The Genetic Simulated Annealing is used to optimize the initial center of clusters. Modified QPSO determines the parameters of RBF. This model is used for prediction of traffic flow and shows better results of comparison to QPSO or QPSO optimized RBF.

18.2   LITERATURE REVIEW

QPSO has shown excellent performance in solving optimization problems of different domains The different domains covering most of the sustainable development issues, shown in Table 18.1, are design problems, networking and communication systems, control systems, planning and scheduling, machine learning, data analytics, real life/engineering optimization problems, system biology, and image processing. The literature exploring the capability of QPSO to solve major issues to maintain sustainability in each of these application domains is reviewed in this section and summarized in Table 18.2. To achieve sustainable development in the manufacturing field, focus has shifted to service manufacturing, and product service systems (PSS) are configured to satisfy customers with a combination of individualized products and service (Cui & Geng, 2021). In sustainable energy management, authors M. Chen et al. (2018) have used QPSO for micro-grid scheduling. Various variants of QPSO optimize the solutions of an extensive range of diverse problems raised as challenges in sustainable development.

Table 18.2 presents the brief overview of the research papers reviewed to identify the improvements proposed by researchers in the previously stated domains, the datasets used and methods to control the parameters and the important observations. We can observe here that improved versions of QPSO have solved the wide range of a diverse set of problems of sustainable development. These are inverse problems of radiation heat transfer and phase change forecasting of electric demand, prediction of remaining lithium-ion battery life, optimization problems in renewable energy, identification problem in auto-docking applications, smart car parking systems, prediction of the operating status of wind turbines, prediction of wheel tread and the wear, environmental economic dispatch problems, forecasting of traffic flow in smart

TABLE 18.1

Literature Review of QPSO in Various Application Domains of Sustainability

Application Domain	References
Design Problems	(Fang et al., 2006a, 2006b, 2006c), (Luitel & Venayagamoorthy, 2008), (Sun et al., 2008), (Wen-bo, 2008), (Wu et al., 2008), (Sabat et al., 2009, 2010), (Shayeghi et al., 2010), (Wei, 2010), (Sarangi et al., 2011), (Chang et al., 2012), (Liu et al., 2012), (Sun et al., 2008), (Wen-bo, 2008), (Wu et al., 2008), (Sabat et al., 2009, 2010), (Shayeghi et al., 2010), (Wei, 2010), (Sarangi et al., 2011), (Chang et al., 2012), (Liu et al., 2012)
Networking and Communication Systems	(Oliveira et al., 2006), (Sun et al., 2006a), (Zhao et al., 2008), (Hongwu, 2009), (Liu & Song, 2009), (Sun et al., 2011), (Gong et al., 2012), (Yu, 2012), (Fang et al., 2015), (Oliveira et al., 2006), (Fang et al., 2015), (Sun et al., 2006a), (Zhao et al., 2006a), (Sun et al., 2006a), (Hongwu, 2009), (Liu & Song, 2009), (Sun et al., 2011), (Gong et al., 2012), (Yu, 2012), (Fang et al., 2015)
Control Systems	(Xi et al., 2007), (Jalilzadeh et al., 2009), (Liu et al., 2009), (Fu et al., 2013), (Shayeghi & Safari, 2013), (Zhang et al., 2015), (Xi et al., 2007), (Jalilzadeh et al., 2009), (Liu et al., 2009), (Fu et al., 2013), (Shayeghi & Safari, 2013), (Zhang et al., 2015)
Planning and Scheduling	(Sun et al., 2006b), (Wen-bo, 2006), (Kong et al., 2007), (Loo & Mastorakis, 2007), (Fu et al., 2009), (Wen-bo, 2009), (Chang et al., 2010), (Zhengchu et al., 2010), (Zheng, 2010), (Chakraborty et al., 2011), (Hosseinnezhad et al., 2011), (Niu et al., 2011), (Chakraborty et al., 2012), (Fu et al., 2012), (Liu et al., 2012), (Niu et al., 2012), (Farzi et al., 2013), (Yang, 2013), (Davoodi et al., 2014), (Lenin & Reddy, 2014), (Li et al., 2014), (Yong-mei & Wan-ye, 2014), (Huang et al., 2016), (Xin-gang et al., 2020), (Xiong et al., 2022), (Li et al., 2020), (Zhang & Hu, 2019), (M. Chen et al., 2018), (Kong et al., 2007), (Loo & Mastorakis, 2007), (Fu et al., 2009), (Wen-bo, 2009), (Chang et al., 2010), (Zhengchu et al., 2010), (Chakraborty et al., 2011), (Niu et al., 2011), (Chakraborty et al., 2011), (Niu et al., 2011), (Hosseinnezhad et al., 2011), (Fu et al., 2012), (Liu et al., 2012), (Niu et al., 2012), (Farzi et al., 2013), (Yang, 2013), (Davoodi et al., 2014), (Lenin & Reddy, 2014), (Li et al., 2020), (Zhang & Hu, 2019), (Xiong et al., 2020), (Li et al., 2020)
Machine Learning and Data Analytics	(Sun et al., 2006c, 2006d), (Jun, 2007), (Li et al., 2007), (Pan et al., 2007), (Wang et al., 2007), (Chen et al., 2008), (Kezhong et al., 2008), (Luo et al., 2008), (Zhou et al., 2008), (Li et al., 2009), (Meshoul & Al-Owaisheq, 2009), (Shi et al., 2009), (Zhao et al., 2009), (Li et al., 2010), (Luitel & Venayagamoorthy, 2010a), (Zhang et al., 2010), (Hamed et al., 2011), (Tang et al., 2011), (Deepa & Kalimuthu, 2012), (Farzi, 2012), (Lei et al., 2012), (Lin et al., 2012), (Sun et al., 2012), (Xin & Xiao-feng, 2012), (Du et al., 2013), (Liu et al., 2013), (Mu et al., 2013), (Zhang et al., 2013), (Bagheri et al., 2014), (Liao & Ding, 2014), (Zhang, 2014), (Liu et al., 2014), (Cui & Geng, 2021), (Jun, 2007), (Li et al., 2007), (Sun et al., 2006c, 2006d), (Li et al., 2008), (Luo et al., 2008), (Kezhong et al., 2008), (Zhou et al., 2008), (Li et al., 2009), (Meshoul & Al-Owaisheq, 2009), (Shi et al., 2009), (Zhao et al., 2009), (Li et al., 2010), (Hamed et al., 2011), (Tang et al., 2011), (Deepa & Kalimuthu, 2012), (Farzi, 2012), (Lei et al., 2012), (Lin et al., 2012), (Sun et al., 2012), (Xin & Xiao-feng, 2012), (Du et al., 2013), (Liu et al., 2013), (Mu et al., 2013), (Zhang et al., 2013), (Bagheri et al., 2014), (Liao & Ding, 2014), (Liu et al., 2014), (Zhang, 2014), (Cui & Geng, 2021)

Real Life/ Engineering Optimization Problems	(Sun et al., 2007), (Wen-bo, 2007a, 2007b), (Wen-hui, 2007), (Chi et al., 2008), (Du & Wei, 2009) (Jeonget al., 2009), (Omkaret al., 2009), (Sun et al., 2009), (Zhao et al., 2009a, 2009b), (Chen & Yang, 2010), (Chunhong et al., 2010), (Coelho, 2010), (Farzi & Baraani-Dastjerdi, 2010), (Jeong et al., 2010), (Li et al., 2010), (Luitel & Venayagamoorthy, 2010b), (Meng et al., 2010), (Tian et al., 2010), (Yin et al., 2010), (Yu et al., 2010), (Zhao et al., 2010), (Kou et al., 2011), (Zhang et al., 2011), (Dawei et al., 2012), (Mariani et al., 2012), (Wu et al., 2012), (Chen et al., 2013), (Ho et al., 2013), (Jamalipour et al., 2013a, 2013b), (Lakshmi & Rao, 2013), (Priyadharshini, 2013), (Chang et al., 2014), (Gainget al., 2014), (Han et al., 2014), (Xi et al., 2015), (Jin and Jin, 2015b), (Haddar et al., 2016), (Prawin et al., 2016), (Yu et al., 2017), (Wu et al., 2019), (Alvarez-Alvarado et al., 2021), (Li et al., 2021), (Ghorpade et al., 2021), (Gu & Ma, 2021), (Wang et al., 2022), (Liu et al., 2019), (Lai et al., 2020), (Zhang et al., 2021), (Xin-gang et al., 2020), (Wei & Wang, 2020), (Rehman et al., 2019), (Islam et al., 2021), (Sun et al., 2007), (Wen-bo, 2007a, 2007b), (Wen-hui, 2007), (Chi et al., 2008), (Du & Wei, 2009) (Jeong et al., 2009), (Omkar et al., 2009), (Sun et al., 2009), (Zhao et al., 2009a, 2009b), (Chen & Yang, 2010), (Chunhong et al., 2010), (Coelho, 2010), (Farzi & Baraani-Dastjerdi, 2010), (Jeong et al., 2010), (Li et al., 2010), (Luitel & Venayagamoorthy, 2010b), (Meng et al., 2010), (Tian et al., 2010), (Yin et al., 2010), (Yu et al., 2010), (Zhao et al., 2010), (Kou et al., 2011), (Zhang et al., 2011), (Daweiet al., 2012), (Mariani et al., 2012), (Wu et al., 2012), (Chen et al., 2013), (Ho et al., 2013), (Jamalipour et al., 2013a, 2013b), (Lakshmi & Rao, 2013), (Priyadharshini, 2013), (Chang et al., 2014), (Gaing et al., 2014), (Han et al., 2014), (Xi et al., 2015), (Jin and Jin, 2015b), (Haddar et al., 2016), (Prawin et al., 2016), (Yu et al., 2017), (Wu et al., 2019), (Alvarez-Alvarado et al., 2021), (Li et al., 2021), (Ghorpade et al., 2021), (Gu & Ma, 2021), (Wang et al., 2022), (Liu et al., 2019), (Lai et al., 2020), (Zhang et al., 2021), (Xin-gang et al., 2020), (Wei & Wang, 2020), (Rehman et al., 2019), (Islam et al., 2021)
System Biology Image Processing	(Cai et al., 2008), (Lu & Wang, 2008), (Lu et al., 2010), (Cai et al., 2008), (Lu & Wang, 2008), (Lu et al., 2010) (Jun, 2008), (Hai-xi, 2009), (Bao & Sun, 2011), (Shabanifard & Amirani, 2011), (Liu et al., 2012), (Su et al., 2013), (Deyet al., 2014), (Jin and Jin, 2015a), (Jun, 2008), (Hai-xi, 2009), (Bao & Sun, 2011), (Shabanifard & Amirani, 2011), (Liu et al., 2012), (Su et al., 2013), (Dey et al., 2014), (Jin and Jin, 2015b), (Latchoumi et al., 2019), (Radha & Gopalakrishnan, 2020), (Shi et al., 2019), (Dashan et al., 2020), (Latchoumi et al., 2019), (Radha & Gopalakrishnan, 2020), (Shi et al., 2019), (Dashan et al., 2020)

TABLE 18.2

Improvements in QPSO

Paper	Dataset	Improvements	Observations
Jun Sun et al., 2004a	Seven benchmark Functions	-	Using Quantum Delta Potential function instead of Quantum Harmonic Oscillator slowed the convergence but improved the global search ability. Parameter selection and control was based on trials, needed an adaptive method for parameter control.
Jun Sun et al., 2004b	Seven benchmark functions	Devised a mechanism to utilize swarm's Mean Best Position to update parameter L.	Since the mean best position is relatively stable during evolutions, different particles convergence goes with that of the entire swarm. This enhances the global search ability of the algorithm.
Jun Sun et al., 2005	Seven benchmark functions	Suggested two methods for parameter control: a. Controlling the value of contraction expansion parameter. b. Setting a threshold on parameter L.	The first approach improves the exploitation ability of the QPSO algorithm. The second approach improves the exploration ability, but increases the convergence rate.
Xu W. & Sun, 2005	Seven benchmark functions	Presented a diversity-guided parameter control method	The method improves the global searching ability of QPSO.
Jun Sun et al., 2006	Four benchmark functions	Proposed two diversity control strategies: a. Controlling the contraction-expansion coefficient β b. Reinitializing the mean best position across the search space	The diversity control methods make the particles explode and stop premature convergence.
Jun Sun et al., 2008	Five benchmark functions	Assigned linearly decreasing weight coefficient to depict the significance of particles in the swarm and utilized it for parameter control.	The weighted mean best position concept used for parameter control yield a balance between the exploration and exploitation in searching.
J Sun et al., 2012	Twelve functions from Suganthan's benchmark suite (six unimodal and six multimodal)	Presented two methods of controlling contraction-expansion coefficient β: a. Utilizing fixed-value b. Linearly decrementing during evolution	When the particle position is updated using the gap between its current and the personal best of a randomly selected particle in the swarm and the β value is decreased linearly from 0.6 to 0.5, the QPSO algorithm yields promising results.

H. Gao et al., 2010	Hundred images from Berkeley segmentation data set	Proposed a cooperative strategy, where each particle is engaged in optimizing an individual dimension of the solution vector	The global searching ability of presented cooperative QPSO is more powerful and faster ensuring stable results for multilevel thresholding based image segmentation problem.
Sun et al., 2010	A first-order, a second-order and a third-order IIR filter in the simulation study	Utilized a mutation operator to enhance the randomness and global search ability	Due to mutation of updated particle positions during iteration, the exploration ability is improved leading to a better searching of the search space.
Liu et al., 2006	Four benchmark functions	Introduced immune operator based on biological immune system mechanism to increase precision and global searching during the latter stages of the search	Immune operator improves the selection process, where particles with low fitness but having favorable evolutionary tendency have chances of getting selected. This eliminates premature convergence. Further, it guides the search process through both vaccination and immune selection leading to convergence precision.
Liu J. et al., 2006	Three benchmark functions	Introduced Simulated Annealing in the selection process of QPSO with a mutation operator	The hybrid method has an increased diversity in population and improved precision in searching.
Liu J. et al., 2006	Seven benchmark test functions	Attempted to solve integer programming using QPSO by truncating the particle positions to nearest integer after each evolution	The interval for creativity coefficient from 1.2 to 0.4 yields better results.
Sun J. et al., 2007a	Five benchmark test functions	Used binary string to represent particle positions, hamming distance to compute distance between particles, and crossover followed by mutation to finally update particle positions	The QPSO with binary encoding and crossover followed by mutation operator yields better quality solutions with lesser time.
Sun J. et al., 2007b	Five benchmark test functions	Introduced Gaussian disturbance on the mean best position to prevent the fitness stagnation during evolution	The consistent Gaussian disturbance in mean best position during evolution helps in maintaining the particle's vigor, making it escape local optima at the later stage of search.

(Continued)

TABLE 18.2 (Continued)

Paper	Dataset	Improvements	Observations
Long, H. et al., 2009	Seven benchmark test functions	Proposed two selection mechanisms for global best position: a. Using tournament selection b. Using roulette-wheel selection	Adopting the selection search in QPSO improves the diversity of the swarm and improves the chance of convergence to global optima.
Wang, X. et al., 2009	Seven benchmark test functions	Utilized variable neighborhood topology model to devise a parallel QPSO algorithm	Subpopulations are employed to search in varied regions of the search space, assuring global convergence.
Z. Chai et al., 2009	Two benchmark functions	Proposed a micro-architecture for suitably implementing QPSO in a low-end FPGA	With the advantages of QPSO, the proposed micro-architecture is reconfigurable to meet diverse requirements of embedded real-time systems implemented on FPGA in terms of computing performance and resource consumption.
J Zhao et al., 2009	Five benchmark test functions	Proposed a normal cloud model-based mutation operator	Randomness and stable tendency of the cloud model increase diversity in the latter period of the search, making it escape from local minima.
Su D. et al., 2009	Five benchmark test functions	Utilized a crossover operator to combine quantum particle's personal best and global best to generate candidate solutions for selection	The modified QPSO has an ability to obtain good solutions avoiding local optima.
J. Sun et al., 2010	2-D zerophase recursive filter examples	Proposed a diversity measure, where diversity is computed from the particle variance and the longest diagonal length. If diversity is low, then the global best position is mutated	The presented method explicitly exerts mutation of global best position by the guidance of diversity measure. It maintains the diversity via mutation at a favorable level for the search, making it a more adaptive and effective approach.
Di Zhou et al., 2010	Six high-dimensional benchmark test functions (three unimodal and three multimodal)	Presented a QPSO parallel algorithm with cooperative search strategy, where sub-swarms with frequent recombination play the role of message passing	The QPSO with cooperative strategy performs well when the frequency of partition and recombination of sub-swarms called combination frequency was set to increase monotonically with iteration for unimodal functions. For multimodal functions it performed well; either it increased or decreased with iteration.

Na Tian et al., 2011	Seven benchmark test functions and an example of the inverse problem	Introduced a perturbation operator having a nonlinear coefficient to ensure global convergence	The random perturbation on each particle exerted during evolution using a nonlinear coefficient maintains the diversity of the swarm particularly at later stages. This leads to a better global search.
Zhao J. et al., 2012	Dynamic environments generated using dynamic function.	Presented a species-based QPSO using a form of speciation for dynamic environment. It allows parallel subpopulations to simultaneously evolve in the search space obtaining multiple optima	The presented QPSO is quite adaptive to the dynamic environment and is effective in tracking multiple optima to obtain good solutions.
Jun Sun et al., 2012	First ten functions in CEC2005 benchmark suite (five unimodal and five multimodal) and Four Gene expression datasets	Proposed a Multi-Elitist strategy to guide QPSO searching in order to escape from local optima	The strategy guides the quantum particles to search in promising regions in the search space maximizing the chance to obtain global optima.
Xi, M & Sun, 2012	Five benchmark functions to be maximized	Proposed a binary QPSO algorithm with new bit crossover operator to compute the personal and global best positions. Then, finally randomize the position transformation	The presented bit crossover operator increases the diversity of populations, thereby improving the global searching ability of binary QPSO.
Sun J. et al., 2013	Fermentation process of the hyaluronic acid production by Streptococcus zooepidemicus	Presented a hybrid of Genetic Programming (GP) and QPSO. First, the parameters of the fermentation process model are obtained by the GP, and then these parameters are tuned by the QPSO algorithm	In the hybrid approach GP and QPSO complement each other and result in a better fitting accuracy.

(Continued)

TABLE 18.2 (Continued)

Paper	Dataset	Improvements	Observations
Zhao J. et al., 2013	Five benchmark test functions	Proposed a binary QPSO algorithm with cooperative approach in which updates of quantum particle's personal best and global best positions are done in each dimension of the position vector	The cooperative approach helps in avoiding the loss of components of position vectors that have moved closer to the global optima during evolution, thereby increasing the convergence speed as well as global searching.
Sheng, X. Y. et al., 2014	Twelve benchmark test functions	Proposed a Q-learning strategy for controlling contraction-expansion parameter of QPSO	The reinforcement strategy enhanced the parameter control process of QPSO, yielding a stable and better performance.
Yumin and Li, 2014	Eight benchmark test functions	Proposed an improved QPSO with combination of rear-end activities and artificial fish clusters for parameter control	The adaptive and dynamic update of control parameter is quite effective for high-dimensional functions also.
Yan-xia JIN et al., 2014	Three benchmark test functions	Utilized logistic chaotic mapping theory in two ways: a. For conducting chaotic search for initial quantum particles b. For chaotic remolding of the locally optimized swarms	Using the traversal of chaotic motion lets the QSO achieve a better diversity in the population. The use of logistic chaotic mapping ensures an even distribution of quantum particles in the initial population well scattered in the search space.
Jiang & Yang, 2014	Five benchmark test functions	Proposed a model-based linear interpolation method for better exploration and exploitation of the search space	Due to the use of linear interpolation, the QPSO algorithm is able to avoid unessential exploration and improve the exploitation leading to better convergence. During iteration, one more function evaluation to generate a trial point is done, which increases the search time.
Sheng, X. et al., 2015	Twelve functions from Suganthan's benchmark suite (six unimodal and six multi-modal)	Presented four adaptive strategies for computing mean best position of quantum particles and the corresponding methods of parameter selection	The coefficient assignment based on non-linear parabolic functions for weighted mean best position calculation proves to be an efficient method. QPSO based on this strategy yields better and stable performance.

Yao, 2014	Travelling Salesman Problem data	Introduced crossover, mutation and inverse operators of genetic algorithm to swap sequence and swap operator	Overcomes the limitations of QPSO such as premature convergence and falling into local optima.
Zhang et al., 2015	Three multi-dimensional optimization Functions: a. Sphere b. Rastrigin c. Schaffer	Improved QPSO algorithm is more robust than the basic PSO and Quantum PSO	IQPSO uses logistic mapping for initialization, random number generation with normal distribution to avoid local optima.
Jin & Jin, 2015a	Image Datasets: a. Corel5k b. Yahoo and Google c. Corel60k	Used the reverse operation in binary encoded particles in IQPSO	The QPSO suffers from premature convergence, and its reason is the lack of diversity in population. Improved QPSO uses the measure method and improvement operation to avoid premature convergence.
Jin & Jin, 2015b	Software metrics (PC1, JM1, KC1, and KC3)	Used Hybrid ANN and QPSO QPSO better controls the selection than PSO	QPSO is used to reduce the dimensionality of software metric space. ANN is used for prediction of fault-proneness.
Haddar et al., 2016	270 benchmarks from OR-library	Improved QPSO with local search method for better exploration and exploitation	Repair operator to add and drop the solution is used to restrict the solution in feasible search space. Static star topology is used for connecting neighbors in a swarm.
Prawin et al., 2016	Nonlinear numerical simulations and breathing crack and chaos problem	Hybrid Dynamic QPSO with reverse path to identify parameters with least error is robust and computationally efficient	Three stages in proposed approach: first to identify the presence and degree of nonlinearity; second to identify the location of nonlinearity, uses reverse path method; third to identify the non-linear parameters
Huang et al., 2016	Demand of electric load data: regional, annual and hourly	Used SVR with chaotic QPSO to improve forecasting accuracy	Quantum mechanics enlarge the search space. Chaotic Cat mapping function breaks the local optima and avoids premature convergence.

(Continued)

TABLE 18.2   (Continued)

Paper	Dataset	Improvements	Observations
Yu et al., 2017	NASA's data of batteries' aging process.	Used QPSO-based particle filter for better performance with global search ability, which has less parameter to control	Particle filter (PF) is a Monte Carlo-based method to generate large number of random particles.
Wu et al., 2019	9 gene expressions and 36 UCI datasets	Hybrid Improved-Binary QPSO improves the accuracy with the remarkable lesser number of features	QPSO guarantees global convergence with fewer parameters to control. Filtering method reduces the search space. It uses crossover and mutation for the computation of local attractor.
Xin-gang et al., 2020	Thermal and electric load data of a city in China	Used Differential Evolution in QPSO	DE Algorithm improves the diversity of population and thus improves the search capability of QPSO.
Alvarez-Alvarado et al., 2021	24 benchmark functions	QPSO with Lorentz and Rosen—Morse potential field (QPSO-LR, QPSO-RM) and Coulomb-like Square Root (QPSO-CS)	Explored the movement of particle in three different fields. Probability density function determines the field space for better exploration and exploitation.
Li et al., 2021	EDock dataset	Diversity-controlled Lamarckian QPSO is used to control the diversity	Novel method of diversity control increases the search ability and prevents the issue of premature convergence.
Ghorpade et al., 2021	IEEE CEC2019: benchmark function and data collected from IOT nodes	Enhanced Differential Crossover QPSO (EDCQPSO) is used here	Differential operator improves the diversity in population. Crossover operator allows information interchange among particles to guide
Gu & Ma, 2021	Collected data of vibrational signals	QPSO with United Least Squares Support Vector Machine (ULSSVM)	QPSO optimizes the kernel and penalty parameters of LSSVM. ULSSVM performs better than time based LSSVM and space based LSSVM.
Wang et al., 2022	Four regression datasets from UCI repository	Chaotic QPSO optimized Derived ELM solved this non-linear problem	Piecewise Logistic chaotic sequences in QPSO enhance the search ability and optimizes the parameters of DELM.
Xiong et al., 2022	Intel XScale and Transmeta Crusoe processors	Directed Acyclic Graph (DAG) with QPSO is used to solve scheduling problem	DAG-QPSO-I and II converges faster than Simulated Annealing and PSO-based algorithm.

Lai et al., 2020	27 benchmark instances of multidimensional Knapsack problem	Diversity preserving QPSO and local optimization method is used to solve this NP Hard problem	Diversity preserving property manages the diversity of population. Local optimization method emphasizes on exploiting the search space.
Latchoumi et al., 2019	2 datasets from UCI repository: liver disorder, Wisconsin Breast Cancer Diagnosis (WBCD)	Weighted QPSO with smooth Support Vector Machine (SSVM) is used to improve the accuracy in identifying defects in medical data	WQPSO is proposed for clustering and SSVM is used for classification. WQPSO is improved with quantum theory and weighted mean of the best features.
Zhang et al., 2021	Traffic flow data collected from of Yuanda road and Beijing 4th Ring Road	Genetic Simulated Annealing (SA) in QPSO is used to optimize the parameters of radial basis function (RBF) neural network	Genetic SA optimizes initial cluster. Suggested Modified PSO-RBF has a simple structure and provides better accuracy than RBF or QPSO-RBF.
Radha & Gopalakrishnan, 2020	Magnetic image resonance (MRI) images	Improved QPSO with intelligent fuzzy based clustering is used for segmentation	IQPSO with Fuzzy-Based Enhanced Level Set Method (LSM) optimizes the opening contours.
Shi et al., 2019	High-resolution computed tomography (HRCT) images	QPSO with Random Forest (RF) is used as wrapper method for early pattern recognition	In QPSO-RF QPSO is used for best feature selection from re-sampled data for building a RF classifier.
Liu et al., 2019	CEC2014 test functions.	Teamwork evolutionary strategy in QPSO (TEQPSO) is used to balance between local and global search	Cross sequential quadratic programming performs the local search and Gaussian chaotic mutation operators identify new regions in the search space in TEQPSO.
Xin-gang et al., 2020	Standard IEEE 30 bus system and 10-Unit system.	Differential Evolution-Crossover QPSO (DECQPSO) is used for fuel cost and pollution emission optimization	Differential evolution and crossover operators in QPSO algorithm enhance the global search ability and also diversity in a particle's generation. It overcomes the convergence issue of QPSO algorithm.

(Continued)

TABLE 18.2 (Continued)

Paper	Dataset	Improvements	Observations
Wei & Wang, 2020	100-digit challenge problem in IEEE CEC 2019	Annealing Krill QPSO (AKQPSO) is used to optimize the problems of 100 digit challenges.	PSO is optimized using quantum behavior and the Krill Herd algorithm is optimized using Simulated Annealing algorithm.
Dashan et al., 2020	Japanese female facial expression (JAFFE) dataset and Facial Emotion Recognition (FER-2013) dataset.	Hybrid approach of Deep Belief Network (DBN) and QPSO is used to identify emotions on human faces	Here QPSO is used to optimize the parameters of a Deep Belief Network. It is better than choosing its parameters manually.
Rehman et al., 2019	Six standard benchmark functions	An enhancement factor (EF) is added in QPSO to enhance the global search efficiency	Modified QPSO randomly select the best particle, an enhancement factor is used for enhancing the global search ability.
Li et al., 2020	Synthetic data for virtual tasks allocation.	Stability QPSO (SQPSO) and Establishment QPSO (EQPSO) is used for better coalition scheduling	SQPSO adds high rewards, benefits, stability and accelerate searching. EQPSO reduces the time cost for coalition formation for effective scheduling.
Zhang & Hu, 2019	Dataset for Flexible Job Scheduling Problem (FJSP).	Improved Hybrid QPSO based on Levy flights is used for efficient scheduling and quality results	In the suggested approach, quantum probability maps the sequence of the process with the particle's position. It then uses local, global and probability random selection of machines for processes.
Islam et al., 2021	Data from 11 level Cascade H-Bridge of Multilevel Inverter.	Modified QPSO (MQPSO) is used to identify proper angles for efficient generation of PWM signals	MQPSO enforces selection of fundamental harmonics, eliminating the lower harmonics and reducing the total harmonic distance.

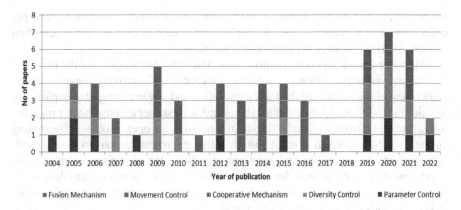

FIGURE 18.2 Applications of QPSO in different domains

transportation systems etc. Improved versions of QPSO have also optimized the solution for segmentation of medical images, prediction of fibrosis in CT scanned images, job scheduling problems, visual features selection (VFS) for automatic image annotation (AIA), prediction of software fault proneness, multidimensional Knapsack NP hard problem and many others.

Figure 18.2 accumulates the contributions of different variants of QPSO in various domains of sustainable development to create better-quality solutions within the acceptable range. The statistics shown here present the extent of utilizations of control mechanisms every year. A variant of QPSO depends on the control mechanism used to modify its characteristics. An improvement in the control mechanism of one type, to enhance the searching capability of QPSO also indicates the improvement in controls of other types. QPSO variants are showing excellent performances in solving unimodal/multimodal, high dimensional and NP-hard optimization problems. There is still scope of improvements in its behavioral control to enrich the performances and to apply it in the thrust areas of achieving sustainability. These areas must include medical domain, biological systems, ecological systems, various design problems etc. where it can perform a crucial role in benefit of society.

18.3 CONCLUSION AND FUTURE WORK

Complex real-world problems seeking sustainability or sustainable development are expressed as the optimization problems characterized by their nature of modality (unimodal/multi-modal), high-dimension (fixed/variable) and non-smooth optimization problems. Different optimization problems have different challenges that are reflected in their fitness landscapes. Introduction of nature-inspired algorithms has made it possible to solve such challenging problems by providing approximate solutions; but there is no single algorithm that can be applied successfully to all these problems. PSO is an efficient swarm intelligent algorithm that has been upgraded by utilizing the characteristics of quantum mechanics. The Quantum-PSO algorithm has gained popularity among researchers since its invention due to its global searching ability and ease of control. It has been applied to solve optimization problems aiming to achieve sustainability in varied domains. Further, researchers have put much effort into suggesting strategies for the improvements in its performance when applied on different types of optimization problems.

This paper provides a comprehensive overview of such strategies and the improvements obtained thereby. Few of such strategies focus on improving the exploitation ability; few of them improve the local exploration and few impose global exploration abilities in traditional QPSO. Although several such strategies have been proposed, we can observe from the study that it still remains an open problem. Depending on the characteristics and need of sustainability, the researcher has to develop suitable methods to obtain optimal solutions improving the objectives of sustainable development. This study provides a glimpse of different strategies used to develop an improved QPSO algorithm and propels the researchers to explore their applicability in various fields of sustainable development.

REFERENCES

Alvarez-Alvarado, M. S., Alban-Chacón, F. E., Lamilla-Rubio, E. A., Rodríguez-Gallegos, C. D., & Velásquez, W. (2021). Three Novel Quantum-Inspired Swarm Optimization Algorithms Using Different Bounded Potential Fields. *Scientific Reports*, 11(1), 1–22.

Bagheri, A., Peyhani, H.M., & Akbari, M. (2014). Financial Forecasting Using ANFIS Networks with Quantum-Behaved Particle Swarm Optimization. *Expert Systems with Applications*, 41, 6235–6250.

Bao, Y., & Sun, J. (2011). Image Registration with a Modified Quantum-Behaved Particle Swarm Optimization. *2011 10th International Symposium on Distributed Computing and Applications to Business, Engineering and Science*, 202–206.

Cai, Y., Sun, J., Wang, J., Ding, Y., Tian, N., Liao, X., & Xu, W. (2008). Optimizing the Codon Usage of Synthetic Gene with QPSO Algorithm. *Journal of Theoretical Biology*, 254(1), 123–127.

Chai, Z., Sun, J., Cai, R., & Xu, W. (2009). Implementing Quantum-Behaved Particle Swarm Optimization Algorithm in FPGA for Embedded Real-Time Applications. *2009 Fourth International Conference on Computer Sciences and Convergence Information Technology*, 886–890.

Chakraborty, S., Senjyu, T., Saber, A., Yona, A., & Funabashi, T. (2012). A Novel Particle Swarm Optimization Method Based on Quantum Mechanics Computation for Thermal Economic Load Dispatch Problem. *IEEJ Transactions on Electrical and Electronic Engineering*, 7, 461–470.

Chakraborty, S., Senjyu, T., Yona, A., Saber, A., & Funabashi, T. (2011). Solving Economic Load Dispatch Problem with Valve-Point Effects Using a Hybrid Quantum Mechanics Inspired Particle Swarm Optimisation. *IET Generation Transmission & Distribution*, 5, 1042–1052.

Chang, C., Tsai, J., & Pei, S. (2012). A Quantum PSO Algorithm for Feedback Control of Semi-Autonomous Driver Assistance Systems. *2012 12th International Conference on ITS Telecommunications*, 553–557.

Chang, C., Tsai, J., & Pei, S. (2014). Quantum Particle Swarm Optimisation Algorithm for Feedback Control of Semi-Autonomous Driver Assistance Systems. *IET Intelligent Transport Systems*, 8, 608–620.

Chang, J., An, F., & Su, P. (2010). A Quantum-PSO Algorithm for No-Wait Flow Shop Scheduling Problem. *2010 Chinese Control and Decision Conference*, 179–184.

Chen, J., & Yang, D. (2010). Constrained Handling in Multi-Objective Optimization Based on Quantum-Behaved Particle Swarm Optimization. *2010 Sixth International Conference on Natural Computation*, 8, 3887–3891.

Chen, J., Zhen, Y., Yang, D., & Province, Z. (2013). *Fast Moving Object Tracking Algorithm Based on Hybrid Quantum PSO*. Control Engineering of China.

Chen, M., Ruan, J., & Xi, D. (2018). Micro Grid Scheduling Optimization Based on Quantum Particle Swarm Optimization (QPSO) Algorithm. *2018 Chinese Control and Decision Conference (CCDC)*, 6470–6475. https://doi.org/10.1109/CCDC.2018.8408267

Chen, W., Sun, J., Ding, Y., Fang, W., & Xu, W. (2008). *Clustering of Gene Expression Data with Quantum-Behaved Particle Swarm Optimization*. IEA/AIE.

Chen, Z. (2011). Improved PSO and its application to SVM parameter optimization. *Jisuanji Gongcheng yu Yingyong (Computer Engineering and Applications)*, 47(10), 38–40.

Chi, Y., Dong, Y., Xia, K., & Shi, J. (2008). Continuous Attribute Discretization Based on Quantum PSO Algorithm. *2008 7th World Congress on Intelligent Control and Automation*, 6187–6191.

Chunhong, C., Limin, W., & Wenhui, L. (2010, October). The Geometric Constraint Solving Based on the Quantum Particle Swarm. In *International Conference on Rough Sets and Knowledge Technology* (pp. 582–587). Springer.

Coelho, L. (2010). Gaussian Quantum-Behaved Particle Swarm Optimization Approaches for Constrained Engineering Design Problems. *Expert Systems with Applications*, 37, 1676–1683.

Cui, Z., & Geng, X. (2021). Product Service System Configuration Based on a PCA-QPSO-SVM Model. *Sustainability*, 13(16), 9450. https://doi.org/10.3390/su13169450

Davis, L. (1987). *Genetic Algorithms and Simulated Annealing*. London: Pitman; Los Altos, Calif.: Morgan Kaufmann Publishers.

Davoodi, E., Hagh, M.T., & Zadeh, S.G. (2014). A Hybrid Improved Quantum-Behaved Particle Swarm Optimization-Simplex method (IQPSOS) to Solve Power System Load Flow Problems. *Applied Soft Computing*, 21, 171–179.

Dawei, Y., Bin, C., Xianrong, Y., & Yubo, F. (2012). Aeroengine Deviation Parameters Nonlinear Estimation Using Improved Quantum-Behaved PSO Algorithm. *Proceedings of the 31st Chinese Control Conference*, 2013–2018.

Deepa, S., & Kalimuthu, M. (2012). An Optimization of Association Rule Mining Algorithm Using Weighted Quantum Behaved PSO. *International Journal of Power Control Signal and Computation (IJPCSC)*, 3(1), 80–85.

Dey, S., Bhattacharyya, S., & Maulik, U. (2014). Quantum Behaved Multi-Objective PSO and ACO Optimization for Multi-level Thresholding. *2014 International Conference on Computational Intelligence and Communication Networks*, 242–246.

Du, J., & Wei, L. (2009). Quantum Behaved Particle Swarm Optimization for Origin—Destination Matrix Prediction. *2009 2nd International Conference on Power Electronics and Intelligent Transportation System (PEITS)*, 1, 133–136.

Du, Z., Zhu, Y., & Liu, W. (2013). Combining Quantum-Behaved PSO and K2 Algorithm for Enhancing Gene Network Construction. *Current Bioinformatics*, 8, 133–137.

El Dahshan, K. A., Elsayed, E. K., Aboshoha, A., & Ebeid, E. A. (2020). Recognition of Facial Emotions Relying on Deep Belief Networks and Quantum Particle Swarm Optimization. *International Journal of Intelligent Engineering and Systems*, 13(4), 90–101.

Fang, W., Sun, J., Wu, X., & Palade, V. (2015). Adaptive Web QoS Controller Based on Online System Identification Using Quantum-Behaved Particle Swarm Optimization. *Soft Computing*, 19, 1715–1725.

Fang, W., Sun, J., & Xu, W. (2006a). *Design IIR Digital Filters Using Quantum-Behaved Particle Swarm Optimization*. ICNC.

Fang, W., Sun, J., & Xu, W. (2006b). Design of Two-Dimensional Recursive Filters by Using Quantum-Behaved Particle Swarm Optimization. *2006 International Conference on Intelligent Information Hiding and Multimedia*, 240–243.

Fang, W., Sun, J., & Xu, W. B. (2008). FIR Filter Design Based on Adaptive Quantum-Behaved Particle Swarm Optimization Algorithm. *Systems Engineering and Electronics*, 30(7), 1378–1381.

Fang, W., Sun, J., & Xu, W. (2009). A New Mutated Quantum-Behaved Particle Swarm Optimizer for Digital IIR Filter Design. *EURASIP Journal on Advances in Signal Processing*, 1–7.

Fang, W., Sun, J., Xu, W., & Liu, J. (2006c). FIR Digital Filters Design Based on Quantum-Behaved Particle Swarm Optimization. *First International Conference on Innovative Computing, Information and Control—Volume I (ICICIC'06)*, 1, 615–619.

Farzi, S. (2012). Training of Fuzzy Neural Networks via Quantum-Behaved Particle Swarm Optimization and Rival Penalized Competitive Learning. *International Arab Journal of Information Technology*, 9, 306–313.

Farzi, S., & Baraani-Dastjerdi, A. (2010). Leaf Constrained Minimal Spanning Trees Solved by Modified Quantum-Behaved Particle Swarm Optimization. *Artificial Intelligence Review*, 34, 1–17.

Farzi, S., Shavazi, A.R., & Pandari, A. (2013). Using Quantum-Behaved Particle Swarm Optimization for Portfolio Selection Problem. *International Arab Journal of Information Technology*, 10, 111–119.

Fu, Y., Ding, M., & Zhou, C. (2012). Phase Angle-Encoded and Quantum-Behaved Particle Swarm Optimization Applied to Three-Dimensional Route Planning for UAV. *IEEE Transactions on Systems, Man, and Cybernetics—Part A: Systems and Humans*, 42, 511–526.

Fu, Y., Ding, M., Zhou, C., Cai, C., & Sun, Y. (2009). Path Planning for UAV Based on Quantum-Behaved Particle Swarm Optimization. *International Symposium on Multispectral Image Processing and Pattern Recognition*.

Fu, Y., Ding, M., Zhou, C., & Hu, H. (2013). Route Planning for Unmanned Aerial Vehicle (UAV) on the Sea Using Hybrid Differential Evolution and Quantum-Behaved Particle Swarm Optimization. *IEEE Transactions on Systems, Man, and Cybernetics: Systems*, 43, 1451–1465.

Fu, Y. G., Zhou, C. P., & Ding, M. Y. (2010). 3-D Route Planning Based on Hybrid Quantum-Behaved Particle Swarm Optimization. *Journal of Astronautics*, 31(12), 2657–2664.

Gaing, Z., Lin, C., Tsai, M., Hsieh, M., & Tsai, M. (2014). Rigorous Design and Optimization of Brushless PM Motor Using Response Surface Methodology with Quantum-Behaved PSO Operator. *IEEE Transactions on Magnetics*, 50, 1–4.

Gao, H., Xu, W., Sun, J., & Tang, Y. (2010). Multilevel Thresholding for Image Segmentation Through an Improved Quantum-Behaved Particle Swarm Algorithm. *IEEE Transactions on Instrumentation and Measurement*, 59, 934–946.

Ghorpade, S. N., Zennaro, M., Chaudhari, B. S., Saeed, R. A., Alhumyani, H., & Abdel-Khalek, S. (2021). Enhanced Differential Crossover and Quantum Particle Swarm Optimization for IoT Applications. *IEEE Access*, 9, 93831–93846.

Gong, L., Sun, J., Xu, W., & Xu, J. (2012). Research and Simulation of Node Localization in WSN Based on Quantum Particle Swarm Optimization. *2012 11th International Symposium on Distributed Computing and Applications to Business, Engineering & Science*, 144–148.

Gu, X., & Ma, X. (2021). Operating State Prediction for Wind Turbine Generator Bearing Based on ULSSVM and QPSO. *Journal of Vibro Engineering*, 23(7).

Haddar, B., Khemakhem, M., Hanafi, S., & Wilbaut, C. (2016). A Hybrid Quantum Particle Swarm Optimization for the Multidimensional Knapsack Problem. *Engineering Applications of Artificial Intelligence*, 55, 1–13.

Hai-xi, L. (2009). *Image Color Segmentation Based on Quantum-Behaved Particle Swarm Optimization Data Clustering*. Microcomputer Information.

Hamed, H. N. A., Kasabov, N., & Shamsuddin, S. M. (2011). *Quantum-Inspired Particle Swarm Optimization for Feature Selection and Parameter Optimization in Evolving Spiking Neural Networks for Classification Tasks.* InTech.

Han, P., Yuan, S., & Wang, D. (2014). Thermal System Identification Based on Double Quantum Particle Swarm Optimization. In *Intelligent Computing in Smart Grid and Electrical Vehicles* (pp. 125–137). Springer.

Ho, S., Yang, S., Ni, G., & Huang, J. (2013). A Quantum-Based Particle Swarm Optimization Algorithm Applied to Inverse Problems. *IEEE Transactions on Magnetics,* 49, 2069–2072.

Hongwu, L. (2009). A Discrete Quantum-Behaved PSO and Its Multiuser Detection Application. *2009 IEEE International Conference on Intelligent Computing and Intelligent Systems,* 3, 566–569.

Hosseinnezhad, V., Hagh, M.T., & Babaei, E. (2011). Quantum Particle Swarm Optimization for Economic Dispatch Problem with Valve-Point Effect. *2011 19th Iranian Conference on Electrical Engineering,* 1–4.

Huang, M. L. (2016). Hybridization of Chaotic Quantum Particle Swarm Optimization with SVR in Electric Demand Forecasting. *Energies,* 9(6), 426.

Islam, J., Meraj, S. T., Negash, B., Biswas, K., Alhitmi, H. K., Hossain, N. I., . . . & Watada, J. (2021). Modified Quantum Particle Swarm Optimization for Selective Harmonic Elimination (SHE) in a Single-Phase Multilevel Inverter. *International Journal of Innovative Computing, Information and Control,* 17(3), 959–971.

Jalilzadeh, S., Shayeghi, H., Safari, A., & Masoomi, D. (2009). Output Feedback UPFC Controller Design by Using Quantum Particle Swarm Optimization. *2009 6th International Conference on Electrical Engineering/Electronics, Computer, Telecommunications and Information Technology,* 1, 28–31.

Jamalipour, M., Gharib, M., Sayareh, R., & Khoshahval, F. (2013a). PWR Power Distribution Flattening Using Quantum Particle Swarm Intelligence. *Annals of Nuclear Energy,* 56, 143–150.

Jamalipour, M., Sayareh, R., Gharib, M., Khoshahval, F., & Karimi, M. (2013b). Quantum Behaved Particle Swarm Optimization with Differential Mutation Operator Applied to WWER-1000 in-Core Fuel Management Optimization. *Annals of Nuclear Energy,* 54, 134–140.

Jeong, Y., Park, J., Jang, S., & Lee, K.Y. (2009). A New Quantum-Inspired Binary PSO for Thermal Unit Commitment Problems. *2009 15th International Conference on Intelligent System Applications to Power Systems,* 1–6.

Jeong, Y., Park, J., Jang, S., & Lee, K. (2010). A New Quantum-Inspired Binary PSO: Application to Unit Commitment Problems for Power Systems. *IEEE Transactions on Power Systems,* 25, 1486–1495.

Jiang, S., & Yang, S. (2014). An Improved Quantum-Behaved Particle Swarm Optimization Algorithm Based on Linear Interpolation. *2014 IEEE Congress on Evolutionary Computation (CEC),* 769–775.

Jin, C., & Jin, S. W. (2015a). Automatic Image Annotation Using Feature Selection Based on Improving Quantum Particle Swarm Optimization. *Signal Processing,* 109, 172–181.

Jin, C., & Jin, S. W. (2015b). Prediction Approach of Software Fault-Proneness Based on Hybrid Artificial Neural Network and Quantum Particle Swarm Optimization. *Applied Soft Computing,* 35, 717–725.

Jin, Y., Xue, J., & Shi, Z. (2014). An Improved Algorithm of Quantum Particle Swarm Optimization. *JSoftware,* 9, 2789–2795.

Jun, S. (2008). Image Enhancement Based on Quantum-Behaved Particle Swarm Optimization. *Journal of Computer Applications,* 28(1).

Kennedy, J., & Eberhart, R. (1995, November). Particle Swarm Optimization. In *Proceedings of ICNN'95-International Conference on Neural Networks* (Vol. 4, pp. 1942–1948). IEEE.

Kezhong, L., Kangnian, F., & Guangqian, X. (2008). A Hybrid Quantum-Behaved Particle Swarm Optimization Algorithm for Clustering Analysis. *2008 Fifth International Conference on Fuzzy Systems and Knowledge Discovery*, 1, 21–25.

Kong, X., Sun, J., Ye, B., & Xu, W. (2007). *An Efficient Quantum-Behaved Particle Swarm Optimization for Multiprocessor Scheduling*. International Conference on Computational Science.

Kou, P., Zhou, J., Wang, C., Xiao, H., Zhang, H., & Li, C. (2011). Parameters Identification of Nonlinear State Space Model of Synchronous Generator. *Engineering Applications of Artificial Intelligence*, 24, 1227–1237.

Lai, X., Hao, J. K., Fu, Z. H., & Yue, D. (2020). Diversity-Preserving Quantum Particle Swarm Optimization for the Multidimensional Knapsack Problem. *Expert Systems with Applications*, 149, 113310.

Lakshmi, K., & Rao, A.R. (2013). Optimal Design of Laminate Composite Isogrid with Dynamically Reconfigurable Quantum PSO. *Structural and Multidisciplinary Optimization*, 48, 1001–1021.

Latchoumi, T. P., Ezhilarasi, T. P., & Balamurugan, K. (2019). Bio-Inspired Weighed Quantum Particle Swarm Optimization and Smooth Support Vector Machine Ensembles for Identification of Abnormalities in Medical Data. *SN Applied Sciences*, 1(10), 1–10.

Lei, X., Huang, X., Shi, L., & Zhang, A. (2012). Clustering PPI Data Based on Improved Functional-Flow Model Through Quantum-Behaved PSO. *International Journal of Data Mining and Bioinformatics*, 6(1), 42–60.

Lenin, K., & Reddy, B. (2014). Quantum Particle Swarm Optimization Algorithm for Solving Optimal Reactive Power Dispatch Problem. *International Journal of Information Engineering and Electronic Business*, 6, 32–37.

Li, C., Long, H., Ding, Y., Sun, J., & Xu, W. (2010). *Multiple Sequence Alignment by Improved Hidden Markov Model Training and Quantum-Behaved Particle Swarm Optimization*. LSMS/ICSEE.

Li, C., Sun, J., Li, L. W., Wu, X. J., & Palade, V. (2021). An Effective Swarm Intelligence Optimization Algorithm for Flexible Ligand Docking. *IEEE/ACM Transactions on Computational Biology and Bioinformatics*, 1–14.

Li, M., Liu, C., Li, K., Liao, X., & Li, K. (2020). Multi-Task Allocation with an Optimized Quantum Particle Swarm Method. *Applied Soft Computing*, 96, 106603.

Li, P., & Xu, W. (2007). Solve Traveling Salesman Problems Based on QPSO. *Computer Engineering and Design*, 28(19), 160–162.

Li, S., Wang, R., Hu, W., & Sun, J. (2007). *A New QPSO Based BP Neural Network for Face Detection*. ICFIE.

Li, S., Zhao, D., Zhang, X., & Wang, C. (2010). Reactive Power Optimization Based on an Improved Quantum Discrete PSO Algorithm. *2010 5th International Conference on Critical Infrastructure (CRIS)*, 1–5.

Li, T., Zhang, J., Wang, S., & Lv, Z. (2014). Research on Route Planning Based on Quantum-Behaved Particle Swam Optimization Algorithm. *Proceedings of 2014 IEEE Chinese Guidance, Navigation and Control Conference*, 335–339.

Li, X., Zhou, L., & Liu, C. (2009). Model Selection of Least Squares Support Vector Regression Using Quantum-Behaved Particle Swarm Optimization Algorithm. *2009 International Workshop on Intelligent Systems and Applications*, 1–5.

Liao, X., & Ding, Q. (2014, May). Application of LS-SVM in the Short-Term Power Load Forecasting Based on QPSO. In *International Conference on Logistics Engineering, Management and Computer Science (LEMCS 2014)* (pp. 225–228). Atlantis Press.

Lin, X., Sun, J., Palade, V., Fang, W., Wu, X., & Xu, W. (2012). *Training ANFIS Parameters with a Quantum-Behaved Particle Swarm Optimization Algorithm*. ICSI.

Liu, C., Li, D., and Du, Y. (2005). Some statistical analysis of the normal cloud model. *Information and Control*, 34, 236–239.

Liu, F., Duan, H., & Deng, Y. (2012). A Chaotic Quantum-Behaved Particle Swarm Optimization Based on Lateral Inhibition for Image Matching. *Optik*, 123, 1955–1960.

Liu, G., Chen, W., Chen, H., & Xie, J. (2019). A Quantum Particle Swarm Optimization Algorithm with Teamwork Evolutionary Strategy. *Mathematical Problems in Engineering*, 2019, Article ID 1805198.

Liu, H., & Song, G. (2009). A Multiuser Detection Based on Quantum PSO with Pareto Optimality for STBC-MC-CDMA System. *2009 IEEE International Conference on Communications Technology and Applications*, 652–655.

Liu, J., Sun, J., & Xu, W. (2006). Quantum-Behaved Particle Swarm Optimization for Integer Programming. In King, I., Wang, J., Chan, L. W., & Wang, D. (Eds.), *Neural Information Processing*. ICONIP 2006. Lecture Notes in Computer Science (Vol. 4233). Springer. https://doi.org/10.1007/11893257_114

Liu, J., Sun, J., & Xu, W. (2006). Improving Quantum-Behaved Particle Swarm Optimization by Simulated Annealing. In Huang, D. S., Li, K., & Irwin, G. W. (Eds.), *Computational Intelligence and Bioinformatics*. ICIC 2006. Lecture Notes in Computer Science (Vol. 4115). Springer. https://doi.org/10.1007/11816102_14

Liu, J., Sun, J., Xu, W. B. & Kong, X. H. (2006). Quantum-Behaved Particle Swarm Optimization Based on Immune Memory and Vaccination. *2006 IEEE International Conference on Granular Computing*, 453–456. https://doi.org/10.1109/GRC.2006.1635838

Liu, J., Wu, Q., & Zhu, D. (2009). Thruster Fault-Tolerant for UUVs Based on Quantum-Behaved Particle Swarm Optimization. In *Opportunities and Challenges for Next-Generation Applied Intelligence* (pp. 159–165). Springer.

Liu, L., Xia, K., Zhou, Y., & Bai, J. (2014). Oil Layer Recognition by Support Vector Machine Based on Quantum-Behaved Particle Swarm Optimization. *The Journal of Information and Computational Science*, 11, 1511–1518.

Liu, P., Leng, W., & Fang, W. (2013). Training ANFIS Model with an Improved Quantum-Behaved Particle Swarm Optimization Algorithm. *Mathematical Problems in Engineering*, 1–10.

Liu, R., Zhang, P., & Jiao, L. (2014). Quantum Particle Swarm Optimization Classification Algorithm and Its Applications. *International Journal of Pattern Recognition and Artificial Intelligence*, 28.

Liu, T., Pan, L., Meng, Q., & Li, Y. (2012). Parameter Identification of Chaotic System Based on Quantum PSO Algorithm. *Proceedings of 2012 2nd International Conference on Computer Science and Network Technology*, 1982–1985.

Liu, Z., Zhang, W., & Wang, Z. (2012, July). Optimal Planning of Charging Station for Electric Vehicle Based on Quantum PSO Algorithm. In *Proceedings of the CSEE* (Vol. 32, No. 22, pp. 39–45).

Long, H., Sun, J., Wang, X., Lai, C. H., & Xu, W. (2009). Using Selection to Improve Quantum-Behaved Particle Swarm Optimisation. *International Journal of Innovative Computing and Applications*, 2, 100–114.

Loo, C. K., & Nikos, M. E. (2007, March). Quantum Potential Swarm Optimization of PD Controller for Cargo Ship Steering. In *11th WSEAS International Conference on Applied Mathematics*, 224–230.

Lu, K., Li, H., & Wang, R. (2010). Optimization of Feeding Rate for Alcohol Fermentation by Quantum-Behaved Particle Swarm Optimization. *2010 8th World Congress on Intelligent Control and Automation*, 4677–4680.

Lu, K., & Wang, R. (2008). Application of PSO and QPSO Algorithm to Estimate Parameters from Kinetic Model of Glutamic Acid Batch Fermetation. *2008 7th World Congress on Intelligent Control and Automation*, 8968–8971.

Luitel, B., & Venayagamoorthy, G. (2008). Particle Swarm Optimization with Quantum Infusion for the Design of Digital Filters. *2008 IEEE Swarm Intelligence Symposium*, 1–8.

Luitel, B., & Venayagamoorthy, G. (2010a). Quantum Inspired PSO for the Optimization of Simultaneous Recurrent Neural Networks as MIMO Learning Systems. *Neural Networks: The Official Journal of the International Neural Network Society*, 23(5), 583–586.

Luitel, B., & Venayagamoorthy, G. (2010b). Particle Swarm Optimization with Quantum Infusion for System Identification. *Engineering Applications of Artificial Intelligence*, 23, 635–649.

Luo, Z., Zhang, W., Li, Y., & Xiang, M. (2008). SVM Parameters Tuning with Quantum Particles Swarm Optimization. *2008 IEEE Conference on Cybernetics and Intelligent Systems*, 324–329.

Mariani, V., Duck, A., Guerra, F., Coelho, L., & Rao, R. (2012). A Chaotic Quantum-Behaved Particle Swarm Approach Applied to Optimization of Heat Exchangers. *Applied Thermal Engineering*, 42, 119–128.

Mauryan, K.S., Thanushkodi, K., & Sakthisuganya, A. (2012). Reactive Power Optimization Using Quantum Particle Swarm Optimization. *Journal of Computer Science*, 8, 1644–1648.

Meng, K., Wang, H., Dong, Z., & Wong, K. (2010). Quantum-Inspired Particle Swarm Optimization for Valve-Point Economic Load Dispatch. *IEEE Transactions on Power Systems*, 25, 215–222.

Meshoul, S., & Al-Owaisheq, T. (2009, October). QPSO-MD: A Quantum Behaved Particle Swarm Optimization for Consensus Pattern Identification. In *International Symposium on Intelligence Computation and Applications* (pp. 369–378). Springer.

Millonas, M. (1994). Swarms, Phase Transitions, and Collective Intelligence. In *Proceedings of Artificial Life*. Springer.

Mu, L., Zhao, M., & Zhang, C. (2013). Quantum Particle Swarm Optimisation Based on Chaotic Mutation for Automatic Parameters Determination of Pulse Coupled Neural Network. *International Journal of Computing Science and Mathematics*, 4, 354–362.

Nishant, R., Kennedy, M., & Corbett, J. (2020). Artificial Intelligence for Sustainability: Challenges, Opportunities, and a Research Agenda. *International Journal of Information Management*, 53, 102104.

Niu, Q., Zhou, Z., & Zeng, T. (2011). *A Hybrid Quantum-Inspired Particle Swarm Evolution Algorithm and SQP Method for Large-Scale Economic Dispatch Problems*. ICIC.

Niu, Q., Zhou, Z., Zhang, H., & Deng, J. (2012). An Improved Quantum-Behaved Particle Swarm Optimization Method for Economic Dispatch Problems with Multiple Fuel Options and Valve-Points Effects. *Energies*, 5, 1–19.

Oliveira, L. D., Ciriaco, F., Abrão, T., & Jeszensky, P. (2006). Particle Swarm and Quantum Particle Swarm Optimization Applied to DS/CDMA Multiuser Detection in Flat Rayleigh Channels. *2006 IEEE Ninth International Symposium on Spread Spectrum Techniques and Applications*, 133–137.

Omkar, S., Khandelwal, R., Ananth, T.S., Naik, G., & Gopalakrishnan, S. (2009). Quantum Behaved Particle Swarm Optimization (QPSO) for Multi-Objective Design Optimization of Composite Structures. *Expert Systems with Applications*, 36, 11312–11322.

Pan, G., Xia, K., Dong, Y., & Shi, J. (2007). An Improved LS-SVM Based on Quantum PSO Algorithm and Its Application. *Third International Conference on Natural Computation (ICNC 2007)*, 2, 606–610.

Prawin, J., Rao, A. R. M., & Lakshmi, K. (2016). Nonlinear Parametric Identification Strategy Combining Reverse Path and Hybrid Dynamic Quantum Particle Swarm Optimization. *Nonlinear Dynamics*, 84(2), 797–815.

Priyadharshini, P. (2013). ATC Enhancement with Facts Devices Using Quantum Inspired PSO. *International Journal of Computer Applications*, 71, 39–45.

Radha, R., & Gopalakrishnan, R. (2020). A Medical Analytical System Using Intelligent Fuzzy Level Set Brain Image Segmentation Based on Improved Quantum Particle Swarm Optimization. *Microprocessors and Microsystems*, 79, 103283.

Rehman, O. U., Yang, S., Khan, S., & Rehman, S. U. (2019). A Quantum Particle Swarm Optimizer with Enhanced Strategy for Global Optimization of Electromagnetic Devices. *IEEE Transactions on Magnetics*, 55(8), 1–4.

Sabat, S.L., Coelho, L., & Abraham, A. (2009). MESFET DC Model Parameter Extraction Using Quantum Particle Swarm Optimization. *Microelectronics Reliability*, 49, 660–666.

Sabat, S.L., Udgata, S.K., & Murthy, K. (2010). Small Signal Parameter Extraction of MESFET Using Quantum Particle Swarm Optimization. *Microelectronics Reliability*, 50, 199–206.

Sarangi, A., Mahapatra, R.K., & Panigrahi, S. (2011). DEPSO and PSO-QI in Digital Filter Design. *Expert Systems with Applications*, 38, 10966–10973.

Shabanifard, M., & Amirani, M. (2011). A Modified Quantum-Behaved Particle Swarm Optimization Algorithm for Image Segmentation. *2011 19th Iranian Conference on Electrical Engineering*, 1–6.

Shayeghi, H., & Safari, A. (2013). Optimal Design of UPFC Based Damping Controller Using PSO and QPSO. In *Analysis, Control and Optimal Operations in Hybrid Power Systems* (pp. 157–186). Springer.

Shayeghi, H., Shayanfar, H., Jalilzadeh, S., & Safari, A. (2010). Tuning of Damping Controller for UPFC Using Quantum Particle Swarm Optimizer. *Energy Conversion and Management*, 51, 2299–2306.

Sheng, X. Y., Sun, J., & Xu, W. B. (2014). Quantum-Behaved Particle Swarm Optimization Using Q-Learning. *Applied Mechanics and Materials*, 556–562, 3965–3971. https://doi.org/10.4028/www.scientific.net/amm.556-562.3965

Sheng, X. Y., Xi, M., Sun, J., & Xu, W. (2015). Quantum-Behaved Particle Swarm Optimization with Novel Adaptive Strategies. *Journal of Algorithms and Computational Technology*, 9, 143–161.

Shi, Y., Wong, W. K., Goldin, J. G., Brown, M. S., & Kim, G. H. J. (2019). Prediction of Progression in Idiopathic Pulmonary Fibrosis Using CT Scans at Baseline: A Quantum Particle Swarm Optimization-Random Forest Approach. *Artificial Intelligence in Medicine*, 100, 101709.

Shi, Z., Li, Y., Song, Y., & Yu, T. (2009). Fault Diagnosis of Transformer Based on Quantum-Behaved Particle Swarm Optimization-Based Least Squares Support Vector Machines. *2009 International Conference on Information Engineering and Computer Science*, 1–4.

Su, D., Xu, W., & Sun, J. (2009). Quantum-Behaved Particle Swarm Optimization with Crossover Operator. *2009 International Conference on Wireless Networks and Information Systems*, 399–402.

Su, X., Fang, W., Shen, Q., & Hao, X. (2013). An Image Enhancement Method Using the Quantum-Behaved Particle Swarm Optimization with an Adaptive Strategy. *Mathematical Problems in Engineering*, 1–14.

Sun, J., Xu, W., & Feng, B. (2004, December). A Global Search Strategy of Quantum-Behaved Particle Swarm Optimization. In *IEEE Conference on Cybernetics and Intelligent Systems, 2004* (Vol. 1, pp. 111–116). IEEE.

Sun, J., Xu, W., & Ye, B. (2006). Quantum-Behaved Particle Swarm Optimization Clustering Algorithm. In Li, X., Zaïane, O.R., & Li, Z. (Eds.), *Advanced Data Mining and Applications*. ADMA 2006. Lecture Notes in Computer Science (Vol. 4093). Springer. https://doi.org/10.1007/11811305_37

Su, X., Zhao, J., & Sun, J. (2009). Online System Identification Based on Quantum-Behaved Particle Swarm Optimization Algorithm. *2009 International Conference on Web Information Systems and Mining*, 475–479.

Sun, J., Chen, W., Fang, W., Wu, X., & Xu, W. (2012). Gene Expression Data Analysis with the Clustering Method Based on an Improved Quantum-Behaved Particle Swarm Optimization. *Engineering Applications of Artificial Intelligence*, 25, 376–391.

Sun, J., Fang, W., Chen, W., & Xu, W. (2008). Design of Two-Dimensional IIR Digital Filters Using an Improved Quantum-Behaved Particle Swarm Optimization Algorithm. *2008 American Control Conference*, 2603–2608.

Sun, J., Fang, W., & Xu, W. (2010). A Quantum-Behaved Particle Swarm Optimization with Diversity-Guided Mutation for the Design of Two-Dimensional IIR Digital Filters. *IEEE Transactions on Circuits and Systems II: Express Briefs*, 57, 141–145.

Sun, J., Fang, W., Wang, D., & Xu, W. (2009). Solving the Economic Dispatch Problem with a Modified Quantum-Behaved Particle Swarm Optimization Method. *Energy Conversion and Management*, 50, 2967–2975.

Sun, J., Fang, W., Wu, X., Palade, V., & Xu, W. (2012). Quantum-Behaved Particle Swarm Optimization: Analysis of Individual Particle Behavior and Parameter Selection. *Evolutionary Computation*, 20(3), 349–393. https://doi.org/10.1162/EVCO_a_00049

Sun, J., Fang, W., Wu, X., Xie, Z., & Xu, W. (2011). QoS Multicast Routing Using a Quantum-Behaved Particle Swarm Optimization Algorithm. *Engineering Applications of Artificial Intelligence*, 24, 123–131.

Sun, J., Feng, B., & Xu, W. (2004). Particle Swarm Optimization with Particles Having Quantum Behavior. In *Proceedings of the 2004 Congress on Evolutionary Computation*. IEEE Cat. No.04TH8753 (Vol. 1, pp. 325–331). IEEE.

Sun, J., Lai, C.H., Xu, W., Ding, Y., & Chai, Z. (2007). A Modified Quantum-Behaved Particle Swarm Optimization. In Shi, Y., van Albada, G. D., Dongarra, J., & Sloot, P. M. A. (Eds.), *Computational Science—ICCS 2007*. ICCS 2007. Lecture Notes in Computer Science (Vol. 4487). Springer. https://doi.org/10.1007/978-3-540-72584-8_38

Sun, J., Liu, J., & Xu, W. (2006a). *QPSO-Based QoS Multicast Routing Algorithm*. SEAL.

Sun, J., Liu, J., & Xu, W. (2007). Using Quantum-Behaved Particle Swarm Optimization Algorithm to Solve Non-Linear Programming Problems. *International Journal of Computer Mathematics*, 84, 261–272.

Sun, J., Palade, V., Wang, Z., & Wu, X. (2013). Hybrid Approach of Genetic Programming and Quantum-Behaved Particle Swarm Optimization for Modeling and Optimization of Fermentation Processes. In *Combinations of Intelligent Methods and Applications* (pp. 117–136). Springer.

Sun, J., Wu, X., Fang, W., Ding, Y., Long, H., & Xu, W. (2012). Multiple Sequence Alignment Using the Hidden Markov Model Trained by an Improved Quantum-Behaved Particle Swarm Optimization. *Information Sciences*, 182, 93–114.

Sun, J., Xu, W., & Fang, W. (2006, May). Quantum-Behaved Particle Swarm Optimization Algorithm with Controlled Diversity. In *International Conference on Computational Science* (pp. 847–854). Springer.

Sun, J., Xu, W., & Fang, W. (2006b). *Solving Multi-period Financial Planning Problem Via Quantum-Behaved Particle Swarm Algorithm*. ICIC.

Sun, J., Xu, W., & Fang, W. (2006e). Quantum-Behaved Particle Swarm Optimization Algorithm with Controlled Diversity. In Alexandrov, V. N., van Albada, G. D., Sloot, P. M. A., & Dongarra, J. (Eds.), *Computational Science—ICCS 2006*. ICCS 2006. Lecture Notes in Computer Science (Vol. 3993). Springer. https://doi.org/10.1007/11758532_110

Sun, J., Xu, W., & Feng, B. (2004). A Global Search Strategy of Quantum-Behaved Particle Swarm Optimization. *IEEE Conference on Cybernetics and Intelligent Systems*, 1, 111–116.

Sun, J., Xu, W., & Feng, B. (2005). Adaptive Parameter Control for Quantum-Behaved Particle Swarm Optimization on Individual Level. *2005 IEEE International Conference on Systems, Man and Cybernetics*, 4, 3049–3054.

Sun, J., Xu, W., & Liu, J. (2006c). *Training RBF Neural Network Via Quantum-Behaved Particle Swarm Optimization*. ICONIP.

Sun, J., Xu, W., & Ye, B. (2006d). *Quantum-Behaved Particle Swarm Optimization Clustering Algorithm*. ADMA.

Tang, Q., Zhao, L., Qi, R., Cheng, H., & Qian, F. (2011). Tuning the Structure and Parameters of a Neural Network by Using Cooperative Quantum Particle Swarm Algorithm. *Applied Mechanics and Materials*, 48–49, 1328–1332.

Tian, N., Sun, J., Xu, W., & Lai, C. (2011). An Improved Quantum-Behaved Particle Swarm Optimization with Perturbation Operator and Its Application in Estimating Groundwater Contaminant Source. *Inverse Problems in Science and Engineering*, 19, 181–202.

Tian, N., Xu, W., Sun, J., & Lai, C. (2010). Estimation of Heat Flux in Inverse Heat Conduction Problems Using Quantum-Behaved Particle Swarm Optimization. *Journal of Algorithms and Computational Technology*, 4, 25–46.

Wang, H., Yang, S., Xu, W., & Sun, J. (2007). Scalability of Hybrid Fuzzy C-Means Algorithm Based on Quantum-Behaved PSO. *Fourth International Conference on Fuzzy Systems and Knowledge Discovery (FSKD 2007)*, 2, 261–265.

Wang, L., Liu, L., Qi, J., & Peng, W. (2020). Improved Quantum Particle Swarm Optimization Algorithm for Offline Path Planning in AUVs. *IEEE Access*, 8, 143397–143411.

Wang, M., Jia, S., Chen, E., Yang, S., Liu, P., & Qi, Z. (2022). Research and Application of Neural Network for Tread Wear Prediction and Optimization. *Mechanical Systems and Signal Processing*, 162, 108070.

Wang, X., Sun, J., & Xu, W. (2009). A Parallel QPSO Algorithm Using Neighborhood Topology Model. *2009 WRI World Congress on Computer Science and Information Engineering*, 4, 831–835.

Wei, C. L., & Wang, G. G. (2020). Hybrid Annealing Krill Herd and Quantum-Behaved Particle Swarm Optimization. *Mathematics*, 8(9), 1403.

Wei, S. (2010). Parameter Optimization of PID Controller Based on Quantum-Behaved Particle Swarm Optimization Algorithm with Weight Coefficient. *Computer Engineering and Applications*, 10, 256–270.

Wen-bo, X. (2009). Path Planning for Mobile Robot Based on Binary Quantum-Behaved Particle Swarm Optimization. *Journal of System Simulation*, 17, 5516–5523.

Wen-hui, J. (2007). Application of Quantum-Behaved PSO Algorithm with Mutation Operation System Parameters Identication. *Computer Engineering and Applications*, 29, 222–224.

Wu, Q., Ma, Z., Fan, J., Xu, G., & Shen, Y. (2019). A Feature Selection Method Based on Hybrid Improved Binary Quantum Particle Swarm Optimization. *IEEE Access*, 7, 80588–80601.

Wu, R., Wang, J., Xia, K., & Yang, R. (2008). Optimal Design on CMOS Operational Amplifier with QPSO Algorithm. *2008 International Conference on Wavelet Analysis and Pattern Recognition*, 2, 821–825.

Wu, X., Zhang, B., Wang, K., Li, J., & Duan, Y. (2012). A Quantum-Inspired Binary PSO Algorithm for Unit Commitment with Wind Farms Considering Emission Reduction. *IEEE PES Innovative Smart Grid Technologies*, 1–6.

Wu, X., Zhang, B., Yuan, X., Li, G., Luo, G., & Zhou, Y. (2013). Solutions to Unit Commitment Problems in Power Systems with Wind Farms Using Advanced Quantum-Inspired Binary PSO. *Proceedings of the CSEE*, 33(4), 45–52.

Xi, M., & Sun, J. (2012). A Modified Binary Quantum-Behaved Particle Swarm Optimization Algorithm with Bit Crossover Operator. *Advanced Materials Research*, 591–593, 1376–1380.

Xi, M., Qi, Z., Zou, Y., Raghavan, G., & Sun, J. (2015). Calibrating RZWQM2 Model Using Quantum-Behaved Particle Swarm Optimization Algorithm. *Computers and Electronics in Agriculture*, 113, 72–80.

Xi, M., Sun, J., & Xu, W. (2007). Parameter Optimization of PID Controller Based on Quantum-Behaved Particle Swarm Optimization Algorithm. *Complex Systems and Applications-Modelling*, 14, 603–607.

Xi, M., Sun, J., & Xu, W. (2008). An Improved Quantum-Behaved Particle Swarm Optimization Algorithm with Weighted Mean Best Position. *Applied Mathematics and Computation*, 205(2), 751–759. ISSN 0096-3003. https://doi.org/10.1016/j.amc.2008.05.135

Xin, J., & Xiao-feng, H. (2012). A Quantum-PSO-Based SVM Algorithm for Fund Price Prediction. *Journal of Convergence Information Technology*, 7, 267–273.

Xin-gang, Z., Ji, L., Jin, M., & Ying, Z. (2020). An Improved Quantum Particle Swarm Optimization Algorithm for Environmental Economic Dispatch. *Expert Systems with Applications*, 152, 113370.

Xin-gang, Z., Ze-qi, Z., Yi-min, X., & Jin, M. (2020). Economic-Environmental Dispatch of Microgrid Based on Improved Quantum Particle Swarm Optimization. *Energy*, 195, 117014.

Xiong, W., Guo, B., & Yan, S. (2022). Energy Consumption Optimization of Processor Scheduling for Real-Time Embedded Systems Under the Constraints of Sequential Relationship and Reliability. *Alexandria Engineering Journal*, 61(1), 73–80.

Xu, M., & Xu, W. B. (2008). Parameter Estimation of Complex Functions Based on Quantum-Behaved Particle Swarm Optimization Algorithm. *DCABES 2008 Proceedings*, Vols I and II, 591–596.

Xu, W., & Sun, J. (2005). Adaptive Parameter Selection of Quantum-Behaved Particle Swarm Optimization on Global Level. In Huang, D. S., Zhang, X. P., & Huang, G. B. (Eds.), *Advances in Intelligent Computing*. ICIC 2005. Lecture Notes in Computer Science (Vol. 3644). Springer. https://doi.org/10.1007/11538059_44

Xue, Y. C., Sun, J., & Xu, W. B. (2006). QPSO Algorithm for Rectangle-Packing Optimization. *J Comput Appl*, 9, 2068–2070.

Yang, S., & Wang, M. (2004, June). A Quantum Particle Swarm Optimization. In *Proceedings of the 2004 Congress on Evolutionary Computation*. IEEE Cat. No. 04TH8753 (Vol. 1, pp.320–324). IEEE.

Yao, W. Q. (2014). Genetic Quantum Particle Swarm Optimization Algorithm for Solving Traveling Salesman Problems. In *Fuzzy Information & Engineering and Operations Research & Management* (pp. 67–74). Springer.

Yin, Q., Li, W., Zhang, X., & Huo, F. (2010). Continuous Quantum Particle Swarm Optimization and Its Application to Optimization Calculation and Analysis of Energy-Saving Motor Used in Beam Pumping Unit. *2010 IEEE Fifth International Conference on Bio-Inspired Computing: Theories and Applications (BIC-TA)*, 1231–1235.

Yong-mei, W., & Wan-ye, Y. (2014). The Planning of Distribution Generation (DG) Based on Multi-Objective Quantum Particle Swarms Optimization (QPSO). *International Journal of Advanced Pervasive and Ubiquitous Computing (IJAPUC)*, 6(1).

Yu, G., Huang, Y., & Huang, L. (2010). T-S Fuzzy Control for Magnetic Levitation Systems Using Quantum Particle Swarm Optimization. *Proceedings of SICE Annual Conference 2010*, 48–53.

Yu, J., Mo, B., Tang, D., Liu, H., & Wan, J. (2017). Remaining Useful Life Prediction for Lithium-Ion Batteries Using a Quantum Particle Swarm Optimization-Based Particle Filter. *Quality Engineering*, 29(3), 536–546.

Yu, Z. (2012). Computer Network Attack Detection Based on Quantum Pso and Relevance Vector Machine. *International Journal on Advances in Information Sciences and Service Sciences*, 4, 268–273.

Yumin, D., & Li, Z. (2014). Quantum Behaved Particle Swarm Optimization Algorithm Based on Artificial Fish Swarm. *Mathematical Problems in Engineering*, 1–10.

Zhang, B., Qi, H., Sun, S. C., Ruan, L. M., & Tan, H. P. (2015). Solving Inverse Problems of Radiative Heat Transfer and Phase Change in Semitransparent Medium by Using Improved Quantum Particle Swarm Optimization. *International Journal of Heat and Mass Transfer*, 85, 300–310.

Zhang, D., Wang, J., Fan, H., Zhang, T., Gao, J., & Yang, P. (2021). New Method of Traffic Flow Forecasting Based on Quantum Particle Swarm Optimization Strategy for Intelligent Transportation System. *International Journal of Communication Systems*, 34(1), e4647.

Zhang, J., Jia, F. Y., & Tang, L. (2011). An Improved Quantum-Behaved PSO Algorithm for a New Process Optimization Problem Based on Mechanism Model of LBE Converter. *2011 International Symposium on Advanced Control of Industrial Processes (ADCONIP)*, 204–209.

Zhang, K. (2014). A Network Traffic Prediction Model Based on Quantum Inspired Pso and Wavelet Neural Network. *Mathematical & Computational Applications*, 19, 218–229.

Zhang, K., Liang, L., & Huang, Y. (2013). A Network Traffic Prediction Model Based on Quantum Inspired PSO and Neural Network. *2013 Sixth International Symposium on Computational Intelligence and Design*, 2, 219–222.

Zhang, Q., & Hu, S. (2019, February). An Improved Hybrid Quantum Particle Swarm Optimization Algorithm for FJSP. *Proceedings of the 2019 11th International Conference on Machine Learning and Computing*, 246–252.

Zhang, Q., Lei, X., Huang, X., & Zhang, A. (2010). An Improved Projection Pursuit Clustering Model and Its Application Based on Quantum-Behaved PSO. *2010 Sixth International Conference on Natural Computation*, 5, 2581–2585.

Zhao, D., Xia, K., Wang, B., & Gao, J. (2008). An Approach to Mobile IP Routing Based on QPSO Algorithm. *2008 IEEE Pacific-Asia Workshop on Computational Intelligence and Industrial Application*, 1, 667–671.

Zhao, J., Sun, J., & Palade, V. (2012). Tracking Multiple Optima in Dynamic Environments by Quantum-Behavior Particle Swarm Using Speciation. *International Journal of Swarm Intelligence Research*, 3, 55–76.

Zhao, J., Sun, J., & Xu, W. (2009a). Application of Online System Identification Based on Improved Quantum-Behaved Particle Swarm Optimization. *2009 Second International Symposium on Computational Intelligence and Design*, 2, 186–189.

Zhao, J., Sun, J., & Xu, W. (2009d). Quantum-Behaved Particle Swarm Optimization with Normal Cloud Mutation Operator. *2009 International Conference on Computational Intelligence and Software Engineering*, 1–4.

Zhao, J., Sun, J., & Xu, W. (2013). A Binary Quantum-Behaved Particle Swarm Optimization Algorithm with Cooperative Approach. *International Journal of Computer Science Issues (IJCSI)*, 10(1), 112.

Zhao, J., Sun, J., Chen, W., & Xu, W. (2009b). Tracking Extrema in Dynamic Environments with Quantum-Behaved Particle Swarm Optimization. *2009 WRI Global Congress on Intelligent Systems*, 2, 103–108.

Zhao, J., Sun, J., Xu, W., & Zhou, D. (2009c). Structure Learning of Bayesian Networks Based on Discrete Binary Quantum-Behaved Particle Swarm Optimization Algorithm. *2009 Fifth International Conference on Natural Computation*, 6, 86–90.

Zhao, X., Sun, J., & Xu, W. (2010). Application of Quantum-behaved Particle Swarm Optimization in Parameter Estimation of Option Pricing. *2010 Ninth International Symposium on Distributed Computing and Applications to Business, Engineering and Science*, 10–12.

Zheng, T. (2010). Minimizing Total Flowtime in Flow Shop Scheduling by a Quantum-Inspired Swarm Evolutionary Algorithm. *2010 International Conference on Electronics and Information Engineering*, 1, V1-351–V1-355.

Zhengchu, W., Mu-xun, Z., Xiufeng, L., Chun, F., & Feixiang, J. (2010). A Quantum Particle Swarm Optimization for Solving the Capacitated Vehicle Routing Problem. *2010 8th World Congress on Intelligent Control and Automation*, 3281–3285.

Zhou, D., Sun, J., & Xu, W. (2010). An Advanced Quantum-Behaved Particle Swarm Optimization Algorithm Utilizing Cooperative Strategy. *Third International Workshop on Advanced Computational Intelligence*, 344–349.

Zhou, L., Yang, H., & Liu, C. (2008). QPSO-Based Hyper-Parameters Selection for LS-SVM Regression. *2008 Fourth International Conference on Natural Computation*, 2, 130–133.

Index

Printed in the United States
by Baker & Taylor Publisher Services